T0315078

The Science of Stem Cells

The Science of Stem Cells

Jonathan M. W. Slack

WILEY Blackwell

Registered Offices
John Wiley & Sons, Inc., 111 River Street, Hoboken, NJ 07030, USA

Editorial Office
111 River Street, Hoboken, NJ 07030, USA

For details of our global editorial offices, customer services, and more information about Wiley products visit us at www.wiley.com.

Wiley also publishes its books in a variety of electronic formats and by print-on-demand. Some content that appears in standard print versions of this book may not be available in other formats.

Library of Congress Cataloguing-in-Publication Data

Names: Slack, J. M. W. (Jonathan Michael Wyndham), 1949– author.
Title: The science of stem cells / Jonathan M. W. Slack.
Description: Hoboken, NJ : Wiley, 2018. | Includes bibliographical references and index. |
Identifiers: LCCN 2017028793 (print) | LCCN 2017030609 (ebook) | ISBN 9781119235231 (pdf) | ISBN 9781119235255 (epub) | ISBN 9781119235156 (hardback)
Subjects: LCSH: Stem cells. | BISAC: SCIENCE / Life Sciences / Cytology.
Classification: LCC QH588.S83 (ebook) | LCC QH588.S83 .S575 2017 (print) | DC 616.02/774–dc23
LC record available at https://lccn.loc.gov/2017028793

Cover Design: Wiley
Cover Image: Gist F. Croft, Alessia Deglincerti, Ali Brivanlou, The Rockefeller University

About the cover: Optical section through a day 12 human embryo which has developed in vitro. The epiblast (OCT4: green) has formed an amnion-like cavity. The trophectoderm has become cytotrophoblast (KERATIN 17: lilac) and multinucleated synciotiotrophoblast (hCG beta, acqua). DNA is stained red, showing cell nuclei. Image kindly supplied by Dr Ali H Brivanlou, Rockefeller University.

Set in 10/12pt Warnock by SPi Global, Pondicherry, India

10 9 8 7 6 5 4 3 2 1

Contents

Preface   *xi*
About the Companion Website   *xiii*

1   What is a Stem Cell?   *1*
Stem Cell Markers   *3*
Label-Retention   *4*
The Niche   *5*
Asymmetric Division and Differentiated Progeny   *6*
Clonogenicity and Transplantation   *6*
In Vivo Lineage Labeling   *7*
Conclusions   *9*
Further Reading   *10*

2   Characterizing Cells   *13*
Histological and Anatomical Methods   *13*
  Histological Sections   *13*
    Fixation   *13*
    Sectioning   *14*
    Staining   *14*
  Electron Microscopy   *15*
  Fluorescence Microscopy   *16*
  Wholemounts   *17*
  Immunostaining   *17*
  In Situ Hybridization   *18*
Other Methods   *19*
  RNAseq   *19*
  Laser Capture Microdissection   *19*
  Flow Cytometry   *20*
Dividing Cells   *21*
  The Cell Cycle   *21*
  Studying Cell Turnover   *24*
  Reporters for the Cell Cycle   *26*
  Identification of Very Slow Cell Turnover   *26*
  Classification of Cell Types by Proliferative Behavior   *28*
  Cell Death   *28*
Further Reading   *30*

3   Genetic Modification and the Labeling of Cell Lineages   *31*
Introducing Genes to Cells   *31*
   Transfection and Electroporation   *31*
   Gene Delivery Viruses   *33*
   Controlling Gene Expression   *35*
     Tet System   *35*
     Cre System   *35*
     Inhibiting Gene Activity   *37*
     CRISPR-Cas9   *37*
Transgenic Mice   *38*
   Animal Procedures   *38*
   Modification of Embryonic Stem Cells   *40*
   Types of Transgenic Mice   *41*
Cell Lineage   *42*
   Cell Lineage, Fate Maps, Clonal Analysis   *43*
   Use of CreER for Lineage Analysis   *44*
   Retroviral Barcoding   *46*
   Clonal Analysis in Humans   *47*
   Further Reading   *47*

4   Tissue Culture, Tissue Engineering and Grafting   *49*
Simple Tissue Culture   *51*
   Media   *51*
   Contamination   *53*
   Growth in Culture   *54*
   Cryopreservation and Banking   *55*
   GMP Cultivation   *56*
Complex Tissue Culture   *56*
   Induced Differentiation   *56*
   Three Dimensional Cell Culture   *57*
   Artificial Organs and Organoids   *59*
Grafting   *60*
   The Immune System   *61*
     T Cells   *61*
     The Major Histocompatibility Complex   *62*
     T and B Cell Responses   *63*
   Reactions to a Graft   *64*
   Immunosuppressive Drugs   *65*
   Animal Experiments Involving Grafting   *66*
Further Reading   *67*

5   Early Mouse and Human Development   *69*
Gametogenesis   *70*
   Germ Cells   *70*
   Mitosis and Meiosis   *70*
   Primordial Germ Cells (PGCs)   *72*
   Spermatogenesis   *73*
   Oogenesis   *73*
Fertilization   *76*

Early Development *77*
  Preimplantation Phase *77*
  Implantation Period – Mouse *80*
  Implantation Period – Human *82*
  Ethical and Legal Issues Concerning the Early Human Conceptus *85*
  Sex Determination *86*
  X-Inactivation *87*
  Imprinting *87*
  Cloning by Nuclear Transplantation (SCNT) *89*
Further Reading *90*

6 Pluripotent Stem Cells *93*
Mouse Pluripotent Stem Cells *93*
  Mouse Embryonic Stem Cells *93*
    Differentiation of Mouse ES cells *95*
  Mouse iPS Cells *97*
Human Pluripotent Stem Cells *101*
  SCNT-Derived Embryonic Stem Cells *102*
  Ethical Issues Concerning Human ES Cells *102*
Pluripotent Stem Cells from Postnatal Organisms *103*
Applications of Pluripotent Stem Cells *104*
Further Reading *105*

7 Body Plan Formation *107*
Embryological Concepts *107*
  Developmental Commitment *107*
  Embryonic Induction *109*
  Symmetry Breaking *110*
Key Molecules Controlling Development *111*
  Genes Encoding Developmental Commitment *111*
  Inducing Factors *112*
    Wnt System *112*
    FGF System *113*
    Nodals and BMPs *114*
    Notch System *114*
    Hedgehog System *115*
    Growth Promoting Pathways *115*
    Retinoic Acid *115*
Body Plan Formation *116*
  General Body Plan *116*
  Gastrulation *116*
  Embryo Folding *120*
Further Reading *123*

8 Organogenesis *125*
Nervous System *125*
  The Brain *126*
  Regional Specification of the CNS *128*
    Rostrocaudal *128*

Mediolateral *130*
Dorsoventral *131*
The Eye *131*
The Neural Crest *132*
Epidermis *134*
Hair Follicles *135*
Mammary Glands *136*
Somitogenesis *137*
The Somite Oscillator and Gradient *138*
Subdivision of the Somites *139*
Myogenesis *140*
The Kidney *140*
Blood and Blood Vessels *142*
Blood *142*
Blood Vessels *143*
The Heart *145*
The Gut *146*
Regional Specification of the Endoderm *148*
The Intestine *149*
The Pancreas *150*
The Liver *150*
Further Reading *151*

9 Cell Differentiation and Growth *155*
Organs, Tissues and Cell Types *155*
Epithelia *156*
Connective Tissues *156*
Cell Differentiation *158*
Regulation of Gene Activity *158*
Lateral Inhibition *161*
Asymmetrical Cell Division *162*
Neurogenesis and Gliogenesis *164*
Neurons and Glia *164*
Neurogenesis *166*
Gliogenesis *168*
Postnatal Cell Division *169*
Adult Neurogenesis *169*
Neurospheres *171*
Skeletal and Cardiac Muscle *171*
Skeletal Muscle *171*
Development of Skeletal Muscle *172*
Muscle Satellite Cells *173*
Cardiac Muscle *175*
Endodermal Tissues *176*
Cell Differentiation in the Pancreas *176*
Cell Differentiation in the Intestine *178*
Cell Differentiation in the Liver *179*
Hepatocytes and Cholangiocytes *180*
Liver Growth and Regeneration *181*

Transdifferentiation and Direct Reprogramming of Cell Type   *183*
Differentiation Protocols for Pluripotent Stem Cells   *184*
Further Reading   *185*

10 Stem Cells in the Body   *189*
The Intestinal Epithelium   *189*
   Intestinal Stem Cells   *191*
      In Vitro Culture   *193*
   Clonality of Intestinal Crypts   *193*
The Epidermis   *195*
   Hair Follicles   *197*
   Cornea and Limbus   *199*
   Mammary Glands   *200*
      Mammary Stem Cells   *203*
The Hematopoietic System   *204*
   Analysis by Transplantation and in Vitro Culture   *204*
   Hematopoiesis in the Steady State   *207*
      The Hematopoietic Niche   *209*
Spermatogenesis   *211*
Further Reading   *213*

11 Regeneration, Wound Healing and Cancer   *217*
Planarian Regeneration   *217*
   Neoblasts   *218*
Amphibian Limb Regeneration   *220*
   The Regeneration Blastema   *220*
   Pattern Formation in Regeneration   *222*
Mesenchymal Stem Cells   *223*
Mammalian Wound Healing   *225*
   Soft Tissue Wounds   *225*
   Healing of Bone Fractures   *225*
   Spinal Cord Injuries   *227*
Regeneration and Repair   *228*
Cancer   *228*
   Genetic Heterogeneity of Cancer   *230*
Cancer Stem Cells   *233*
Further Reading   *236*

Index   *239*

Preface

This book originates from a widespread perception that many students studying stem cell biology, and even many junior workers in stem cell research labs, lack essential knowledge of the scientific underpinnings of the subject. This can lead to undesirable consequences, most notably the tendency for clinics to offer to patients "miracle cells" whose injection can cure all ills rather in the manner of a medieval elixir. Such unrealistic attitudes are also, unfortunately, highly prevalent among the general public. Excellent educational work is done by bodies such as the International Society for Stem Cell Research, which has an accurate patient website open to all. However, a correct perception among the public about the capabilities of stem cell therapy cannot be expected until the practitioners themselves have a clear idea of what sort of cells they are working with and what these cells can, and cannot, be expected to do. This book seeks to improve the situation by exploring the scientific basis of stem cell biology in a concise and accessible manner. It is designed to be suitable for all students studying stem cell biology at undergraduate or graduate school level.

The book deals with basic science and so does not cover the current clinical applications of stem cells. I considered that to include clinical material would make the book too long, lose focus, and cause it rapidly to become out of date. However, because of the inevitable demand for such information the book has an online supplement which summarizes briefly the state of play in each

clinical application of stem cells to date, and this may be found at www.wiley.com/go/slack/thescienceofstemcells.

The "parent science" of stem cell biology is, to a large extent, developmental biology. Embryonic stem cells were discovered by developmental biologists, and the methods used for controlling their differentiation rely heavily on our knowledge of the normal mechanisms of regional specification and cell differentiation in the embryo. So developmental biology is necessarily an important part of this book. However, this is by no means just another textbook of developmental biology: it is primarily a book about stem cells and the concepts, technology, and experimental facts needed to understand them properly.

Because the focus is on the discussion of approaches and concepts the level of molecular detail has been kept fairly basic. Modern molecular life sciences are all very fact-heavy and comprehension can be obscured by too many facts. However, where readers require more detail on specific topics than is given here, it can always be acquired by reading recent review articles or key primary papers, a selection of which are cited at the end of each chapter. Although I have been sparing with molecular detail, I have listed where possible the major gene products that are indicative of the particular cell types which are of interest to stem cell biologists, and also some of the key physiological properties of these cell types. Familiarity with these criteria for cell identification is very important when assessing the results of experiments involving the directed differentiation of pluripotent stem cells or the direct reprogramming of one cell type to another.

A further type of detail that is worth paying careful attention to is the difference between species. Most of the experimental work in this area has been done on the mouse, but stem cell biology is inevitably oriented toward developing eventual applications for human patients. I have been careful where possible to distinguish normal events in human and mouse, and also occasional results from other model organisms such as the zebrafish, so that students are not misled by an exclusive focus on the mouse. In keeping with normal convention the names of genes are italicized when they refer specifically to the gene or the RNA, and are in normal type when they refer to the protein.

The central message of the book is that there is nothing magic about stem cells. In fact, it turns out that stem cell behavior is more important than the stem cells themselves. Certain cell populations in the body may adopt a stem cell type of behavior under particular circumstances, depending on their developmental history and their environment. So being able to work with stem cells successfully means being aware of how cells behave in different contexts and understanding how to characterize and manipulate them properly.

In the long term, stem cell biology does have huge potential for generating novel therapies for many common and recalcitrant diseases, and this potential will be realized most easily when all students and practitioners can become real masters of the science of stem cells.

Jonathan M. W. Slack
Bath 2017

About the Companion Website

This book is accompanied by a companion website:

www.wiley.com/go/slack/thescienceofstemcells

which summarizes the current clinical applications of stem cells.

1

What is a Stem Cell?

In the popular media and even in some medical circles, stem cells are presented as miracle cells that can do anything. When administered to a patient with some serious disease they will rebuild the damaged tissues and make the patient young again. Alas, in reality there are no such cells. However, there are cells that exhibit stem cell behavior and the future of regenerative medicine will undoubtedly be built on a good scientific understanding of their properties. In this chapter these properties are briefly outlined, and in the remainder of the book each of them will be underpinned by an explanation of the relevant areas of science and technology.

A list of characteristics of stem cell behavior that is generally agreed upon is the following:

- Stem cells reproduce themselves.
- Stem cells generate progeny destined to differentiate into functional cell types.
- Stem cells persist for a long time.
- Stem cell behavior is regulated by the immediate environment (the niche).

This is shown diagrammatically in Figure 1.1. The first two items on the list indicate the key abilities of self-renewal and of generation of differentiated progeny. As will be explained below, these abilities may be shown at a cell population level rather than by every single stem cell at every one of its divisions. Also, the second item indicates "destined to differentiate" meaning that cell division may continue for a while before differentiation, but not indefinitely. Cells derived from stem cells

that proliferate for a limited number of cycles are called progenitor cells or transit amplifying cells. The third item on the list means that if the stem cell population is one of those that exists in tissue culture then it should be capable of indefinite growth, while if it is part of an organism it should be very long lasting, normally persisting for the whole life of the organism. The fourth characteristic indicates that all stem cells exist in a specific microenvironment that controls their program of division and differentiation. This may seem at first sight only to apply to stem cells within the body and not to those grown in vitro, but in order to get them to grow, the cells in vitro are always provided with specialized medium ingredients that, in effect, mimic the components normally provided in the niche.

This fourfold definition involves not just intrinsic properties of stem cells, but also properties that depend on aspects of their environment such as the lifespan of the animal, the nature of the niche, or the composition of the culture medium. This emphasizes the fact that the goal of stem cell biology is understanding the *behavior* and not just the intrinsic nature of stem cells. To achieve this, the characteristics of the stem cell environment are just as important as the properties of the stem cells themselves. Moreover, understanding stem cell behavior means understanding various aspects of cell and developmental biology which are not always familiar to workers in stem cell laboratories.

The above definition is of value in indicating the special characteristics of stem cell

The Science of Stem Cells, First Edition. Jonathan M. W. Slack.
© 2018 John Wiley & Sons, Inc. Published 2018 by John Wiley & Sons, Inc.
Companion website: www.wiley.com/go/slack/thescienceofstemcells

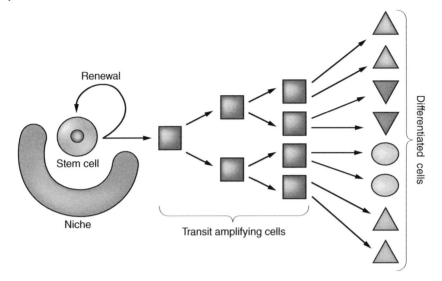

Figure 1.1 A consensus diagram showing stem cell behavior. (Modified from Slack, J.M.W. (2013). Essential Developmental Biology, 3rd edn. Reproduced with the permission of John Wiley and Sons.)

behavior, but is also helpful in indicating what is not stem cell behavior. For example, most of the cells in the body that are dividing are not stem cells. In particular cells in the embryo that differentiate after a certain period of time, such as the earliest cells formed by division of the fertilized egg, are not stem cells. Nor are differentiated cells that divide during postnatal life to generate more of themselves, such as hepatocytes or tissue-resident macrophages. A common term found in the literature is "stem/progenitor cell". This is a singularly unhelpful designation as it conflates two entirely different cell behaviors. Progenitor cells are precisely those that differentiate into functional cell types after a finite period of multiplication. They include the transit amplifying cells that arise from stem cells (Figure 1.1) and also cells of the embryo and of the growing individual that are destined to differentiate after a certain time.

Real stem cells comprise two fundamentally different types: pluripotent stem cells that exist only in vitro, and tissue-specific stem cells that exist in vivo in the postnatal organism. Pluripotent stem cells comprise embryonic stem cells (ESC) and induced pluripotent stem cells (iPSC). There are various subdivisions that will be considered later, but the essential features of these cells are first that they can be propagated without limit in vitro, and second that, under appropriate culture conditions, they are able to give rise to a wide variety of cell types, perhaps all the cell types in the normal organism except for the trophectoderm of the placenta. By contrast, tissue-specific stem cells exist within the body and generate progeny to repopulate the tissue in question. Well-studied tissue-specific stem cells include those of the hematopoietic (blood-forming) system, the epidermis, the intestinal epithelium and the spermatogonia of the testis. Under normal circumstances, tissue-specific stem cells do not produce cells characteristic of other tissue types. There are also some well-characterized stem cells that do not undergo continuous division, but seem to be kept in reserve to deal with tissue regeneration when required. A good example is the muscle satellite cells, which are normally quiescent but are able to be mobilized to divide and fuse to form new myofibers following injury. This type of stem cell behavior is sometimes called facultative.

Many criteria for identifying stem cells have been proposed and used. These are briefly listed here and the concepts and technologies will be developed in later chapters of the book.

Stem Cell Markers

Very often a cell is said to be a stem cell because it expresses one or more gene products associated with stem cells. However, there is no molecular marker that identifies all stem cells and excludes all non-stem cells. Those components required for general cell metabolism and cell division are certainly found in all stem cells, but they are also found in many other cell populations as well.

Pluripotent stem cells (ESC and iPSC) express an important network of transcription factors which are necessary for maintenance of the pluripotent state (see Chapter 6). Transcription factors are the class of proteins that control the expression of specific genes. A key member of the pluripotency group is the POU-domain transcription factor OCT4 (also known as OCT3 and POU5F1). The presence of OCT4 is certainly necessary for the properties of pluripotent stem cells. However it is not expressed in any type of tissue-specific stem cells except at a low level in spermatogonia.

A component that might be expected to be found in all stem cells is the telomerase complex. At the end of each chromosome is a structure called the telomere, made up in vertebrate animals of many repeats of the simple sequence TTAGGG. Because of the nature of DNA replication, the double helix cannot be copied right up to the end, so a part of the telomere is lost in each cell cycle. After enough cycles, the erosion of chromosome ends activates the system which senses DNA double-stranded breaks and causes death of the cell. This process is an important reason for the limited survival time of most primary tissue culture cell lines, which undergo senescence after a certain number of population doublings in vitro. Obviously

there must be a mechanism for repairing telomeres in vivo, and this is provided by the telomerase complex, of which the most important components are an RNA-dependent DNA polymerase called TERT, and an RNA called TERC which contains the template CCCTAA for the telomere sequence. High levels of telomerase are found in germ cells, ensuring the survival of full length chromosomes for the next generation. Telomerase is also upregulated in permanent ("transformed") tissue culture cell lines and in most cancers. However most types of somatic cell have little or no telomerase. Tissue-specific stem cells do contain some telomerase; generally enough to maintain cell division for a normal lifetime, but not enough to fully reverse the erosion of the telomeres. In situations such as repeated transplantation of hematopoietic stem cells from one mouse to another, there is an upper limit to the number of possible transplants and this is determined at least partly by telomere erosion. The presence of telomerase can be considered to be a stem cell marker, although it is also found in permanent tissue culture lines, early embryos and most cancers.

In human or animal tissues, various markers have been advanced as characteristic of all stem cells. For example the cell surface glycoprotein CD34 is found on human hematopoietic stem cells (HSCs) and can be used to enrich them from bone marrow by fluorescence-activated cell sorting (FACS). However it is also found on other cell types, such as capillary endothelial cells, and it is unclear whether it is actually necessary for the stem cell behavior of the HSC. In fact, since it is not found on mouse HSC, which are generally similar in behavior to human HSC, it is probable that it is not necessary. CD34 is not found on human embryonic stem cells or on most epithelial stem cell types, indicating that it is not a generic stem cell marker. A molecular marker which is known to be required for stem cell function is LGR5. This is an accessory receptor for the Wnt family of signaling molecules (see Chapter 7) and is found on stem cells in the intestine, hair follicle, mammary gland and stomach. These types of stem cell all depend on Wnt signaling from their environment for continued cell division, so the presence of the LGR5 is really necessary. However it is not found on other types of stem cell, so is also not a universal marker.

An interesting type of marker is that offered by dye exclusion, in particular exclusion of the Hoechst 33342 dye. This is a bisbenzimide dye, excited by UV light to emit a blue fluorescence. It is widely used as a DNA-binding reagent, but it is also actively pumped out of some cell types. If a subgroup of cells has lost more dye than the rest of the population, then it appears in flow cytometry as a cluster of cells showing less blue fluorescence than average. This is called a side population. The side population is enriched for stem cells in some situations, especially in murine bone marrow where it provides a similar degree of enrichment of hematopoietic stem cells to FACS using a panel of cell surface markers (Figure 1.2). The dye exclusion property is due to the activity of cell membrane transporter molecules including the P-glycoprotein (MDR1) and transporters of the ABC class. Dye exclusion is indicative of an increased capacity for export of all

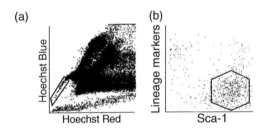

(a) Hoechst Blue / Hoechst Red
(b) Lineage markers / Sca-1

Figure 1.2 Flow cytometry plots showing a side population of cells active in Hoechst dye exclusion. (a) Whole mouse bone marrow, the boxed region is the side population. (b) Side population cells refractionated with regard to differentiated lineage markers, absent from stem cells, and Sca-1, a cell surface marker present on stem cells. (From: Goodell, M.A., Brose, K., Paradis, G., Conner, A.S. and Mulligan, R.C. (1996) Isolation and functional properties of murine hematopoietic stem cells that are replicating in vivo. *Journal of Experimental Medicine* 183, 1797–1806. Reproduced with the permission of The Rockefeller University.)

hydrophobic small molecules, many of which are toxic to cells. Although useful to the investigator, it is unlikely that this capacity is really important for stem cell behavior. For example, mouse embryonic stem cells show dye exclusion while human embryonic stem cells do not.

In summary, there is no single gene product which is found in all stem cells and not in any non-stem cell. Many so-called stem cell markers are probably not necessary for stem cell behavior. Of those gene products which are necessary for stem cell behavior, some, such as the cell division machinery and telomerase, are found in stem cells and in some non-stem cells. Others, such as OCT4 or LGR5, are found in some, but not all, types of stem cell.

Label-Retention

When a cell population is exposed to a DNA precursor, such as the nucleoside bromodeoxyuridine (BrdU), which is metabolized by cells in the same way as thymidine, all cells undergoing DNA synthesis will incorporate it into their DNA and so become labeled. The BrdU in the cell nuclei can be detected by immunostaining. After the BrdU supply is withdrawn, so long as cell division is continuing, then the level of BrdU in the DNA will halve with every subsequent S phase and become undetectable to immunostaining after about six divisions. If a cell divides slower than average, it will retain detectable BrdU for longer. This label-retaining behavior is often considered to be a characteristic of stem cells. In Figure 1.C.1 is shown an image of a hematopoietic stem cell (HSC) visualized with an antibody to the cell surface marker CD150. It retains a DNA precursor (EdU) label from a pulse given 30 days previously. Likewise, muscle satellite cells, that enable muscle regeneration following damage, are usually in a quiescent state. This relatively quiescent behavior is considered necessary to maintain regenerative function of some types of stem cell over a lifetime. If the mechanisms of quiescence are disturbed in mice by knocking out key components, then hematopoietic stem cells or muscle satellite cell populations have been shown to become exhausted during the lifetime of the animal, because they are dividing too much. Relative quiescence also serves to protect the stem cells against the oxidative damage which results from continuous growth with its associated oxidative metabolism.

Slow division is the cause of label retention in stem cells, but it must be remembered that not all label retention is due to stem cell behavior. In particular differentiation to a completely non-dividing (post-mitotic) cell type leads to permanent label retention. This property has been used especially to establish the differentiation time of neurons in embryonic development, and the final mitosis is often referred to as the cell birthday. Moreover, label retention is by no means universal among stem cells. For example, it is not shown by intestinal or epidermal stem cells. It is also, of course, not shown by the pluripotent stem cells (ES or iPS cells) which undergo rapid division in culture.

The Niche

The concept of a stem cell niche arose in the 1970s to explain the fact that the spleen colony-forming cells from the bone marrow had a lesser differentiation potency than hematopoietic stem cells in vivo (see Chapter 10). The idea is that stem cells require continuous exposure to signals from surrounding cells in order to maintain their stem cell behavior. This was first proved experimentally using the fruit fly *Drosophila*. In the *Drosophila* ovary there are female germ cells which lie in contact with somatic cells called cap cells. These secrete a TGFβ-like molecule called Decapentaplegic (Dpp). Dpp maintains the stem cells in mitosis. But as they divide, some of the stem cell progeny become displaced from contact with the cap cells, and are then exposed to less of the Dpp. This fall in Dpp lifts a repression on the oocyte maturation process and enables

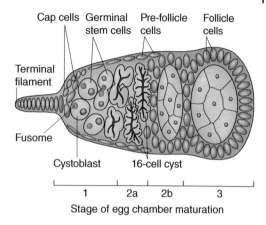

Figure 1.3 The stem cell niche in the Drosophila ovary. Female germ cell stem cells require continued contact with cap cells to remain stem cells. Once they lose contact with cap cells they differentiate into a cyst of one oocyte and 15 nurse cells. (Slack, J.M.W. (2009) Essential Developmental Biology, 2nd edn. Reproduced with the permission of John Wiley and Sons.)

the daughter cell to differentiate to a cystoblast. This then undergoes a fixed differentiation program, dividing four times to generate a post-mitotic complex of one oocyte and 15 supporting nurse cells. This situation illustrates the behavior of a niche very nicely. The stem cells continue to divide so long as they are in contact with the niche, and they differentiate when they are no longer in contact. If a stem cell is removed experimentally, its position may be taken by a progeny cell which would normally have differentiated, but because of its renewed occupancy of the niche it remains a dividing stem cell.

Probably all the stem cells types in the mammalian body exist within specific niches like this which control their behavior. For example the intestinal stem cells lie adjacent to Paneth cells which supply WNT, and spermatogonial stem cells lie adjacent to Sertoli cells that supply them with glial derived neurotrophic factor (GDNF). In both cases the signaling molecules are needed to maintain the stem cells in mitosis, and removal from the niche brings an end to cell division unless the factors are provided experimentally. In the bone marrow, there has been controversy about the exact nature of the niche, but hematopoietic stem cells are often found adjacent to blood vessels, as shown in Figure 1.C.1.

Asymmetric Division and Differentiated Progeny

It is often thought that all stem cells must undergo asymmetric divisions, with one daughter being a stem cell and the other destined to differentiate. This does sometimes occur, but it is also possible for stem cells to have a less rigid program of cell division with some divisions producing two stem cells, some two progenitor cells, and some producing one of each. Statistically a steady state requires that the stem cell number remains constant, although there may be occasions where it needs expanding, such as during normal growth of the organism or following injury. In the intestine for example it has been shown by cell labeling and by direct visualization that symmetric divisions predominate (see Chapter 10).

By definition, stem cells must produce differentiated progeny, but how many differentiated cell types do they actually produce? The answer is very variable and depends on the tissue concerned. In the intestine, stem cells produce absorptive, goblet, tuft and Paneth cells, together with several types of enteroendocrine cells. In the bone marrow, the hematopoietic stem cells produce all the cell types of the blood and immune system. At the other end of the scale, the spermatogonia of the testis produce only sperm. Epidermal stem cells are often said to produce only keratinocytes, but they can also form a type of neuroendocrine cell called the Merkel cell, responsible for touch sensitivity. The examples of both the intestine and the epidermis indicate that neuroendocrine cells can arise from epithelial stem cells quite distinct from the central or peripheral nervous systems, but they are not indicative of a wider potency enabling other tissue types to be formed.

Around the year 2000 there was a rash of papers indicating that hematopoietic stem cells (HSCs) were able to repopulate many, or perhaps all, other tissue types in the body following transplantation. This phenomenon was known as "transdifferentiation" (a term more usefully reserved for changes of differentiation type between fully differentiated cells). However further investigation showed that the phenomenon could mostly be explained by donor cells lodging within other tissues but not actually differentiating into them, or by cell fusion with cells of other tissues. Unfortunately the idea of very wide plasticity of tissue-specific stem cells became established in many people's minds at this time and has helped promote the present worldwide industry of "stem cell therapy" much of which has no scientific rationale or real clinical effectiveness.

Clonogenicity and Transplantation

It is often supposed that stem cells are those which grow rapidly and can form large clones in vitro. This perception came from early studies on epidermal stem cells, where the proportion of such cells (holoclones) does correlate well with the estimated proportion of stem cells in the basal layer (Figure 1.4). Sometimes, "spheres", such as neurospheres or mammospheres, which contain both stem cells and their differentiated progeny, can be grown from tissue samples in suspension. However, stem cell behavior depends both on the cells and their environment, and it is well-known that cell behavior can be greatly changed by the environment of in vitro culture. For example, neurospheres can be grown from parts of the central nervous system in which there are no stem cells in vivo. As another well-known example, cells of the mammalian embryo epiblast, which rapidly develop into other cell types in vivo, can give rise to pluripotent embryonic stem cells in vitro, which continue to divide indefinitely in an appropriate medium.

Transplantation behavior also looms large in thinking about stem cells. The ability to

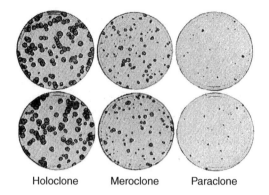

Holoclone   Meroclone   Paraclone

Figure 1.4 Clones of epidermal cells growing in culture. In this study the clones were classified as holoclones (large), meroclones (medium) and paraclones (small). The holoclones were considered to arise from stem cells. (From Barrandon, Y. and Green, H. (1987) Three clonal types of keratinocyte with different capacities for multiplication. Proceedings of the National Academy of Sciences of the United States of America 84, 2302–2306. Reproduced with the permission of Proceedings of the National Academy of Sciences of the United States of America.)

rescue irradiated animals with bone marrow transplants was the original discovery that, decades later, led to the identification of hematopoietic stem cells. It is generally felt that a hematopoietic stem cell is defined by the ability to repopulate the entire blood and immune system of an irradiated host. This is certainly an important property, although a single-minded focus upon it has tended to obscure the important distinction between cell behavior in an extreme regeneration situation and that in normal homeostasis. Some authors even suggest that a cell is not a bona fide stem cell unless it, as a single cell, can repopulate an entire tissue following transplantation. While this has been done a few times and may be a theoretical possibility for all stem cell systems, there are always practical limits to transplantation. All adult vertebrate animals have highly sophisticated immune systems that, as a by-product to their role in defending against infection, cause the rejection of cell and tissue grafts from other individuals. This is a very complex subject, but in general grafting between

adults is only possible between genetically identical individuals (e.g. identical twins or inbred mouse strains), or following immunosuppression with drugs, or by using highly immunodeficient strains of animal as hosts.

A type of stem cell defined almost entirely by transplantation is the so-called cancer stem cell. These are subsets of cells from tumors, isolated using various stem cell markers, which will generate tumors in immunodeficient hosts following grafting, under conditions where the majority of cells from the same tumor do not. Cancer stem cells are discussed in Chapter 11.

In Vivo Lineage Labeling

This is the most reliable method for establishing the existence of stem cell behavior in vivo because it can provide direct visualization. So far it has only been widely used in mice, but the wide availability of CRISPR-Cas9 technology will soon make it available for other organisms as well. The principle is to use a DNA recombinase enzyme (Cre) to impart a permanent genetic label to a cell in vivo that expresses a particular gene, or, more precisely, has a particular promoter highly active. The label is subsequently heritable on cell division and is unaffected by any differentiation events occurring in the progeny cells. A modification of the Cre recombinase to make it activatable by estrogen-like hormones (CreER) has been widely used in mice and enables the labeling to be initiated at a specific time. This method is described in Chapter 3. Once it has been labeled a stem cell will produce a sector of labeled tissue in which all its dividing and differentiated progeny carry the label. The labeled sector will grow initially as cells divide and mature and will eventually reach a steady state in which addition of new labeled cells is balanced by the removal of dead ones. This pattern should then remain unchanged in the long term. An example is shown in Figure 1.5, showing intestinal stem cells labeled using the *Lgr5* promoter. These cells reside in the

(a)   (b)   (c)

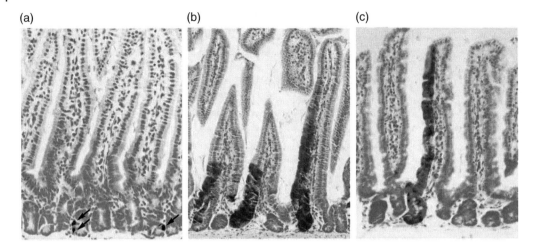

Figure 1.5 Descendants of stem cells in the mouse intestine visualized by the CreER method. The stem cells express a protein, LGR5 whose promoter is used for labeling. (a) The mice were labeled 1 day previously, (b) 5 days previously and in (c) 60 days previously. The initial label is in the LGR5 positive cells themselves (arrows); subsequently, ribbons of descendant cells up the crypts and villi become labeled. (Originally from: Barker, N. et al. (2007) Identification of stem cells in small intestine and colon by marker gene Lgr5. Nature 449, 1003–1007. Reproduced with the permission of Nature Publishing Group.)

intestinal crypts and generate a file of cells up the crypt and onto the neighboring villus. Near the villus tip the cells die and are then lost into the intestinal lumen. Some other examples of stem cell labeling are shown in Figures 10.C.1–10.C.6.

Because it can provide visualization of individual stem cell domains, cell lineage labeling has provided the data for an influential model of stem cell behavior which may be called stochastic. Here the idea is that the stem cells have a certain chance of dividing to form two stem cells, two transit amplifying cells or one of each (Figure 1.6). If the outcome is 50% of new stem cells and 50% of transit amplifying cells, then this gives quantitatively the same outcome as a situation of obligatory asymmetric division in which every division yields one stem cell daughter and one transit amplifying daughter. However it gives different predictions about the behavior of labeled stem cell clones. In the situation of the obligatory asymmetric division, labeled stem cell clones will each comprise one stem cell plus all their descendants. When the steady state has been reached, the labeled clones should persist for life and stay the same size. However in the stochastic model, clones may be lost if their stem cell divides to form two transit amplifying cells. They may also increase in size if the stem cell divides to form two stem cells. This situation has been modeled mathematically and it predicts that the number of labeled clones should steadily decline with time while the size of clones should become progressively more disparate, with the average size increasing. This means that the proportion of the tissue occupied by the labeled clones remains constant, but the number of labeled clones becomes progressively fewer and their size more varied. In fact this behavior is precisely what is observed when lineage labeling data are analyzed quantitatively, at least for the epidermis, spermatogonia and intestinal epithelium. In particular there is a property called "scaling behavior" which means that the frequency distribution of labeled clone sizes, divided by the average clone size, stays the same over time. Under such circumstances, which may turn out to be the norm for mammalian stem cell systems, the key stem cell properties of self-renewal, persistence and differentiation are still maintained, but they exist at a cell population level rather than as the properties of a single cell.

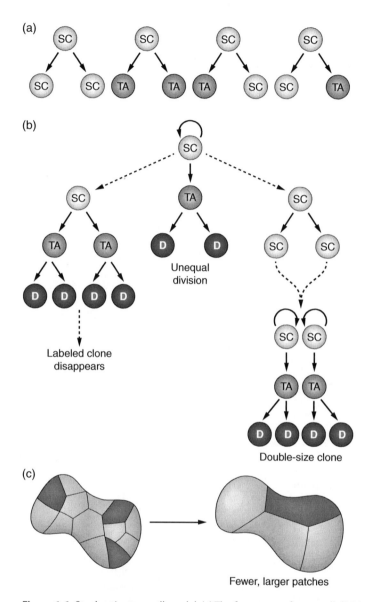

Figure 1.6 Stochastic stem cell model. (a) The four types of stem cell division. (b) Disappearance of labeled clone, and doubling of size of labeled clone. (c) Tendency of labeled clones to become fewer but larger with time. SC: stem cell, TA: transit-amplifying cell, D: differentiated cell.

Conclusions

The above brief discussion indicates that stem cells carry no universal molecular marker, that they are not all quiescent, that they are not necessarily the same as transplantable cells, and even that they may not be definable at the individual cell level at all but only at a population level.

Of course, what is considered to be a stem cell or not to be a stem cell all depends on the definition employed. The definition given at the start of this chapter comprises four properties:

- Stem cells reproduce themselves.
- Stem cells generate progeny destined to differentiate into functional cell types.

- Stem cells persist for a long time.
- Stem cell behavior is regulated by the immediate environment (the niche).

Together, these properties do make up a consensus view of stem cells acceptable to most scientists today. However, some will disagree with one or another element of the set. It is always possible to change the definition and thereby change what counts as a stem cell and what does not. For example, there are some cells at the periphery of the developing mammalian kidney that are sometimes called stem cells because they reproduce themselves and generate new kidney tubules. But these cells only persist for a few cell cycles in late gestation so do not satisfy the third criterion given here. They could be counted as stem cells only if this property were abandoned. As another example, some will insist that a stem cell must be able to give rise to more than one type of differentiated progeny, in which case the spermatogonia have to be removed from the list of tissue-specific stem cells since they produce only one type of differentiated cell: the sperm.

Perhaps because of the high profile of stem cell research generally, some attention has been given to the stem cell concept by philosophers of science. In a recent book, Laplane considers the various proposed attributes of stem cells and classifies these as categorical, dispositional, relational and system-based. A categorical property is essential and intrinsic, for example the presence of OCT4 in pluripotent stem cells. A dispositional property is a property revealed under particular conditions. For example intestinal stem cells need a Wnt signal from their niche in order to proliferate, so this property depends not just on the nature of the stem cell, but also on something else. A relational property is said to exist if it depends entirely on something else: for instance a hypothetical niche which would make any cell whatever that occupied it into a stem cell. Finally a systemic property belongs, in this context, not to individual cells but to the whole system, for example a tumor in which cells might acquire or lose stem cell behavior in specific circumstances. Laplane concludes that stem cells do comprise a "natural kind" (i.e. a real thing, out there, not just a figment of our imagination), but that they require a complex definition with a general part "stem cells are the cells from which tissues are developed and maintained", and a set of specific parts which in effect list the various properties discussed in this chapter. Apparently philosophers will accept that a natural kind may properly be defined by a set of attributes which are both categorical and dispositional, and of which not all need apply in every instance.

Debates such as these may cause momentary panic: how can such an important part of modern life as the stem cell become so intangible when examined critically? The contribution of philosophers may be considered negative if it just makes familiar entities disappear. However the philosophical view is worthwhile if it makes us examine the concepts critically. In the case of stem cell biology what emerges from a critical evaluation is that we should think not about stem cells as such but about *stem-type behaviors* that may be shown by various cell populations in specific circumstances. Defining stem cells is slippery and difficult, but defining stem cell behavior is relatively easy and stem cell behavior is real and important. In order to manipulate it in practical situations we need to understand the complete context, and for this an approach based on the underpinning sciences, such as cell and developmental biology, is really necessary.

Further Reading

Challen, G.A. and Little, M.H. (2006) A side order of stem cells: the SP phenotype. Stem Cells 24, 3–12.

Clevers, H., Loh, K.M. and Nusse, R. (2014) An integral program for tissue renewal and regeneration: Wnt signaling and stem cell control. Science 346.

Flores, I., Benetti, R. and Blasco, M.A. (2006) Telomerase regulation and stem cell behaviour. Current Opinion in Cell Biology 18, 254–260.

Fuchs, E. and Horsley, V. (2011) Ferreting out stem cells from their niches. Nature Cell Biology 13, 513–518.

Hsu, Y.-C. (2015) Theory and practice of lineage tracing. Stem Cells 33, 3197–3204.

Klein, A.M. and, Simons, B.D. (2011) Universal patterns of stem cell fate in cycling adult tissues. Development 138, 3103–3111.

Knoblich, J.A. (2008) Mechanisms of asymmetric stem cell division. Cell 132, 583–597.

Lander, A.D. (2009) The "stem cell" concept: is it holding us back? Journal of Biology 8, 70.

Lander, A.D., Kimble, J., Clevers, H., Fuchs, E., et al. (2012) What does the concept of the stem cell niche really mean today? BMC Biology 10, 19.

Laplane, L. (2016) Cancer Stem Cells. Philosophy and Therapies. Harvard University Press, Cambridge, Mass.

Magavi, S.S.P. and Macklis, J.D. (2002) Identification of newborn cells by BrdU labeling and immunocytochemistry in vivo. In: T. Zigova, P.R.S. and J.R. Sanchez-Ramos (eds). Neural Stem Cells: Methods and Protocols. Humana Press Inc., Totowa, NJ. Methods in Molecular Biology, Vol. 198, pp. 283–290.

Morrison, S.J. and Spradling, A.C. (2008) Stem cells and niches: mechanisms that promote stem cell maintenance throughout life. Cell 132, 598–611.

Rumman, M., Dhawan, J. and Kassem, M. (2015) Concise review: quiescence in adult stem cells: biological significance and relevance to tissue regeneration. Stem Cells 33, 2903–2912.

Schofield, R. (1983) The stem cell system. Biomedicine and Pharmacotherapy 37, 375–380.

Sidney, L.E., Branch, M.J., Dunphy, S.E., Dua, H.S. and Hopkinson, A. (2014) Concise review: evidence for CD34 as a common marker for diverse progenitors. Stem Cells 32, 1380–1389.

Snippert, H.J., van der Flier, L.G., Sato, T., van Es, J.H., et al. (2010) Intestinal crypt homeostasis results from neutral competition between symmetrically dividing Lgr5 stem cells. Cell 143, 134–144.

Spradling, A., Drummond-Barbosa, D. and Kai, T. (2001) Stem cells find their niche. Nature 414, 98–104.

Wagers, A.J. and Weissman, I.L. (2004) Plasticity of adult stem cells. Cell 116, 639–648.

Yanger, K. and Stanger, B.Z. (2011) Facultative stem cells in liver and pancreas: Fact and fancy. Developmental Dynamics 240, 521–529.

2

Characterizing Cells

One of the fundamental aspects of stem cell biology is the ability to characterize and identify cells: not just the stem cells themselves but also the cells that make up their environment and the cells they turn into. This chapter introduces the methods for doing this. Cells in the body exist as numerous functional types. These are known as differentiated cells and examples would be hepatocytes in the liver, keratinocytes in the epidermis or neurons in the brain. There are also present in the body some immature and some undifferentiated cells. Some cells divide; others do not; others divide only occasionally, for example in case of injury to the tissue. A few of the dividing cell populations qualify as stem cells according to the definition of the previous chapter, but the majority are progenitor cells, also called transit amplifying cells in the adult tissue context.

To characterize cells we need to know both their gene expression pattern and their renewal characteristics. The methods for doing this have become quite sophisticated and are briefly outlined here. This chapter mostly deals with cells in the body, although many of the techniques can also be used with tissue culture cells. In Chapter 4 we shall consider cells in tissue culture more closely. For the moment it may just be noted that cells placed into culture often change their morphology, gene expression pattern and behavior compared with the situation in vivo.

Histological and Anatomical Methods

Histological Sections

A crude, but surprisingly effective, method of cell type identification is visual examination of cells after staining of tissue sections with suitable dyes (Figure 2.C.1). This is effectively the science of histology which arose in the late nineteenth century and has since formed the basis of the classification of tissue and cell types. It serves as the principal tool for pathologists diagnosing the type and severity of tumors and other tissue-based diseases. The components of this technology are: fixation, sectioning, staining and microscopy.

Fixation

Fixation is necessary to stabilize the tissue and render it robust enough to be able to be cut into sections a few microns thick. The most common fixatives are aldehydes which harden the specimen by chemical combination with amino groups in proteins and nucleic acids, causing the formation of chemical cross links between adjacent molecules. Formalin is a 40% w/v solution of the gas formaldehyde (HCHO) and is usually used in neutral buffered solution at 10% v/v (4% w/v of HCHO). Paraformaldehyde is a solid polymerized form of formaldehyde and in solution it depolymerizes back to formaldehyde so a 4% w/v solution of paraformaldehyde is very similar to 4%

The Science of Stem Cells, First Edition. Jonathan M. W. Slack.
© 2018 John Wiley & Sons, Inc. Published 2018 by John Wiley & Sons, Inc.
Companion website: www.wiley.com/go/slack/thescienceofstemcells

formaldehyde. Glutaraldehyde is OHC-(CH$_2$)$_3$-CHO and because it has two aldehyde groups it produces a denser and more durable form of cross-linking. It is often used for electron microscopy. Other types of fixative in common use are various acids, organic solvents and inorganic salts. Acids, such as picric, hydrochloric or trichloroacetic acid, and solvents, such as ethanol or acetone, work mostly by precipitation of macromolecules. This means that they retain less material than aldehyde fixation and they are used mostly for fixation of solid specimens that are not due for sectioning (wholemounts), where the ability of stains or antibodies to be able to penetrate to the interior is important. Inorganic salts, including mercuric chloride and potassium dichromate, are found in some fixatives. They work by a variety of methods including cross linkage of SH groups by mercuric chloride and oxidation of amines and other groups by potassium dichromate.

Sectioning

Once fixed, a specimen needs to be embedded in a suitable material for sectioning. Although various other materials have been introduced in recent decades, paraffin wax is still very widely used. This is because, unlike almost all other embedding agents, it will form a ribbon when sectioned. This means that each section adheres to the previous one and so a whole row of adjacent sections can easily be picked up and mounted on a microscope slide. It also means that it is easy to cut sections all the way through a small specimen so that all parts of it can be examined. Paraffin wax is only soluble in very apolar organic solvents such as xylene or the less toxic proprietary solvent Histoclear. So before a specimen can be infiltrated with wax it needs to have all traces of water removed. This is achieved by treating the specimen with a series of ethanol solutions, typically 50%, 70%, 90%, 95% and 100%. The gradual succession of alcohols avoids the specimen being damaged by the strong mixing forces generated when pure water and ethanol are mixed. The time in each solution depends on

the size of the specimen; large specimens needing longer times to reach equilibrium. 100% ethanol is miscible with xylene or Histoclear and so the specimen can then be equilibrated in one of these, and then in paraffin wax, which needs to be heated above the melting point to make it liquid (usually 56–58 °C). After a couple of changes of molten paraffin wax, and when the specimen is fully equilibrated, it is allowed to set. The paraffin is then trimmed round the specimen so that it is enclosed in a block. Such blocks are extremely stable and can be successfully sectioned and stained after storage for many decades. Nowadays the procedures of dehydration and embedding are usually carried out by automated tissue processing machines.

For sectioning the block is mounted on a machine called a microtome. This works by passing the block across a very sharp blade and advancing a few microns between each cut. The ribbon of sections hangs down from the blade and can be collected and placed on a glass slide. Paraffin wax sections are usually cut at 6–8 microns thickness, which is thin enough to give good transparency and provides good access to stains or other reagents. Depending on the type of specimen, it may need to be correctly oriented when mounted on the microtome. For example, a sample of skin would normally be cut transversely and viewed so that the outer surface is on top. A sample of intestine would also be cut transversely and viewed with the luminal surface on top. Before anything can be done with the sections they need to be dewaxed and rehydrated. This is achieved by passing the slides, with sections attached, through a reverse series of solvents: Histoclear, then 100%, 95%, 90% and 70% ethanol, and then water. Again, the graduated series of ethanols is used to avoid tissue damage by severe mixing forces.

Staining

Tissue sections are transparent and need to be stained before they can be viewed under the microscope (Figure 2.C.1). One of the most popular general stain combinations has long been hematoxylin and eosin. Hematoxylin is normally used by overstaining and then partially destaining in an acid solution. This stains cell nuclei blue/purple and leaves most other components unstained. Eosin is a general "plasma stain" that stains everything red/orange. The result is that the nuclei are blue/purple and the cytoplasm red/orange. Masson's trichrome is a complex mixture of stains that gives a multi-color result and is often used for connective tissues. Nuclei are dark brown, cytoplasm is pink, high concentrations of protein such as muscle fibers, are red, and collagen fibers are blue or green, depending on the precise recipe. Periodic acid–Schiff stain reacts with carbohydrates to give a bright pink color and is very useful for visualizing glycogen granules. Alcian blue combines with acidic polysaccharides and may be used at different pHs to stain different components: the lower the pH the more selective is the stain for highly acidic groups such as the sulfates found in the glycosaminoglycans of cartilage.

Once stained the sections need to be mounted and made permanent. The transparency of tissue sections depends on the difference of refractive index between the section and the surrounding medium. This is quite high in water (refractive index 1.33) but is reduced to almost nothing in proprietary mounting media such as DPX, which is a solution of distyrene and tricresyl phosphate in xylene and has a refractive index of 1.52. DPX is not miscible with water so the stained sections need once again to pass through the sequence of 70%, 90%, 95%, 100% ethanol and Histoclear before they can be covered in DPX. Then a coverslip is placed on top and the DPX soon sets solid to give a robust and permanent slide.

A few stains need to be applied before sectioning. Notable among these is the Golgi silver stain for neural tissues. This stains a small proportion of the cells a black/brown color and leaves the others unstained, hence showing up the neuronal processes of the stained cells very clearly. It is carried out by exposing the fixed specimen to potassium dichromate followed by silver nitrate and is visualized in relatively thick sections.

Paraffin sectioning has its limitations. Principal among these is the fact that the exposure to fixatives, organic solvents and heat destroys many of the antigens that might otherwise be visualized by immunostaining. If the antigen of interest is destroyed by paraffin embedding, then it is necessary to cut frozen sections instead. This is done by freezing the specimen and encasing it in a suitable embedding medium, usually "optimal cutting temperature" (OCT) compound, then cutting sections in a cryostat, which is a microtome enclosed in a refrigerated chamber cooled to about −20 °C. Frozen sections do not form ribbons and, especially if the specimen has not been fixed, they are of lower quality than paraffin sections. On the other hand, all the tissue components are preserved intact. This allows the use of a wide variety of antibodies for immunostaining. In addition, frozen sections need to be used to visualize lipid droplets which would be extracted by the solvents used in paraffin embedding. Lipids can be stained with Sudan black or oil red O. In order to avoid exposure to organic solvents, frozen sections are usually mounted in aqueous mounting media. These contain water soluble components to increase refractive index, such as glycerol or polyvinylpyrrolidone, but these substances have a lower refractive index than DPX and so give less good transparency.

Electron Microscopy

Although the limits of light microscopy have been pushed very successfully and it is now possible to examine living cells at subcellular resolution, the superior resolving power of electron microscopy is still necessary to visualize most cell organelles in any detail. In stem cell biology it can aid in cell type identification by visualizing components such as dense core granules (in beta cells), bile canaliculi (in hepatocytes) or gap junctions (in cardiomyocytes). There are two basic types of instrument: scanning and transmission

(a)   (b)

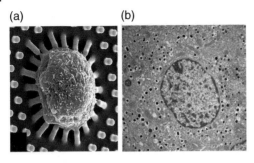

Figure 2.1 (a) Scanning electron micrograph of a "mesenchymal stem cell" cultured on a micropost array. (https://www.google.co.uk/search?q=scanning+electron+microscopy+stem+cells&hl=en&tbm=isch&source=lnms&sa=X&ved=0ahUKEwj_usrJivXMAhXJBsAKHQzpCkkQ_AUIBygB&biw=1366&bih=643&dpr=1#imgrc=gwMf4VCSv6py2M%3A). (b) Transmission electron micrograph of a β-cell from a mouse pancreas. (Author's picture.)

electron microscopes. Scanning electron microscopy gives a pseudo-three dimensional view of whole cells or cell components which have been fixed and "shadowed" with heavy metals. It provides very striking and beautiful images (Figure 2.1a). Transmission electron microscopy examines thin (50 nm) sections (Figure 2.1b). The procedures are somewhat different from those used in paraffin histology. Fixation is normally with glutaraldehyde, followed by osmium tetroxide (OsO_4) which combines with $C=C$ double bonds and so highlights membranes containing unsaturated lipids by introduction of the electron dense osmium atoms. The specimens are then dehydrated and embedded in an epoxy resin such as Araldite or Epon. These are transparent and hard and specimens may be stored indefinitely in the blocks. Sections are cut with an ultramicrotome which has a diamond knife and is capable of cutting very thin sections. For transmission EM these are usually 50 nm thick. They are mounted on a copper grid and placed in the electron microscope for examination. The appearance of an electron microscope section depends on differences of electron density and this depends on the binding pattern of the OsO_4 used for

staining. In addition some "thick" sections of 1 μm are normally cut for light microscopy in order to have an overview of the specimen and to locate areas of interest for viewing in the electron microscope. Very few conventional stains can penetrate epoxy resins, but one that can is Toluidine Blue, so this is often used to stain the survey sections. Toluidine Blue stains most components blue, but highly acidic glycosaminoglycans, such as those found in cartilage matrix, stain pink (metachromatic staining).

Fluorescence Microscopy

Many experiments depend on the ability to visualize specific fluorescence either from fluorescent proteins such as GFP, introduced to the organism under study as a gene, or the fluorescent antibodies used for immunostaining or in situ hybridization (see below). Fluorescence microscopes work by illuminating the specimen with a specific excitation wavelength, selected by a filter. This is chosen to maximize the fluorescence of the fluorophore in question. The light emitted by the fluorophore will be of longer wavelength (redder) than the excitation wavelength and will be viewed through an emission filter to reduce background. Fluorescence microscopy enables the visualization of several different fluorophores in the same specimen by using different excitation and emission filters. Three to four different colors can be visualized in a normal fluorescence microscope. If the microscope enables "spectral analysis", whereby measurements are taken at several wavelengths and a computational separation is made of the emission of different fluorophores, then even more colors can be distinguished. Better resolution can be obtained by use of a confocal microscope in which the specimen is scanned across a particular optical section using a laser. This helps to remove out of focus background signal and provides a sharper image. Even better results can be obtained, at greater expense, using two photon or multi photon microscopes.

Wholemounts

Wholemounts are three dimensional specimens that are rendered transparent for microscopic examination. They are very widely used by developmental biologists for immunostaining and in situ hybridization, as whole embryos at early stages are the right size for this application and it provides a good three dimensional view of the specimen. They are less used in stem cell biology although there have been some fruitful applications, for example to visualize hair follicles (Figure 2.C.2).

For the preparation of wholemounts to be viewed down the microscope, the specimen needs to be small: usually less than 1 mm in linear dimensions. Fixation needs to be quite light as otherwise the fixed specimen may not allow for penetration by antibodies or in situ hybridization probes or even conventional stains. A brief exposure to 4% formaldehyde is often used. Otherwise organic solvents are popular as they extract the lipid content during fixation and much of the protein and nucleic acid redissolves and becomes extracted when the specimen is rehydrated. However, this may mean that antigens of interest are lost. Because of the long time it takes for antibodies or probes to penetrate to the center of a wholemount, the staining procedures can take several days. If the final visualization method is compatible with organic solvents then at the end of the staining procedure the specimen can be mounted in DPX which will render it transparent. Otherwise aqueous media must be used, which often contain high concentrations of glycerol. 80% glycerol has a refractive index of 1.44 which gives less transparency than DPX but is often adequate for a small specimen.

Wholemounts are also used on a larger scale for skeletal staining with alcian blue and alizarin, a combination which renders cartilage blue and bone red. This can be applied to specimens a few cm in size such as late mouse embryos or small fish. It involves prolonged treatment with potassium hydroxide to render the specimen sufficiently transparent. The availability of transparent zebrafish now means that genetically labeled cell grafts can, in favorable circumstances, be visualized in whole fish without such drastic methods of tissue clearing.

Immunostaining

Immunostaining highlights the presence of a particular antigen, usually a protein or carbohydrate, on cells (Figure 2.C.3a–c). The particular part of the target molecule recognized by the antibody is known as the epitope. Immunostaining may be used on sections or whole mounts, or on fixed cells in a tissue culture situation. Fixation and embedding need to be compatible with preservation of the antigen of interest, and retaining sufficient permeability of the specimen to allow penetration by large antibody molecules.

Immunostaining procedures usually involve two steps. The first antibody is a specific antibody recognizing the antigen of interest and the second antibody recognizes and detects the first antibody. First it may be necessary to permeabilize the specimen by treatment with detergents or protease. Then it is "blocked" by exposure to a protein solution that will bind non-specific sites. After that it is exposed to the first antibody for a suitable time. It is washed thoroughly and then exposed to the second antibody. This is directed against the constant region of the first antibody, in other words the conserved region characteristic of the species and antibody class. The second antibody will also be modified to enable detection. This may be by fluorescence, in which case the second antibody will carry a chemical group that fluoresces at a particular wavelength (a fluorophore). It may be by histochemical detection, in which case the second antibody will carry a suitable enzyme such as alkaline phosphatase or horseradish peroxidase for which there are sensitive histochemical methods producing an insoluble and intensely colored precipitate. In either event the signal is visible at the site of antibody binding, which indicates the position of the original antigen. If detection is by histochemistry then the subsequent mounting procedure need not conserve the immune complex and if the colored product is insoluble in organic solvents (such as the brown product of peroxidase oxidation of diaminobenzidine) then the section or whole mount can be dehydrated and mounted in DPX, which gives good transparency. If the colored product is soluble in organic solvents, or if the visualization depends on preservation of the immune complex, as with a fluorescent second antibody, then mounting must be done in an aqueous medium. It is possible to increase the sensitivity of immunostaining by increasing the number of steps to amplify the final signal. For example, the first antibody can be labeled with biotin, and this is followed with avidin or streptavidin, proteins that bind ultra-tightly to biotin. The avidin will have many fluorescent groups or histochemically detectable enzyme molecules attached giving a correspondingly larger signal. However all such methods of amplification depend on increasing the signal/noise ratio as there is no point increasing the signal if the background is also increased so much as to obscure the signal.

Although labs do sometimes still prepare their own antibodies, most reagents for immunostaining are now purchased from lab supply companies. The most critical element for success is the quality of the primary antibody. Unfortunately some primary antibodies from commercial sources do not work well, so it is always worth testing a new antibody on a positive control specimen known to contain the appropriate target molecule. In addition the procedures of immunostaining are capable of generating numerous artifacts and it is important always to perform appropriate controls for nonspecific binding of primary or secondary antibody, and for endogenous signals such as fluorescent components or the presence of endogenous enzymes with similar activity to those used for detection (e.g. peroxidase, alkaline phosphatase).

In Situ Hybridization

In situ hybridization is a method of detecting the location of specific messenger RNA in a specimen (Figure 2.C.3d). It is very widely used in developmental biology, mostly on small whole mounts, although it can also be used on sections. The advantage over immunostaining is that a probe can be prepared against any mRNA, whereas an antibody may or may not exist or be able to be generated against a molecule of interest. On the other hand, in situ hybridization cannot detect molecules that are not the direct result of transcription, in particular carbohydrates whose structure depends on the activity of specific glycosyl transferase enzymes.

Normally an in situ probe is prepared by in vitro transcription of the cloned gene of interest. A modified nucleotide is incorporated into the synthesis which carries a readily detectable chemical modification such as digoxigenin (a plant steroid), biotin, or fluorescein. The probe is partly hydrolysed to give rise to oligonucleotides that can penetrate the specimen. Hybridization is conducted at a suitable temperature and followed by thorough washing. The chemical tag on the probe is detected by immunostaining with an antibody, or, in the case of biotin, with a suitably modified avidin or streptavidin. Depending on the detection method, analysis will be by conventional light microscopy for histochemical detection or fluorescence microscopy for fluorescent antibody detection.

Like immunostaining, in situ hybridization is vulnerable to numerous artifacts and it is important to conduct controls for specificity and sensitivity of the probe. Of course both techniques have a lower limit for detection and some antigens or mRNAs are just too low abundance to be detected.

A recent improvement to in situ hybridization technology is called RNA Scope. This offers to improve the signal to noise ratio substantially and to approach a detection limit of single molecules of mRNA in anatomical specimens. It can be used on fixed cells, wholemounts or sections from formalin-fixed, paraffin-embedded tissues. Here two probes are used which are complementary to adjacent target sequences in the mRNA. Each probe carries a spacer and a tail sequence of 14 nucleotides. Detection depends on a cascade of three further hybridization steps. The procedure involves considerable amplification and typically multiple primary probe pairs will be used providing an even greater signal. Using different colored fluorescent detection probes it is possible to examine several mRNAs in a single specimen.

Other Methods

RNAseq

The most definitive way of identifying a cell is to examine exactly which genes are being expressed and at what relative levels. This can now be done by RNAseq (RNA sequencing). The procedure starts with a sample of RNA from the cells of interest. If it is desired to examine just protein coding gene expression, this is fractionated using polyT-beads to isolate RNA with 3'polyA sequences, i.e. mRNA. This is reverse transcribed to cDNA, the second strand is synthesized, and adapters are added to the ends of all DNA molecules. If necessary, the sample is PCR amplified and then sequenced using a DNA sequencing machine. In the case of stem cell oriented investigations the genome (e.g. human, mouse) will be well-characterized and so software is available to align the sequence reads with the genome. This enables a catalog of expressed genes to be deduced. Unlike previous methods of gene expression analysis, such as microarrays, RNAseq has a very high dynamic range. Since transcripts are literally counted, an accurate measure is obtained of the relative levels of expression of all genes, which will range from many thousands of transcripts per cell down to very few. RNAseq is also useful for identifying different splice forms of mRNA and the expression of alternate alleles. If non-protein coding RNAs are of interest, they can be enriched at the beginning, for example by using size selection for small RNAs rather than polyA selection.

RNAseq can now be applied even to single cells. The cell is isolated by fluorescence activated cell sorting (FACS) or by laser dissection (for these methods, see below). The procedure is the same as for bulk samples except that extensive PCR amplification is essential. Because of the tiny amount of starting material in one cell there is a greater danger of bias during reverse transcription and PCR amplification. Nonetheless single cell sequencing can prove very useful for purposes such as deducing a sequence of differentiation states in a mixed population of cells. Of course, the need to isolate single cells means that positional information is lost.

Laser Capture Microdissection

Until recently it has been difficult to combine the high resolution of histology with the sophisticated analysis of gene expression offered by modern molecular techniques such as microarrays, genomic sequencing or RNAseq. However it is now possible through the method of laser capture microdissection (Figure 2.2). This enables individual cells or small groups of cells to be picked out from a section or a tissue culture plate, and transferred to a tube for molecular analysis. There are several designs of equipment, but they all have in common the fact that the region of interest can be cut out using a laser which is controlled by the investigator viewing the specimen down the microscope. Following this, some procedures involve flipping the excised tissue fragment into a tube, others attaching it to an adhesive surface, but whatever the method, the tissue is isolated and available for analysis. It does need to be borne in mind that the amount of material in a few cells is very small so nucleic acid

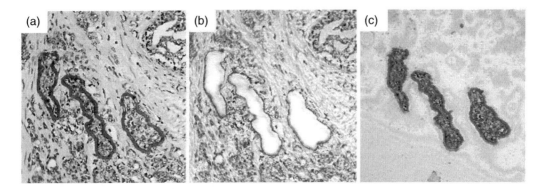

Figure 2.2 Laser capture microdissection: mouse prostate gland. (a) Laser outline of cells to be collected. (b) Remaining cells after laser capture. (c) Cells collected. (From Garnis, C., Buys, T.P.H. and Lam, W.L. (2004) Genetic alteration and gene expression modulation during cancer progression. Molecular Cancer 3.9.)

analysis is likely to require considerable amplification by PCR, and this may introduce errors and artifacts.

Flow Cytometry

Flow cytometry is a method of viewing cells one by one in suspension. It can be used for counting cells with different characteristics or, with somewhat more complex apparatus, it can be used for fractioning different types of living cell into separate tubes.

In flow cytometry a suspension of cells is passed one at a time through a flow cell illuminated by a laser. The simplest devices examine just the light scattering properties of cells. "Forward scatter" roughly corresponds to cell size, and "side scatter" roughly corresponds to cell granularity (Figure 2.3). However, usually instruments have more than one laser and several detectors to examine specific fluorescence from antibodies that have been attached to the cells by similar methods to those used in immunostaining. Propidium iodide is a DNA intercalating fluororescent dye often used to check cell viability, as it does not penetrate live cells. If the instrument is just an analytical one rather than a cell sorter, then the cells are discarded after counting.

Flow cytometry has always found most of its applications in hematology, as cells of the blood are already in suspension and are

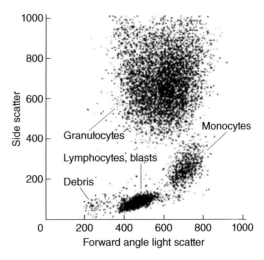

Figure 2.3 Flow cytometry of white blood cells, showing separate clusters of monocytes, granulocytes and lymphocytes. (From: Introduction to Flow Cytometry. AbCAM. http://docs.abcam.com/pdf/protocols/Introduction_to_flow_cytometry_May_10.pdf. Provided courtesy of AbCAM. Copyright©2016 AbCAM.)

robust enough to tolerate being squirted through a narrow orifice. A large number of antibodies have been identified which distinguish difficult classes of lymphocyte and other cells of the blood. Many of these belong to the so called "cluster of differentiation" (CD) antigen series, which are all cell surface molecules. The majority of CD antigens have specific immune functions, but they are not necessarily confined to the immune system and several are well-known by other names, for instance CD29 is integrin β1, CD71 is the transferrin receptor, CD143 is angiotensin converting enzyme, CD331 is FGF receptor 1. Flow cytometry can also successfully be applied to tissue culture cells or cells from intact tissues, provided that they can be dissociated to a single cell suspension by suitable enzyme treatment, and without losing the cell surface markers used for identification. Flow cytometry may be carried out either on live cells, stained with antibodies to cell surface components, or on live or fixed cells which express fluorescent reporter proteins, or on fixed cells, stained with antibodies to intracellular components. As with immunostaining, careful controls need to be carried out to allow for background signals, and for artifacts due to cross talk between fluorescence channels.

If the machine is a cell sorter, (fluorescence-activated cell sorter = FACS, Figure 2.4) then the signals from the detectors can be used to control where each cell is sent. The stream of cells is converted to a stream of tiny droplets, each containing one cell, and a small electric charge is imparted to each droplet by a drop charge device, depending on the settings programmed by the operator and on the readings from the detectors. This electric charge will determine which direction is taken when the cell droplet passes through the detector plates between which there is an electric field. Ideally the original mixed cell population can be fractionated into several tubes each containing a pure population based on the preset criteria. Successful cell sorting is an art that requires considerable experience, with careful attention to the nozzle size, flow characteristics, staining, gate settings and appropriate controls.

Dividing Cells

The Cell Cycle

Many cells in the adult body are quiescent. Some are permanently post-mitotic, such as neurons and multinucleate muscle fibers. Others are capable of division but do so only rarely. Some, including stem cells but also progenitor cells, divide regularly. The situation is very different in vitro because most tissue culture cells are in a situation of continuous exponential growth.

The normal mitotic cell cycle has four phases (Figure 2.5a). The first is G1 (gap phase 1) during which growth occurs. Then there is S (synthesis) phase during which the DNA is replicated. This means that each chromosome, which is a double-stranded DNA molecule, becomes copied into two identical chromosomes, called chromatids, each a double-stranded DNA molecule. Because of the semi-conservative nature of DNA replication, each chromatid has one old strand and one new strand in its double helix. After DNA synthesis there follows the G2 phase (gap phase 2), during which cells have double the normal DNA content and in which there is further growth, and finally the M (mitosis) phase. During mitosis the chromosomes condense and become visible. The centrosome divides, the nuclear membrane is lost, and a mitotic spindle is assembled from microtubules which radiate from the two halves of the centrosome. During metaphase the chromosomes line up in the center of the mitotic spindle. Kinetochore microtubules attach to the chromosomal centromeres and at anaphase the chromatid pairs are pulled apart so that one set heads for each daughter cell. The events of chromosome segregation are closely followed by the actual division of the cell (cytokinesis) which comprises the events required to separate one cell into two. This is usually symmetrical but may on occasion be unsymmetrical, especially for divisions of stem cells. Quiescent cells are said to be in a state called G0, which is not part of the cell division cycle.

An analysis of cells with respect to DNA content may be performed by flow cytometry after fixing/permeabilizing and staining with a DNA intercalating fluorochrome such as propidium iodide (Figure 2.5b).

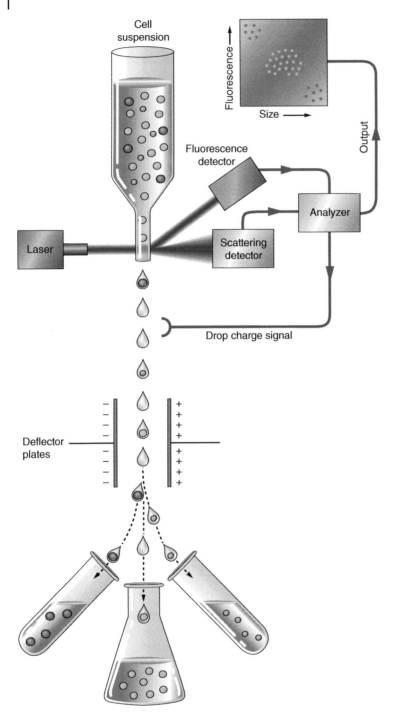

Figure 2.4 Fluorescence activated cell sorter (FACS). This shows a hypothetical separation of three cell types differing in size and also in fluorescence, with the smaller cells being more brightly fluorescent. The drop charge signal, and hence the destination tube, depends on the forward scatter (size) and the fluorescence measured by the detectors. (Slack, J.M.W. (2013). Essential Developmental Biology, 3rd edn. Reproduced with the permission of John Wiley and Sons.)

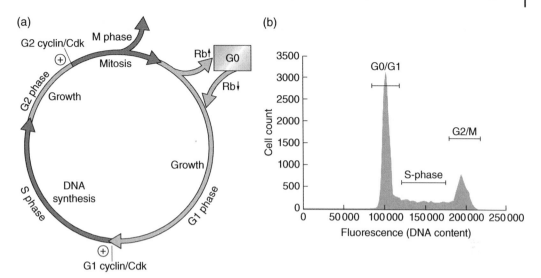

Figure 2.5 The cell cycle. (a) Diagram of the cycle. Entry to S phase and M phase is controlled by the checkpoints shown as ⊕. Cdk = cyclin-dependent kinase, Rb = retinoblastoma protein. (Modified from Slack, J.M.W. (2013). *Essential Developmental Biology*, 3rd edn. Reproduced with the permission of John Wiley and Sons.) (b) Flow cytometry of DNA content in a dividing cell population. Clear peaks are apparent for the unreplicated (G1) and replicated (G2) cells. The cells undergoing S phase have an intermediate DNA content. (http://uic.igc.gulbenkian.pt/fc-protocols.htm.)

The proportions of cells in G1 and G2 are given by the peaks of 1x and 2x DNA content, and the cells in S phase are represented by an intermediate DNA content indicative of partial replication. This method is more useful for tissue culture cells than for actual tissues because the cells need to be dissociated into a single cell suspension, and the value of the analysis is greatest when the cell population is homogenous rather than composed of mixed types which are likely to have different cycles.

Control of the cell cycle is based both on an intrinsic oscillatory mechanism and on external signals. The intrinsic mechanism is similar in all eukaryotic organisms, and ensures that the events of the cycle occur in the correct order and that division is closely linked to growth so that cells divide when they have doubled in volume. In fact there are cases where this does not happen, most obviously in the situation of early embryos where there is no growth and cell divisions are cleavage divisions in which the cell volume is halved each cell cycle. Cell growth and division may also be disengaged in the case of multinuclearity or polyploidy, whereby cells can increase both their size and DNA content without dividing. This is common in mammalian hepatocytes and cardiomyocytes.

The intrinsic cell cycle mechanism depends on proteins called cyclins, which bind to and activate cyclin dependent kinases (CDKs). The cycle contains checkpoints which can be passed only if the appropriate complex of a cyclin and CDK in present. For entry into mitosis, this complex is called M-phase promoting factor, or maturation promoting factor (both names conveniently abbreviating to MPF) and phosphorylates and activates various cell components required for mitosis. For entry into S phase, a different cyclin–CDK complex activates the enzymes of DNA replication. Following each phase the cyclins are rapidly destroyed, enabling the cycle to progress to the next stage.

The external signals controlling cell division are very varied, but often include growth factors that stimulate tyrosine kinase receptors on the cell surface. These usually connect to the ERK signal transduction pathway (shown in Figure 7.4), and will mobilize cells from the G0 to the G1 state by phosphorylating the retinoblastoma (RB) protein.

Phosphorylation inactivates RB, and dissociates it from a complex with the transcription complex E2F-DP, enabling this to activate transcription of various components needed to enter the G1 state. RB is also involved in the transition from G1 to S phase. Tissue culture cells are well-supplied with growth factors in their medium, especially platelet-derived growth factor (PDGF) in the serum. Similarly, dividing cell populations in vivo usually have local sources of growth factors, especially Wnt factors, provided by other cells in their immediate environment.

Studying Cell Turnover

It is possible to identify dividing cells in histological sections by immunostaining. The most specific method is to stain for a protein called Ki67 (Figure 2.6a). This coats mitotic chromosomes and is present throughout the cell cycle of dividing cells. Another widely used proliferation marker is phosphohistone H3, the period of phosphorylation being specific to M phase. Dividing cells also express proliferating cell nuclear antigen (PCNA) which is a cofactor of DNA polymerase and is supposed to be present during S phase. However PCNA tends to be detected more widely than this and is not now considered very reliable as a marker of dividing cells.

An extrinsic DNA label that was used for many years was tritiated thymidine (^{3}HTdR). This is metabolized to ^{3}H-TTP and becomes incorporated into DNA during replication. However detection requires autoradiography which takes longer than immunostaining, and easier methods are now available for other DNA precursor labels. Very widely used is bromodeoxyuridine (BrdU) (Figure 2.6b,c). This behaves metabolically like thymidine and can be detected by immunostaining. More recently some other derivatives of uridine have been introduced. Chloro- and iodo-deoxyuridine can with suitable antibodies be distinguished from BrdU, enabling double labeling experiments to be carried out. Ethynyl-dU is visualized by an even simpler detection system based on histochemistry rather than immunostaining. It should be borne in mind that DNA precursors are incorporated into DNA during repair synthesis as well as during replication, so not all DNA labeling necessarily means that cells are in cycle.

The importance of the extrinsic labels is that they can be used for pulse-chase experiments and give a lot more information than the intrinsic markers of dividing cells. Suppose a single dose of BrdU is given and the tissue examined shortly afterwards. This is what is shown in Figure 2.6 and provides what is known as a flash labeling index.

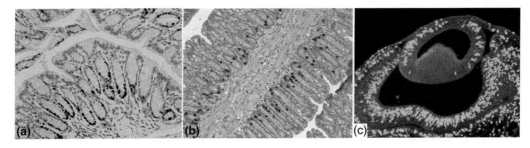

Figure 2.6 Examples of cell division markers. (a) Immunostain of mouse colon with Ki67 antibody. Dividing cells lie in the lower part of the crypts. (b) Mouse colon with a short BrdU label visualized by immunostaining. Fewer cells are labeled than in (a). (c) Chick embryo brain with long BrdU label visualized by immunostaining. Most cells are labeled. (*Sources*: (a) https://media.cellsignal.com/products/images/30670670/32799549/12202_colon_jp.jpg. (b) http://www.abcam.com/brdu-antibody-bu175-icr1-ab6326.html#description_images_7. (c) http://www.sigmaaldrich.com/catalog/product/sigma/b2531?lang=en&region=GB. Reproduced with the permission of Sigma-Aldrich Co. LLC, provided courtesy of Abcam. Reproduced courtesy of Cell Signaling Technology, Inc. (www.cellsignal.com) Ki-67 (D3B5) Rabbit mAb (Mouse Preferred; IHC Formulated) #12202.)

All the cells are labeled that were in S-phase while the BrdU was available in the system. It will only label cells in cycle if they pass through an S-phase, and so underestimates the proportion of cells in cycle. If repeated doses of BrdU are given so that it is continuously available for a long time, then eventually all dividing cells will pass through an S-phase and the percentage of labeled cells will climb to a maximum corresponding to the proportion of cells in cycle, which should theoretically be the same as the Ki67 index.

Suppose a dose of BrdU is given to an animal and then no further label is administered for some time. Because the half-life of an intraperitoneal injection to a rodent is only about 15 minutes, the BrdU will no longer be available after 1–2 hours and so after this there is a period of "chase" with endogenous unlabeled thymidine. If the tissues are examined some time later then any cells labeled with BrdU must be no longer-dividing cells which underwent their final S-phase shortly after the label was given. Completely quiescent tissues will obviously not be labeled, but nor will continuously dividing tissues. This is because the BrdU that is incorporated after the label becomes progressively diluted with each subsequent DNA replication in the absence of label. After 6 further cell cycles, the dilution factor will be 2^6 or 64×, which means the BrdU will no longer be detectable by immunostaining. This type of labeling indicates the final division time of a progenitor and is called the "cell birthday" and has been widely used to study formation of neurons in the central nervous system. Cessation of division is the usual reason for label retention, but it will be apparent that labeling might also be seen in very slowly dividing cells that just happened to pass through one of their very rare S-phases at the time of the label. As mentioned in Chapter 1, because stem cells are sometimes considered to divide slowly, something of a mythology has grown up around label retention along the lines that label retaining cells must be stem cells. Although this may be true on occasion, label retention is certainly not sufficient evidence that a cell is behaving as a stem cell.

There is one other possible reason for label retention. It was argued by Cairns in 1975 that in order to avoid accumulating somatic mutations, stem cells would retain one DNA strand permanently and all newly synthesized strands would be exported to daughter cells. This is sometimes known as the "immortal strand" theory because the retained DNA strand would survive unaltered as long as the stem cell itself, and so potentially as long as the whole organism's lifetime. The Cairns hypothesis seems intrinsically unlikely as it requires that all chromosomes must be lined up with the same orientation on the spindle such that all old strands pass to the parent stem cell and all new ones to the daughter cell. Furthermore it requires the complete suppression of sister chromatid exchange, which often occurs during mitosis, to prevent the sister chromatids becoming mixed up by recombination. Evidence for the Cairns hypothesis has been hard to obtain. It almost certainly does not occur in hematopoietic stem cells or intestinal stem cells, although it may perhaps occur in muscle satellite cells.

The vast majority of observations of dividing cells using intrinsic markers or DNA precursors are comparisons between different situations, for example cells treated or not treated with some factor. This enables us to say that there is probably more (or less) cell production in the treated condition. But what cannot be deduced from measurements of labeling index is the actual rate of production of new cells, because this also requires an estimate of the duration of the cell cycle. For example, if the long term labeling index is 80% and the cell cycle time is 24 hours, then the rate of cell production is 80% of the original population each day. If the cell cycle time is 12 hours, then the rate of cell production is treble this: 240% of the original cell population in 24 hours. There are two doublings over one day so the increase is $3 \times 80\%$. There are several methods of estimating cell cycle time, but they are rarely applied to in vivo situations and are not very accurate, largely because there is usually a considerable variation of cell cycle time within an otherwise homogeneous cell population. A relatively simple method with some popularity involves the administration of two distinguishable labels, such as BrdU and EdU. For it we need to make the basic assumption that the population is in steady state and is completely unsynchronized, so that the proportion of dividing cells in S phase is the same as the proportion of S phase as a fraction of the total cell cycle duration. Suppose BrdU is given for 2 hours and EdU just for the last 30 minutes of this period. Both BrdU and EdU will label all cells currently in S phase. The fraction of cells labeled by BrdU but not by EdU represents those cells which exited S phase during the 1.5 hours before the EdU was administered. If the basic steady state assumption is correct then:

$$\frac{1.5}{\left(\text{S phase duration}\right)} = \frac{\left(\text{proportion BrdU label only}\right)}{\left(\text{proportion BrdU + EdU label}\right)}$$

To give a numerical example: suppose the EdU labels 16.7% and the BrdU alone labels 5%. This means that the S phase duration is $(1.5 \times 0.167)/0.05 = 5$ hours. To find the total cell cycle time we also need to know the fraction of all cells which are in cycle. This can obtained from Ki67 staining or from the plateau reached by prolonged BrdU labeling. Then, using the same basic assumption:

$$\frac{\left(\text{S phase duration}\right)}{\left(\text{total cell cycle duration}\right)} = \frac{\left(\text{proportion EdU label}\right)}{\left(\text{proportion cells in cycle}\right)}$$

Let us say that 50% cells are in cycle. So for our example the total cell cycle time will be:

$$\left(5 \times 0.5\right)/0.167 = 15 \text{ hours.}$$

Despite the huge attention given to cell division in vivo, rather few studies are ever made in which cell production rates are measured. Even fewer are made of the flux to cell death (see below) and in general the impression given by flash labeling studies considerably underestimates the cell production and cell death rates.

Reporters for the Cell Cycle

There are also methods for labeling cells in cycle using genetic reporters. The gene for one of the histones which is a normal component of the nucleosomes, H2B, has been fused to the coding sequence for green fluorescent protein (GFP) and knocked in to mouse cells in a doxycycline repressible form (for methods see Chapter 3). For such mice, removal of doxycycline upregulates expression of *H2B-GFP* and the nuclei become green due to incorporation of the green histone in their chromatin. If doxycycline is restored then *H2B-GFP* is repressed, and as the cells continue dividing they will lose the green color by dilution with unlabeled H2B. This is effectively similar to a pulse-chase situation using a DNA precursor, and has especially been used for studying label-retaining cells.

A recently developed technique is called FUCCI (fluorescent ubiquitination based cell cycle indicator). It depends on the fact that two normal cell cycle components: the licensing factor Cdt1, and its inhibitor, geminin, are degraded at different times in the cell cycle. In FUCCI, transgenic mice are made containing genes for fluorescent proteins carrying target sequences for degradation from either Cdt or geminin. If one is red and the other green this means that cells change from red to green as they pass through the cycle. The transgenic methods of observing the cell cycle have the advantage of being able to be used on living specimens.

Identification of Very Slow Cell Turnover

The analysis of proliferative behavior using DNA precursors can only be informative if there is enough cell division for some S-phase cells to be detectable. If very few or no S-phases are apparent, does this mean that the tissue is completely quiescent or even post-mitotic? This question is harder to answer in the human situation where the lifespan is 80 years or so, compared to a mouse that lives only 3 years. A novel approach has been pioneered by Jonas Frisen and colleagues at the Karolinska Institute. They noted that in the 1950s and early 60s there had been many nuclear weapons tests in the atmosphere leading to considerable radioactive fallout all over the world. Among the isotopes released was ^{14}C. This is in the form of CO_2, so becomes absorbed by plants, and later by animals and humans who eat these plants. Therefore, developing plants and animals become fully labeled with the prevailing atmospheric content of ^{14}C. Because of the test ban treaty of 1963, which confined nuclear bomb tests to underground locations, the ^{14}C level in the atmosphere declined exponentially after this date. The decline is not due to radioactive decay, which is a much slower process, but by removal of ^{14}C from the atmosphere into oceans and soils by the

normal environmental carbon cycle. The question in relation to cell turnover is whether DNA of a person who underwent embryonic development at one level of ^{14}C, retains that content or whether it changes during their lifetime to reflect the new atmospheric level of ^{14}C. In the first case the cells cannot have divided since embryonic life, in the second they must be dividing continuously.

The calibration curve of ^{14}C abundance in the atmosphere was obtained by analysis of tree rings, each of which represents one year of growth of the tree. This confirmed that the recent ^{14}C spike is indeed a unique event in human history (Figure 2.7a). The tissue samples are obtained postmortem from people who have died. Because all tissues contain multiple cell types, and each cell type may have different proliferative behavior it is necessary to isolate nuclei from the cell type of interest. This is done by flow cytometry of dissociated nuclei using a specific nuclear antigen. Then the DNA is purified, and analyzed for its ^{14}C content. This needs very special equipment which is only available in a few centers. Application of the method has shown that neurons of the human cerebral

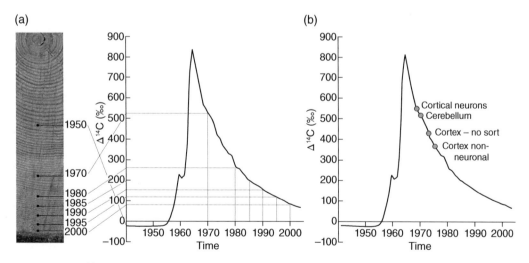

Figure 2.7 The ^{14}C dilution method for estimating the degree of cell turnover in humans over long time periods. (a) Changes of ^{14}C in the atmosphere: the peak is due to nuclear bomb tests. Values are measured from tree ring samples. (b) The ^{14}C abundance of DNA in brain from an individual born in 1967 who died in 2003. The bulk cerebral cortex shows some turnover, but when neuronal and non-neuronal nuclei are separated it can be seen that neuronal turnover is virtually zero. (From Spalding, K.L., Bhardwaj, R.D., Buchholz, B.A., Druid, H. and Frisen, J. (2005) Retrospective birth dating of cells in humans. Cell 122, 133–143. Reproduced with the permission of Elsevier.)

cortex do not turn over at all, even if neurons in some other brain regions do so (Figure 2.7b). Heart muscle turns over very slightly, perhaps half the cardiomyocytes being renewed over a complete human lifespan. Adipose tissue turns over more rapidly, at about 10% per annum. By contrast, well characterized tissue-specific stem cell systems such as the blood turn over very rapidly and show the approximate ^{14}C abundance of the year of death.

Classification of Cell Types by Proliferative Behavior

Claude Leblond, working in Montreal in the 1940s and 1950s, studied mammalian tissues initially by observing the mitotic figures and later by labeling animals with ^{3}HTdR. He realized that some cell populations divided while others did not. Leblond classified cells into three classes: "static", like neurons; "expanding", meaning dividing during growth and subsequently becoming quiescent, like hepatocytes; and "renewal", meaning a situation of continuous cell production, differentiation and death, as found in stem cell systems such as the epidermis or the intestinal epithelium. This classification relates to cell types and not to tissues. So neurons are indeed post-mitotic, but the tissue to which they belong, the CNS, does have some persistent regions of neural stem cells which continue to generate neurons throughout life. Likewise, multinucleate muscle fibers are post-mitotic, but skeletal muscle as a tissue contains muscle satellite cells that can generate new fibers during adult life. The curiously named "expanding" compartment contains many glandular cell types and these often have the capacity to resume division in case of injury so are considered quiescent rather than post-mitotic.

Leblond's classification was based on what can be observed with ^{3}HTdR labeling and so could not address difficult problems such as the degree of cell turnover in cardiac muscle, but it is still of value in emphasizing that different types of cell have very different proliferative behavior in the body and that stem cells are not found in all tissues.

Cell Death

Cell death is just as important as cell production. During development programmed cell death is essential for various events, including the sculpting of the shape of the limbs and the adjustment of cell numbers in spinal ganglia. In adult life it is obvious that in any tissue maintained by stem cells that the rate of cell removal must balance the rate of cell production.

It is usual to distinguish between programmed cell death, which occurs in the above situations and depends on mechanisms within the cell, from necrotic cell death, which arises from damage and involves a chaotic destruction of the cell often accompanied by local inflammation arising from release of bioactive components (Figure 2.8). The usual form of programmed cell death is called apoptosis and involves cell shrinkage and blebbing, fragmentation of the nucleus, degradation of DNA to nucleosome-size fragments and eventual phagocytosis by other cells. Apoptosis is carried out by proteases called caspases which degrade poly-ADPR polymerase (involved in DNA repair), nuclear lamins, and cytoskeletal proteins required to maintain cell shape. Activation of caspases may be initiated by external signals, often via the "death receptor" Fas (= CD95), or by internal DNA damage. A key step is the release of cytochrome c from mitochondria leading to the activation of caspase 9. The signal for eventual phagocytosis of the dead cell is provided by exposure on the surface of the cell of the membrane phospholipid phosphatidylserine. A subset of apoptosis is called anoikis and refers specifically to apoptosis initiated by loss of cell contacts with the extracellular matrix. This is normally prevented by the binding of integrins together with the action of growth factors including insulin and IGF1. The death

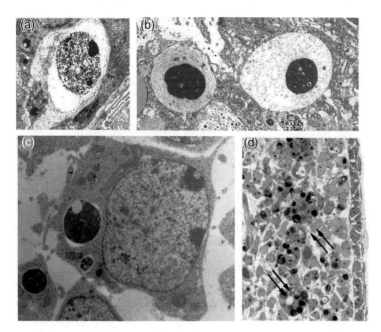

Figure 2.8 Different types of cell death. (a) Necrosis of a mouse prostate cell. The nucleus and cytoplasm are disorganized. (b) Apoptosis of two cells in mouse mammary epithelium. The nuclei are condensed. (c) An apoptotic cell in a mouse embryo phagocytized by a neighboring cell. (d) Apoptosis of cells in the interdigital region of a 12.5 d mouse embryo. The arrows indicate viable cells that have phagocytosed autophagic fragments. (From: Lockshin, R.A. and Zakeri, Z. (2004) Apoptosis, autophagy, and more. The International Journal of Biochemistry and Cell Biology 36, 2405–2419.)

of human embryonic stem cells on dissociation is due to anoikis, and is usually controlled by adding Rho-associated protein kinase inhibitor (ROCK inhibitor) to the medium.

In larger post-mitotic cells with a lot of cell organelles, autophagy can play a large part in cell death. Autophagy refers to destruction of cell components in autophagosomes, which are formed by the combination of an intracellular membrane with a lysosome. It is a normal ongoing mechanism for turnover of cell components, but in some circumstances it can destroy the whole cell and this is referred to a type II programmed cell death, as opposed to conventional apoptosis which is called type I.

Cell death is normally observed by in situ methods, usually immunostaining for one of the caspases, or a method called TdT-mediated dUTP nick end labeling (TUNEL). An aspect of apoptosis is the degradation of chromosomal DNA into nucleosome-size pieces of about 180 bp. In TUNEL these are detected by using the enzyme terminal nucleotidyl transferase to add a biotinylated nucleotide to the end of the DNA fragments and to visualize this with a streptavidin-based detection system. These methods of cell death detection give a measure of the proportion of cells currently undergoing apoptosis. They do not provide a measure of the rate of cell death, which is often underestimated because the duration of visibility of the apoptotic markers is quite short. For example if the observed apoptotic index is 1% and the duration of presence of the apoptotic marker is 2 hours, then the flux to cell death would be (24 × 1%)/2 or 12% per day.

Further Reading

Histological Methods

Bancroft, J.D. and Gamble, M. (2007) Theory and Practice of Histological Techniques, 6th edn. Churchill Livingstone, Philadelphia, PA.

Ross, M.H. and Pawlina, W. (2016) Histology: A Text and Atlas: With Correlated Cell and Molecular Biology, 7th edn. Wolters Kluwer Health, Philadelphia.

Polak, J.M. (2003) Introduction to Immunocytochemistry, 3rd edn. BIOS Scientific Publishers, Oxford.

Polak, J.M. and McGee, J.O'D. (1999) In Situ Hybridization: Principles and Practice. Oxford University Press, Oxford.

Wang, F., Flanagan, J., Su, N., Wang, L.-C., et al. (2012) RNAscope: A novel in situ RNA analysis platform for formalin-fixed, paraffin-embedded tissues. Journal of Molecular Diagnostics 14, 22–29.

Young, B., O'Dowd, G. and Woodford, P. (2014) Wheater's Functional Histology: A Text and Colour Atlas, 6th edn. Elsevier, Churchill Livingstone, Philadelphia.

Other Methods

Espina, V., Wulfkuhle, J.D., Calvert, V.S., VanMeter, A., et al. (2006). Laser-capture microdissection. Nature Protocols 1, 586–603.

Givan, A.L. (2001) Flow Cytometry: First Principles. Wiley-Liss, New York.

Wang, Z., Gerstein, M. and Snyder, M. (2009) RNA-Seq: a revolutionary tool for transcriptomics. Nat Rev Genet. 10, 57–63.

Cell Turnover

Alexiades, M.R. and Cepko, C. (1996) Quantitative analysis of proliferation and cell cycle length during development of the rat retina. Developmental Dynamics 205, 293–307.

Cairns, J. (1975) Mutation, selection and the natural history of cancer. Nature 255, 197–200.

Goodlad, R.A. (2017) Quantification of epithelial cell proliferation, cell dynamics and cell kinetics in vivo. Wiley Interdisciplinary Reviews: Developmental Biology.

Hengartner, M.O. (2000) The biochemistry of apoptosis. Nature 407, 770–776.

Jackson, P.K. (2008) The hunt for cyclin. Cell 134, 199–202.

Lockshin, R.A. and Zakeri, Z. (2004) Apoptosis, autophagy, and more. International Journal of Biochemistry and Cell Biology 36, 2405–2419.

Messier, B. and LeBlond, C.P. (1960) Cell proliferation and migration as revealed by radioautography after injection of thymidine-H3 into male rats and mice. American Journal of Anatomy 106, 247–285.

Zielke, N. and Edgar, B.A. (2015) FUCCI sensors: powerful new tools for analysis of cell proliferation. Wiley Interdisciplinary Reviews: Developmental Biology 4, 469–487.

Spalding, K.L., Bhardwaj, R.D., Buchholz, B.A., Druid, H. and Frisen, J. (2005) Retrospective birth dating of cells in humans. Cell 122, 133–143.

3

Genetic Modification and the Labeling of Cell Lineages

Genetic modification is central to much of modern biomedical research. The principal methods for introducing new genes to cells and to whole organisms are briefly reviewed here, as are the main methods for controlling their expression in time and space. It is also explained how genetic modification methods can be used to enable the tracing of cell lineage, a subject of special interest in stem cell biology. Genes introduced into cells or organisms are often called transgenes.

Introducing Genes to Cells

Transfection and Electroporation

Introducing genes to tissue culture cells is relatively straightforward. The gene of interest needs to be cloned into a mammalian expression vector (Figure 3.1a). This is a plasmid, which contains a powerful mammalian promoter, such as that of cytomegalovirus (CMV), adjacent to a multiple cloning site at which the gene of interest is inserted. Downstream of the cloning site is a polyA addition sequence from the SV40 virus. This causes addition of polyA to the newly transcribed mRNA produced from the transgene and thereby stabilizes the mRNA and enables its transfer to the cytoplasm. The expression plasmid will be prepared by growing it in bacteria, and to make this possible it also contains an origin of DNA replication, to enable replication, and a bacterial antibiotic resistance locus, to enable selection of bacte-

rial cells containing the plasmid. The bacteria containing the plasmid are grown in the presence of the selection antibiotic, then the bacteria are lysed and the plasmid is purified using a proprietary kit. This generates a sample of pure plasmid DNA which can be introduced to the cells of interest.

The process of introducing the DNA is called transfection and is usually done with the aid of a transfection reagent which is a positively charged lipid preparation which binds DNA and generates microscopic particles that can enter the cell by fusing with the plasma membrane. Once inside the target cell the DNA needs to find its way into the nucleus in order to be expressed. It is then transcribed to mRNA in the same way as the endogenous genes, and the mRNA is exported to the cytoplasm where it is translated to protein. Expression of the transgene may be monitored by antibody staining or Western blotting for the protein product, or by the inclusion of another gene in the same plasmid encoding a fluorescent protein such as green fluorescent protein (GFP). Most transfection experiments give transient rather than permanent expression because the plasmid DNA does not become integrated into the cell's chromosomal DNA and it eventually becomes degraded or diluted out by cell growth. However, it is also possible to obtain permanent transfectants by selecting for DNA integration using a second, mammalian, antibiotic resistance gene on the plasmid.

It is often advantageous to be able to express more than one protein from the same

The Science of Stem Cells, First Edition. Jonathan M. W. Slack.
© 2018 John Wiley & Sons, Inc. Published 2018 by John Wiley & Sons, Inc.
Companion website: www.wiley.com/go/slack/thescienceofstemcells

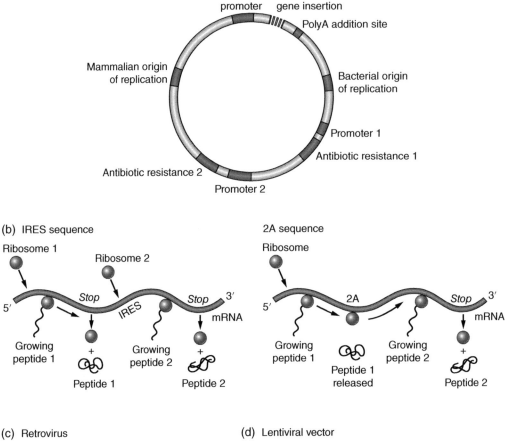

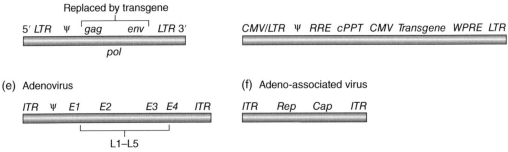

Figure 3.1 (a) Mammalian expression plasmid. (b) IRES and 2A sequences. (c) Retrovirus genome. LTR = long terminal repeat. In a vector, the *gag, pol* and *env* genes are replaced by the gene of interest. (d) Lentiviral vector. LTR = long terminal repeat, RRE = Rep response element, cPPT = polypurine region, CMV = promoter for gene of interest, WPRE = enhancer sequence. (e) Adenovirus genome. ITR = inverted terminal repeat, E1–4 are early expressed genes, L1–5 are late genes. In a first generation vector the E1 gene, needed for replication, is replaced by the insert. (f) AAV genome. ITR = inverted terminal repeat. There are two genes and in a vector both are replaced by the gene of interest.

DNA construct. For example coexpression of a fluorescent protein indicates which cells are actually expressing the transgene. This is usually achieved by encoding them in a single transcription unit (Figure 3.1b). In one method, the two sequences are separated by an internal ribosome entry sequence (IRES), which enables ribosomes to commence translation of the second polypeptide regardless of the presence of the stop codon at the end of the first polypeptide. The second commonly used method is to use a 2A sequence from Newcastle Disease virus. Here the entire transgene is translated into a long polypeptide which then self-cleaves at the 2A sequence into the two components.

For established tissue culture, cell lines in favorable circumstances, very high levels of transfection and of expression can be achieved, but for primary cell lines, grown from animals or from people, the rates are usually much lower. For cells within the tissue of live animals, transfection by plasmids is not efficient enough to be of practical use. An alternative method of gene introduction is electroporation. This involves exposing the cells to the DNA solution and then applying a brief electric pulse which punches small holes in the plasma membrane and introduces some DNA. The holes reseal rapidly but electroporation is still liable to kill a proportion of the cells. Electroporation is used for recalcitrant cell types in culture that cannot readily be transfected, and it is also sometimes used for introducing DNA into animal tissues in vivo, usually for very small specimens such as embryos.

Gene Delivery Viruses

When viruses are used to introduce genes to cells, it is usually called "transduction" rather than "transfection". Viral vectors can often give a higher percentage of cell transduction than simple transfection with DNA because they have specialized systems for penetrating the cell. A large variety of viruses have been used for this purpose and only a few of the most popular are mentioned here. Retroviruses (strictly, γ-retroviruses) are very popular for gene delivery (Figure 3.1c). They are simple RNA viruses encoding just three proteins: a reverse transcriptase (pol), a capsid protein (gag) and an envelope protein (env). The sequences encoding these are flanked by long terminal repeats (LTRs) which act as both the promoter and terminator sequences for transcription of the DNA copy of the viral RNA. The virus particle contains two copies of the RNA plus the reverse transcriptase and capsid proteins, all wrapped in plasma membrane from the host. The envelope protein is inserted into the membrane. Depending on the specificity of the envelope protein, this virus particle can infect a cell and introduce its RNA which becomes reverse transcribed to DNA by the viral reverse transcriptase. The resulting single-stranded DNA is converted to double-stranded DNA which can then be integrated into the host genome. Retroviral DNA integration normally occurs into a region of the genome in which endogenous genes are active. Once integrated, the viral DNA becomes known as a provirus. It functions as a normal gene generating new mRNA copies of itself using the 5′LTR as the promoter.

In a retroviral vector the gene of interest, sometimes accompanied by its own promoter, is cloned between the LTRs, replacing the *gag*, *pol* and *env* genes of the virus. Retroviruses will take inserts up to about 8 kb in size. Because it lacks the key components for replication, the retroviral vector cannot generate further copies of itself, and this is a valuable safety feature for those handling the virus. To produce the viral vector, a packaging cell line is used, usually 293T cells. These are derived human embryo kidney cells and contain an SV40 T gene, which enables replication of plasmids containing an SV40 origin of replication. Into these cells are transfected three different plasmids encoding the following components:

1) The viral vector including the LTRs, the gene of interest and a packaging signal sequence called ψ.

2) A plasmid encoding the *gag* and *pol* genes for capsid assembly and DNA synthesis.
3) A plasmid encoding the envelope protein. This is usually not the original retroviral envelope protein, but instead the Vesicular Stomatitis Virus G protein (VSV-G), which has a broad host range and can infect most types of cell.

Use of a foreign envelope protein to change the host range of a retrovirus is known as "pseudotyping". Unlike the normal virus particles, a retrovirus pseudotyped with VSV-G can be concentrated by pelleting in the ultracentrifuge, which is very useful for purification and storage. Once the three plasmids have been introduced to the packaging cell line, the products will complement each other and the packaging cells then secrete virus particles containing the defective RNA genome inside the plasma membrane envelope containing VSV-G protein.

Following infection, or more accurately transduction, of the target cells, the vector RNA becomes reverse transcribed, made double-stranded, and is often integrated into the genome. The gene of interest is transcribed from the 5′ LTR, or from its own promoter if this has been included in the construct, to produce mRNA which is transported to the cytoplasm and translated into protein. However, after a while, gene expression from integrated retroviruses may be inhibited by "silencing" which involves DNA methylation and other epigenetic changes at the site of integration.

A somewhat more complex type of retrovirus, especially suitable for gene delivery, is the lentivirus (Figure 3.1d). This differs from a simple retrovirus in having three more genes and has the useful property of transducing non-dividing as well as dividing cells. Lentiviral vectors are based on human immunodeficiency virus (HIV) and usually contain a variety of extra safety features to limit the likelihood of production of replication-competent virus by recombination during the packaging process. A mammalian antibiotic selection is often built into the vector to enable selection of cells that have successfully integrated it into the genome, and are resisting silencing.

Adenovirus is a double-stranded DNA virus with a wide host range and the ability to infect both replicating and non-replicating cells. Adenoviral vectors are based on the human adenovirus serotype 5 (Figure 3.1e). They are relatively easy to prepare at high titer and often gives good results in both cell lines and whole animals, although high immunogenicity limits their use in vivo. Adenovirus does not integrate into the genome and so is used to generate high level transient expression. Like the retroviruses it can accommodate up to about 8 kb of cargo. Because the large size of the virus makes in vitro manipulation difficult, adenoviral vectors are prepared by recombination in bacteria between a vector skeleton plasmid lacking the *E1* gene needed for replication, and a shuttle plasmid containing the gene of interest. The resulting replication defective vector is then propagated in the packaging cell line HEK293 which supplies the missing E1 functions. The virus particles are secreted into the medium and are concentrated by cesium chloride banding in the ultracentrifuge, or by using a proprietary purification kit. Adenovirus enters target cells via a receptor called CAR (Coxsackie and Adenovirus Receptor) which is very widespread in mammalian tissues.

A vector with a similar name, but which is actually completely different, is the adeno-associated virus, AAV (Figure 3.1f). This is a single-stranded DNA virus which is much less immunogenic than adenovirus and so is preferred for gene therapy applications. Like retroviruses it is prepared in replication-defective form by cloning the gene of interest into the viral skeleton and transfecting this into the packaging cell line together with additional plasmids encoding the other required functions. The main drawback of AAV is that its cargo capacity is limited to about 4.5 kb. The wild type AAV virus integrates into the host genome at a specific site called AAVS1. However, gene delivery vectors based on AAV do not normally integrate into the host genome, instead persisting as long term as autonomously replicating elements (episomes) in the cell nucleus.

Controlling Gene Expression

There are various ways of controlling the expression of transgenes. The simplest is to drive expression from a tissue-specific promoter, present in the gene delivery vector itself, which is only active in a certain cell type. For example the *transthyretin* promoter is only active in hepatocytes and the *cardiac actin* promoter is only active in cardiac and skeletal muscle cells. However, promoter specificity is not necessarily guaranteed if the environment of the insertion site differs greatly from the endogenous gene locus, so it needs to be checked in each case.

Tet System

It is often advantageous to be able to regulate the time of activity of an introduced gene. There are various ways to do this, but the most popular is based on the tetracycline operon of *E. coli* which has been heavily modified to make it usable in mammalian systems (Figure 3.2a). The external regulator is the tetracycline analog doxycycline, which can be added to the culture medium, or put in the drinking water of laboratory animals, when required. Two DNA components need to be introduced to the cells or animal: the gene of interest controlled by a promoter called the Tet Response Element (TRE), and the gene for the protein that regulates expression in combination with doxycycline. This regulatory protein is the Tet activator (TA), which is greatly modified from the original *E. coli* version. In the Tet-On inducible system, the addition of doxycycline to the cells activates the TA which then activates transcription of the gene of interest at the TRE. There is also a repressible version of the system called Tet-Off, in which the removal of doxycycline activates the TA. In its earlier versions the Tet-Off version was preferred because Tet-On suffered from leakiness due

to residual activity of the TA in the absence of doxycycline. However the system has been gradually improved over the years and Tet-On is now the most popular of the available inducible systems

Cre System

Cre is an endonuclease enzyme from phage P1 which cuts DNA only at specific sites called loxP sites. It is used for a variety of purposes, but here we will focus on its use for regulation of transgene expression. Two components need to be introduced into the cells or animal. First a gene encoding the Cre regulated by a tissue-specific promoter. Second, the gene of interest controlled by a strong ubiquitous promoter, but also with an inhibitory sequence blocking expression which is flanked by loxP sites. The inhibitory sequence is normally a transcriptional stop signal. When the tissue-specific promoter is active, the Cre enzyme is produced. It will then find the loxP sites and cut them, so the inhibitory sequence is excised and the gene of interest is expressed (Figure 3.2b). The key feature of the Cre system which makes it differ from all inducible systems like Tet-On, is that it is based on excision of a DNA sequence which is permanent and will be maintained by the cells regardless of any subsequent cell division or change of differentiation state.

A further level of regulation is provided by use of a modified Cre, called here CreER although other abbreviations are in use, which includes the hormone binding domain from the estrogen receptor, ER. Like several other nuclear hormone receptors, the estrogen receptor is normally sequestered in the cytoplasm, bound to HSP90 protein. When estrogen binds, the HSP90 is displaced and the ER can migrate to the nucleus where it binds DNA and functions as a transcription factor, i.e. a protein that controls gene expression. CreER behaves similarly. Normally it is sequestered in the cytoplasm, but when a suitable estrogen analog (usually tamoxifen or 5-hydroxytamoxifen) is added, it migrates to the nucleus and can cut any loxP sites in

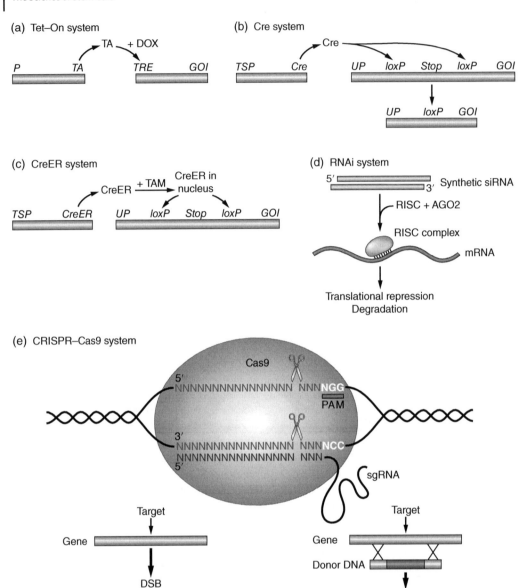

Figure 3.2 (a) Tet system. P = promoter, TA = Tet activator, DOX = doxycycline, TRE = Tet Response Element, GOI = gene of interest. (b) Cre system. TSP = tissue-specific promoter, UP = ubiquitous promoter, GOI = gene of interest. (c) CreER system. TAM = tamoxifen (d) RNAi system. RISC = RNA induced silencing complex, AGO2 = Argonaute 2 (endonuclease). (e) CRISPR-Cas9. PAM = protospacer adjacent motif, sgRNA = single guide RNA, DSB = double-stranded break.

the DNA (Figure 3.2c). The CreER system has been widely used for tracking the cell lineage of tissue-specific stem cells. The method is described in more detail below and several examples are provided in Chapter 10. An equivalent system which is sometimes used, especially where tamoxifen has deleterious side effects, involves controlling *Cre* transcription using the Tet-On system, such that Cre is only produced in the presence of doxycycline.

An alternative to the Cre system, which is very similar in its properties but derived from *Drosophila*, is the FLP/FRT system. Here, FLP is the recombinase and it cuts specifically at FRT sites. It is also possible to use FLP-ER as an inducible form of FLP, activated by treatment with tamoxifen.

Inhibiting Gene Activity

The methods for introducing genes to cells can also be used to inhibit specific cellular components. One way is simply to overexpress a gene that encodes a known inhibitory protein. Alternatively specific inhibitors can be designed. For example transcription factors usually contain a DNA binding domain, which binds specific sites in the genome, and an effector domain, which activates or represses transcription. A construct lacking the DNA binding domain will often act as a dominant negative inhibitor because it will sequester all the normal cofactors needed by the endogenous factor for transcription, and may also form unproductive dimers with the endogenous factor thus inhibiting its activity.

Very commonly used for specific inhibition experiments is RNA interference (RNAi) (Figure 3.2d). This is based on exploiting endogenous mechanisms for handling double-stranded RNA (dsRNA). After introduction to cells, dsRNA becomes cleaved into 21–23 base pair fragments which become incorporated into a silencing complex. This binds to, unwinds, and, if there is a perfect match, cleaves messenger RNAs containing a complementary sequence. Long

molecules of dsRNA can be used for gene inhibition experiments in invertebrate animals and plants but this is not suitable for mammals and other vertebrates as it induces an interferon response that shuts down transcription generally. For mammalian cells it is usual to make the short dsRNA, called siRNA, by chemical synthesis and introduce it to the cells by transfection. Because the siRNA molecules are small, they often give a very high transfection efficiency compared to plasmids with the result that gene expression can be inhibited in most cells in the dish.

dsRNA may also be encoded in DNA plasmids or gene delivery viruses, usually as two inverted complementary sequences that self-assemble by hybridization to a double-stranded hairpin-like structure, called shRNA. In such vectors expression of the RNA may be regulated by Tet or Cre systems as described below, providing additional flexibility in the control of delivery.

CRISPR-Cas9

CRISPR stands for "clustered regularly interspaced short palindromic repeats" but the genome modification system bearing the name does not involve any such repeats. The system is derived from a bacterial mechanism for defense against viruses which involves using RNA to guide nuclease enzymes to the viral DNA. In the molecular biology toolkit CRISPR-Cas9 consists of Cas9, which is an endonuclease, and a guide RNA which binds Cas9 at one end and recognizes a target sequence in the DNA at the other end (Figure 3.2e). So long as the target DNA contains an NGG sequence at the 3′ end of the region of guide RNA homology (the protospacer adjacent motif or PAM), the Cas9 will make a double-stranded break in the DNA. This is repaired by an endogenous process called non-homologous end joining which is very likely to result in small insertions or deletions at the site, causing frameshift mutations and inactivation of the gene. In the presence of an exogenous homologous DNA sequence, there is a good probability that the double-stranded break will provoke recombination which will generate a modified DNA incorporating the exogenous sequence. This adds up to a very simple and reliable system for making site directed changes to genomic DNA. In practice it is just necessary to introduce to the cells the new DNA sequence plus two expression plasmids, one encoding Cas9 and the other encoding the guide RNA under control of a pol-III promoter, which is a promoter suitable for transcription of small non-translated RNAs.

Because of the high efficiency of CRISPR-Cas9, it is possible to alter both maternal and paternal copies of a gene in a cell. This means that "knockout" cell lines lacking a particular gene function can be made relatively easily. It is also possible to target multiple loci simultaneously which may be important at some point in the future for the correction of genetic diseases.

Transgenic Mice

In recent decades transgenic mice have been of huge importance in stem cell biology as well as many other areas of biomedical science. They have played a large role in unravelling developmental mechanisms, and also provided a means for making mouse models of a wide range of human diseases to assist research into pathology and therapy. The term "transgenic" originally referred to mouse lines carrying an extra gene introduced by the experimenter, the gene itself being known as a transgene. However transgenic now often refers to all types of genetic modification including knockouts (specific gene inactivation), conditional knockouts, knock-ins (specific gene replacement) and so on. The techniques for making genetically modified mice have recently been revolutionized by the CRISPR-Cas9 system, although most of the currently available strains were made by previous methods.

The original method for making transgenic mice was to inject DNA into one of the pronuclei of the fertilized egg. This gives random insertions of concatenated repeats and is rarely used today. The standard method for many years has been to make the genetic modifications in mouse embryonic stem cells (ES cells) and to inject the modified stem cells into early mouse embryos where they become incorporated into the inner cell mass and can contribute to the embryo along with cells of the host (Figure 3.3a). The progeny mice then need to be screened for the presence of donor cells (chimerism) and those containing donor cells in the germ line are bred to establish the desired strain. This method has been used very widely and about 25,000 genetically modified mouse strains have so far been produced, including knockouts of about half the protein-coding genes in the mouse genome. But today it is also possible to modify zygotes directly using CRISPR-Cas9, which greatly shortens the overall time required to generate a new strain.

Animal Procedures

Whichever method is used to make the modified zygotes or preimplantation embryos, converting them into actual mice requires that they be inserted into the reproductive tract of a "pseudopregnant" female to allow implantation and normal development to term. Pseudopregnant females are made by mating with vasectomized males, which brings them into the hormonal condition for pregnancy without actually being pregnant. Modified cleavage stage embryos are usually inserted into the oviduct, and modified blastocysts into the uterus, corresponding to the normal migration route of preimplantation embryos down the reproductive tract (see Chapter 5). When the pups are born they need to be screened, usually by polymerase chain reaction (PCR) analysis of DNA from an ear punch or an excised tail tip. If modified ES cells were injected into the blastocysts the strains of the cells and the hosts are usually chosen to differ in coat color allowing a rough preliminary identification of chimerism from the coat color of the pups, but DNA analysis is always necessary as well.

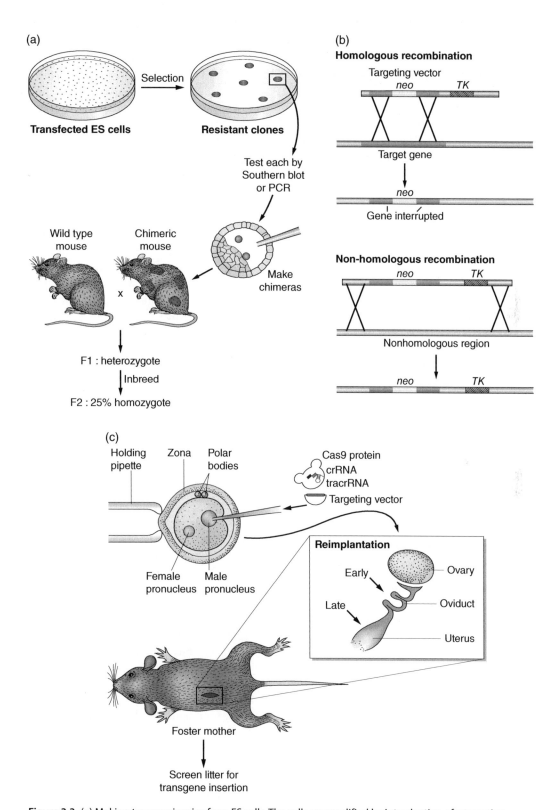

Figure 3.3 (a) Making transgenic mice from ES cells. The cells are modified by introduction of a targeting construct and recombinant clones selected and tested. They are injected into blastocysts which are introduced into the uterus of foster mothers. (b) Positive-negative selection procedure. neo = neomycin resistance gene, TK = thymidine kinase gene. (c) One method of using CRISPR-Cas9 for transgenic modification of zygotes. Here the components are injected into the male pronucleus. (Modified from Slack, J.M.W. (2013) *Essential Developmental Biology*, 3rd edn. Reproduced with the permission of John Wiley and Sons.)

The breeding schedule for transgenic mice will differ depending on the procedure used. The pups arising from the experiment are called founders. DNA injection transgenics usually have a single insertion locus and this behaves in a Mendelian manner. Transgenics made by ES cell injection to blastocysts are usually chimeras containing some donor and some host cells. These are no use unless they have donor cells in the germ line which produce genetically modified gametes. Depending on the level of germ line chimerism they will yield from 0–50% of first generation (F1) offspring containing the genetic modification since the original modification of the ES cells was to just one of the two gene copies. The F1 animals carrying the modification are heterozygous. These can be mated together to generate 25% homozygous mice in the F2 generation, although if the genetic modification is lethal there will be no surviving homozygotes in the F2 and the transgenic line will need to be maintained as heterozygotes. Using CRISPR-Cas9 in zygotes it is possible in principle to obtain founders having a homozygous modification although breeding is still necessary to establish a strain.

Modification of Embryonic Stem Cells

The traditional method for making genetically modified embryonic stem (ES) cells involved introducing the targeting construct to the cells and then using a "positive–negative" selection method with two antibiotics to increase the likelihood of homologous recombination between the construct and the genomic DNA (Figure 3.3b). In other words this should ensure that recombination has occurred at the desired site rather than at some other random site. The clones of ES cells that survive the antibiotic selection are then screened by PCR to ensure that the genetic modification is the expected one and not some unwanted rearrangement. The positive antibiotic selection locus is then removed, for example by using transient expression of Cre to excise a locus flanked by loxP sites. Once the desired clones are obtained the modified ES cells are used for making transgenic mice by the blastocyst injection method.

Using CRISPR-Cas9, the probability of homologous recombination in the ES cells is vastly improved because of the specificity provided by the guide RNA. Homology arms in the targeting construct can be shorter than for conventional homologous recombination (0.3 rather than 2-5 kb). It is also possible to generate homozygous modifications in one step because the method will target the same locus in both of the homologous chromosomes. It is also possible in principle, using different guide RNAs, to modify more than one locus simultaneously. However, whatever procedure is used, antibiotic selection is still necessary to recover the desired recombinant clones and these still need DNA analysis to check the fidelity of the modification and, as far as possible, to ensure a lack of off-target effects from the Cas9 elsewhere in the genome.

Apart from its use to improve the ease of making genetic modifications to ES cells, the CRISPR-Cas system has also been used to make transgenics directly from zygotes (Figure 3.3c). At the time of writing the methods for doing this are still under development. The Cas9 enzyme can be introduced as protein or as RNA. The guide RNA can be in one piece, or two pieces (one to recognize the target and the other to bind the Cas9, as in the natural bacterial immunity system). They can be injected into a pronucleus or into the cytoplasm, or introduced by electroporation. Where homologous recombination is desired, it is best to suppress the normally dominant non-homology end joining process, for example by inclusion of an inhibitor of DNA ligase IV. The potential advantages of this direct procedure is that it can be a much faster method to create a new mouse strain. It may even be possible to obtain homozygous modification from the outset which reduces the amount of subsequent breeding. The method avoids the problem often encountered with ES cells of other spontaneous mutations accumulating in the cell cultures and being incorporated into the transgenic mice. Moreover a similar method can be applied to other animals, especially farm animals, for which the ES cell technology is not currently available. On the other hand, there is always the risk of off-target effects of the Cas9. There is also a risk of mosaicism in the founder mice caused by cutting and recombination events which occur after the first cell cycle. In this case the embryo will be a mixture (mosaic) of cells with and without the modification. Both of these effects require backcrossing to the parent strain and further typing of pups before the desired strain is obtained.

Types of Transgenic Mice

The sophistication of transgenic mouse strains has steadily increased as the technology for making them has improved. Initially genes were overexpressed using either strong ubiquitous promoters or strong tissue-specific promoters. This may produce developmental abnormalities but is usually not very informative about the endogenous gene function. Subsequently, genes were ablated to generate complete loss of function mutants (knockouts). When it was found that knockouts frequently showed no abnormal phenotype they were combined by breeding to create strains containing several knockouts together. In general removal of a whole gene family, for example a complete paralog group of *Hox* genes (see Chapter 7), will generate a deleterious phenotype where removal of a single member of the family does not. This is because there is usually considerable overlap of biological function between members of a gene family (redundancy), so that loss of just one produces little phenotypic effect.

Sometimes knockouts are lethal at an early stage of embryonic development, and this is often because the gene has a function which is required in the placenta. In order to circumvent the early placental requirement and to investigate later functions of the same gene in the embryo, homozygous mutant ES cells can be injected into tetraploid blastocysts, i.e. those composed of cells with four chromosome sets instead of two. This results in the placenta being formed from the tetraploid host while the embryo itself derives from the ES cells and will display a phenotype appropriate to the first embryonic function of the ablated gene. The tetraploid host embryo is made by fusing cells of the two cell stage to a tetraploid single cell by applying a pulse of electric current. On their own, tetraploid embryos can produce a normal placenta but not a viable fetus. When ES cells are injected, these can form a fetus, sustained by the tetraploid placenta.

A more sophisticated type of knockout is the tissue-specific knockout produced using the Cre system. Here a tissue-specific promoter is used to drive expression of Cre, and the target gene is replaced by one flanked with loxP sites ("floxed"). Usually only one of the gene copies is floxed with the other allele being a null mutant. The Cre is expressed only when and where its promoter is active and when the Cre enzyme is produced it excises the floxed gene in those cells. Even this method suffers from the fact that some promoters have an early phase of activity over a very wide domain, causing gene ablation in the whole of this domain. This can be rectified by use of the CreER system. Here the activity of Cre is induced by injection of tamoxifen to the pregnant female and so the knockout can be made to occur in just those cells of the embryo where the Cre promoter is active, and just at the time of induction with tamoxifen (Figure 3.4a). In some cases it may be preferable to drive expression of the Cre using the Tet-On or Tet-Off system (Figure 3.4b).

Exactly the same method of homologous recombination can be used to replace a gene with an altered version, or something completely different. These strains are called "knock-ins". Compared with random insertion of transgenes driven by tissue-specific promoters the knock-in method is more

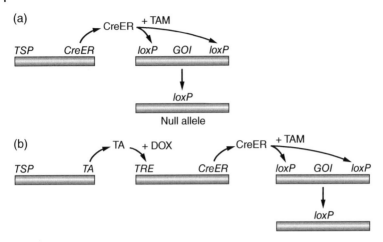

Figure 3.4 Tissue-specific knockout procedures. (a) Using CreER. (b) Using Tet-On and CreER. TSP = tissue-specific promoter, TAM = tamoxifen, GOI = gene of interest, TA = Tet activator, DOX = doxycycline, TRE = Tet Response Element.

reliable because it maintains the normal chromosomal environment and genetic regulation of the modified locus. A knock-in can again be made inducible by the use of the Tet systems.

All of these techniques for genetic modification of mice are used in stem cell biology. But there is one application in particular that has especial relevance: the tracing of cell lineage using the CreER system.

Cell Lineage

Since cells in the body arise by mitotic division, every cell has its own family tree, or lineage, running right back to the fertilized egg. Establishing cell lineage is important in developmental biology because the history of any individual cell tells us which types of progenitor cell were its ancestors. The establishment of each type of progenitor during development represents a developmental decision, usually made in response to an external signaling factor. Cell lineage is also very important in stem cell biology because a stem cell is the precursor for all the cell types of its own tissue. So being able to observe and trace cell lineage is central to defining the existence and properties of different types of stem cell. This section introduces the methods used in mammals for tracing cell lineage. Most depend on genetic modification and are only applicable in mice, although a few depend on naturally existing genetic variation and can be applied to humans. It is conventional to use the term "label" to indicate a substance or gene product introduced into cells for the purpose of tracing cell lineage while the term "marker" refers to an endogenous gene product used to identify cell type.

A common error is to suppose that gene expression patterns remain constant during development, so for example if a cell expresses the intermediate filament protein vimentin, it is considered "of mesenchymal origin". Of course genes may turn on and off in different circumstances and so the presence of a particular marker protein in a cell does not by any means prove that it derived from an ancestor expressing the same marker.

In some invertebrate organisms lineage can be observed directly, and in non-mammalian vertebrates such as frogs and fish which do not grow much during early development it is possible to label cells by injection of passive labels such as fluorescent proteins or enzymes with good histochemical detection methods. Such labels can then be detected at later stages revealing the cell lineage. Passive labels have also been useful for studying the non-growing pre-implantation stages of mammalian development. But following implantation mammalian embryos grow very fast and this means that any passive label is rapidly diluted to invisibility. So after implantation cell lineage in mammals can only be established by the use of genetic labels.

Cell Lineage, Fate Maps, Clonal Analysis

It is useful at this stage to distinguish three related but overlapping concepts: the cell lineage, the fate map and clonal analysis. As mentioned above, the cell lineage is a description of the family history of a cell, maybe back as far as the fertilized egg (Figure 3.5a). In practice it is only possible to know the complete cell lineage for a few types of invertebrate that have a small total cell number and an invariant pattern of cell division and migration. A well-known example is the nematode *Caenorhabditis elegans*, much used as a developmental model organism. To know the cell lineage is important, but on its own the cell lineage does not give any spatial information about where each cell is located and which other cells are its neighbors at different stages of development. This spatial information is given by the fate map, which shows for an early embryo, or a particular region of an embryo, how it moves and changes shape, and what it later becomes in terms of cell differentiation. Unlike nematodes, most animals, including all mammals, have some variation of cell division and cell movement between individuals, so the cell lineage differs a little between individuals. However there is a reproducible overall pattern at the level of tissue regions rather than of single cells and this is what is captured by the fate map. So fate maps are constructed by averaging results from a number of individual embryos, and do not usually have single cell resolution. Fate maps are usually established by prospective labeling of the region of interest with small marks of a passive label, which is located at a later time. An example of a fate map for the mouse embryo egg cylinder, established by injecting small groups of cells with the enzyme horseradish peroxidase, is shown in Figure 3.5b. It is important to remember that a fate map does

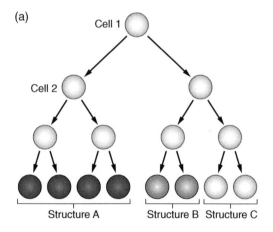

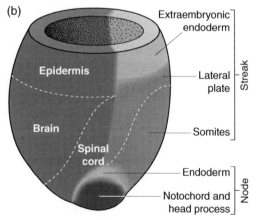

Figure 3.5 (a) Cell lineage diagram demonstrating the principle of clonal analysis. The descendants of cell 1 become three structures so we know that cell 1 is not yet committed to become any one of them. Cell 2 forms only one structure. It may be committed to do so, but this could also be the result of a subsequent signal in this region. (b) Fate map of the egg cylinder stage of a mouse embryo. The boundaries are fuzzier than indicated because there is some variation in cell movements between individual embryos. ((b): Slack, J.M.W. (2013) Essential Developmental Biology, 3rd edn. Reproduced with the permission of John Wiley and Sons.)

not give any information about the developmental commitment of cells at specific developmental stages. For this, additional data are required, such as the effects of moving the cells to a different environment, or observation of expression patterns of key genes involved in commitment.

Clonal analysis focuses on what happens to an individual cell in terms of subsequent differentiation. Unlike the fate map, clonal analysis can tell us something about developmental commitment. If a single cell is labeled and the labeled clone subsequently comprises two cell types, then we can conclude that the decision to form each of these two cell types occurred subsequent to the time of labeling (Figure 3.5a). The converse does not apply. If the entire clone is composed of one cell type, this might be because the cell was already committed to form that type at the time of labeling, but it could also be because the region within which the cell lies was subsequently specified to form that cell type by means of an external signal. Additional data are needed to distinguish these possibilities.

Clonal analysis is particularly useful when analyzing the behavior of stem cells. If a genetic label is applied to a single stem cell the label will later appear as a patch comprising the stem cell itself, its descendant transit amplifying cells, and its descendant differentiated cells. Furthermore this patch will persist long term, as long as the stem cell itself persists. So the clonal labeling indicates directly the two key features of a stem cell: the ability to form the ensemble of cell types in its tissue, and long term persistence.

Clonal analysis can also be carried out in vitro. Here it is usually done by plating individual cells into separate wells and observing what each clone becomes. If one clone can generate more than one differentiated cell type, it is considered to be multipotent. Such assays have been particularly important in establishing the properties of cells in the hematopoietic system (see Chapter 10). Also the clone can be expanded and divided into several wells filled with different media, which may evoke different behaviors. This procedure has been important for establishing the multipotency of mesenchymal stem cells (see Chapter 11). Of course it must always be borne in mind that, because of the range of media that can be employed, cells in tissue culture show a wider range of behaviors than they do in vivo. So clonal analysis in vitro is really what a developmental biologist would call a test of developmental potency, in other words establishing the full range of behaviors of which a cell is capable in different conditions regardless of whether these behaviors are found in vivo or not.

Use of CreER for Lineage Analysis

The CreER system has become the standard method for tracing cell lineage in mammals. In its different guises it can be used either for fate mapping or for clonal analysis. In essence the method imparts a permanent label to cells that have a particular promoter active. If such cells form a coherent patch of tissue, as may often be the case in an early embryo, then labeling it will provide a fate map for the patch by showing where it ends up and what cell types it produces. If just a few of the cells are labeled then individual clones can be visualized and a clonal analysis can be carried out.

The specificity of the method comes from the promoter used to drive the CreER. This needs to be very reliable and it is best to use a knock-in of the *CreER* gene to the locus of interest to achieve this. The label can be a fluorescent protein, such as green fluorescent protein (GFP), or an enzyme, such as β-galactosidase, easily detectable by a histochemical method. The gene for the label protein is controlled by a ubiquitous promoter, often the *Rosa26* promoter whose normal function is to drive expression of a ubiquitous untranslated RNA. There is now available a set of reporter mouse strains which have various different reporter genes knocked in to the *Rosa26* locus. Between the promoter and the coding region is a transcriptional stop sequence flanked by loxP sites. The CreER enzyme is produced from cells in which its promoter is active, and labeling is initiated by dosing the animals with tamoxifen (or 5-hydroxytamoxifen). This activates the CreER as described above and enables excision of the stop sequence and consequent transcription of the reporter. Because the end result is a DNA modification it is permanent and the label is maintained regardless of subsequent cell divisions or cell differentiation events. The operation of the system is shown in Figure 3.6.

The proportion of cells labeled depends on the total tamoxifen dose, and usually multiple injections of tamoxifen are required to get a complete labeling of all cells. By giving a small single dose, it is possible to label just a small proportion of the cells, and, so long as the labeled cells are well separated, this procedure enables clonal analysis. As applied to stem cells it can reveal the domain of the tissue populated by a single stem cell and at

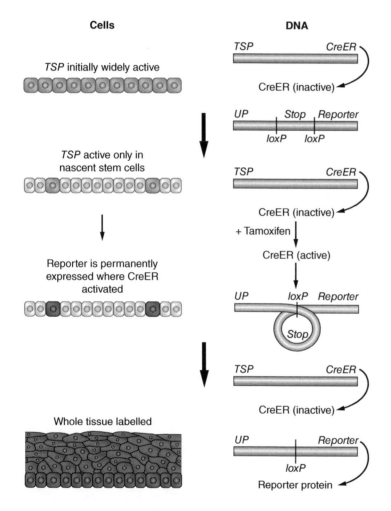

Figure 3.6 The CreER labeling method. This requires the production of mice containing two transgenes. *TSP* = tissue-specific promoter, *UP* = ubiquitous promoter, *Stop* = transcriptional stop sequence. In the scenario shown, *TSP* is initially active over a wide area, but CreER is not activated. At the time of addition of tamoxifen, *TSP* is only active in two nascent stem cells and so only these become modified and express the reporter. Because all descendants have the same modification, subsequently the whole tissue becomes labeled. (Author's figures (modified), first published in Slack, J.M.W. (2008) Origin of Stem Cells in Organogenesis. Science 322, 1498–1501. Reproduced with the permission of The American Association for the Advancement of Science.)

the same time prove its multipotentiality in terms of cell differentiation. More sophisticated versions of the clonal labeling strategy have been derived in which multiple clones are simultaneously labeled with different colors. These methods are generically known as "Brainbow" techniques. They depend on the use of variants of the loxP site such that alternative excision events can occur in the reporter locus, leading to the expression of different colored proteins depending on the particular excision. There are many different variants of Brainbow but one is shown in Figure 3.7 and an actual example of Brainbow labeling in Figure 3.C.1.

The CreER lineage labeling method has been hugely important in stem cell research. But like all methods, it has its problems. The biggest is that the technique depends absolutely on the fidelity of the promoter used to drive the *CreER*. Suppose that in a small region (region 1) of cells a promoter is active at a level of 100 while in another region of cells which is 10 times larger, (region 2) it is active at a level of 10. Let us say that the detection level for activity by in situ hybridi-

zation is 20, so activity can only be seen in region 1. In the CreER procedure an equal number of labeled clones are likely to come from the two regions because the product of promoter activity and cell number is the same. But it may erroneously be concluded that they all come from region 1 in which promoter activity is visible, rather than from region 2 in which it is not visible.

Retroviral Barcoding

A completely different method for clonal analysis has recently been used in the hematopoietic system. This involves making a "barcode" library in a retro- or lentivirus and using this to transduce the cell population under study. A DNA barcode is a short sequence ensemble, usually made by random nucleotide synthesis in vitro, cloned into an integrating viral vector. It is flanked by primer sequences enabling easy sequencing of the barcode from an individual provirus. The library should be designed such that the barcodes are approximately equally represented and that the number of barcodes is at

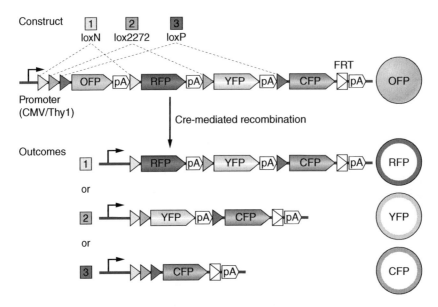

Figure 3.7 One relatively simple method for "Brainbow" labeling. The transgenic mouse contains the construct shown, which has three pairs of different types of loxP site. Depending on which excision event is brought about by the Cre, different colors are expressed. OFP (orange fluorescent protein) is initially expressed in all cells, but is lost in the recombinant clones. (Slack, J.M.W. (2013) Essential Developmental Biology, 3rd edn. Reproduced with the permission of John Wiley and Sons.)

least ten times the number of the stem cells in the experimental sample. The idea is that each stem cell will be transduced with a virus carrying a different barcode and that, following integration, this will be inherited on cell division and persist permanently regardless of the pathways of differentiation followed. The presence and abundance of any barcode can subsequently be measured in any experimental sample, taken in vitro or in vivo, simply by DNA sequencing. This method does not allow for fate mapping as there is no spatial visualization, but it is in principle a very sensitive method for counting clones in a complex cell mixture.

Clonal Analysis in Humans

None of the above techniques for cell lineage tracing are applicable to humans because they all depend on the introduction of transgenic constructs, which would be unethical under most circumstances. However there are some naturally occurring genetic markers in humans that can be useful in this context. Probably the most useful is the gene on the mitochondrial DNA encoding cytochrome c oxidase (CCO). Mitochondrial DNA suffers a high rate of mutation because mitochondria are constantly generating superoxides and other mutagenic free radicals, and because their DNA repair systems are much less effective than those of nuclear DNA. Although there are several copies of the mitochondrial DNA in each mitochondrion, and there are many mitochondria in each cell, it is nonetheless possible by somatic mutation and genetic drift to lose all copies of *CCO* from occasional cells by middle age. This is of particular importance for studying stem cells. Because it takes decades to establish a cell completely

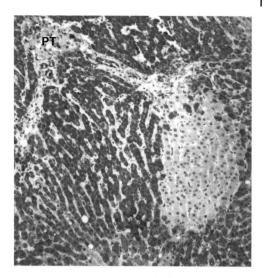

Figure 3.8 Clone of cells (light colored) in the human liver marked by a loss of function mutation of the mitochondrial gene for cytochrome c oxidase. The shape of the clone suggests that it derives from a stem cell in the periportal region. PT = portal triad (From: Walther, V. and Alison, M.R. (2016) Cell lineage tracing in human epithelial tissues using mitochondrial DNA mutations as clonal markers. Wiley Interdisciplinary Reviews: Developmental Biology 5, 103–117. Reproduced with the permission of John Wiley and Sons.)

null for *CCO*, in the renewal tissues that are constantly undergoing cell turnover, any such cells must be stem cells. The shape and composition of the clone indicates the domain of a single stem cell (Figure 3.8). Because mitochondrial DNA is present in many copies per cell, it is easy to check whether a mutant patch is genuinely a single clone by using laser capture dissection to isolate the patch and sequencing the *CCO* locus after PCR amplification. A monoclonal patch will contain only a single mutation, whereas a polyclonal patch will almost certainly contain more than one different mutation.

Further Reading

Buckingham, M.E. and Meilhac, S.M. (2011) Tracing Cells for Tracking Cell Lineage and Clonal Behavior. Developmental Cell 21, 394–409.

Büning, H., Perabo, L., Coutelle, O., Quadt-Humme, S. and Hallek, M. (2008) Recent developments in adeno-associated virus vector technology. Journal of Gene Medicine 10, 717–733.

Chen, Y., Cao, J., Xiong, M., Petersen, A.J., et al. (2015) Engineering human stem cell lines with inducible gene knockout using CRISPR/Cas9. Cell Stem Cell 17, 233–244.

Cockrell, A.S. and Kafri, T. (2007) Gene delivery by lentivirus vectors. Molecular Biotechnology 36, 184–204.

Danthinne, X. and Imperiale, M. (2000) Production of first generation adenovirus vectors: a review. Gene Therapy 7, 1707–1714.

Dykxhoorn, D.M. and Lieberman, J. (2005) The silent revolution: RNA interference as basic biology, research tool, and therapeutic. Annual Review of Medicine 56, 401–423.

Grosselin, J., Sii-Felice, K., Payen, E., Chretien, S., Tronik-Le Roux, D. and Leboulch, P. (2013) Arrayed lentiviral barcoding for quantification analysis of hematopoietic dynamics. Stem Cells 31, 2162–2171.

Horii, T. and Hatada, I. (2016) Production of genome-edited pluripotent stem cells and mice by CRISPR/Cas. Endocrine Journal 63, 213–219.

Hsu, Y.-C. (2015) Theory and practice of lineage tracing. Stem Cells 33, 3197–3204.

Kretzschmar, K. and Watt, Fiona, M. (2012) Lineage tracing. Cell 148, 33–45.

Kumar, P. and Woon-Khiong, C. (2011) Optimization of lentiviral vectors generation for biomedical and clinical research purposes: contemporary trends in technology development and applications. Current Gene Therapy 11, 144–153.

Low, B.E., Kutny, P.M. and Wiles, M.V. (2016) Simple, efficient CRISPR-Cas9-mediated gene editing in mice: strategies and methods. Methods in Molecular Biology (Clifton, N.J.) 1438, 19–53.

Primrose, S.M. and Twyman, R.M. (2006) "Gene transfer to animal cells". Chapter 12 in Principles of Gene Manipulation and Genomics, 7th edn. Wiley-Blackwell, Malden, MA.

Richier, B. and Salecker, I. (2015) Versatile genetic paintbrushes: Brainbow technologies. Wiley Interdisciplinary Reviews-Developmental Biology 4, 161–180.

Schonig, K., Bujard, H. and Gossen, M. (2010) the power of reversibility: regulating gene activities via tetracycline-controlled transcription. pp. 429–453 in Wassarman, P.M. and Soriano, P.M. (eds), Methods in Enzymology, Vol 477: Guide to Techniques in Mouse Development, Part B: Mouse Molecular Genetics, 2nd edn.

Sun, Y., Chen, X. and Xiao, D. (2007) Tetracycline-inducible expression systems: new strategies and practices in the transgenic mouse modeling. Acta Biochimica et Biophysica Sinica 39, 235–246.

Walther, V. and Alison, M.R. (2016) Cell lineage tracing in human epithelial tissues using mitochondrial DNA mutations as clonal markers. Wiley Interdisciplinary Reviews: Developmental Biology 5, 103–117.

4

Tissue Culture, Tissue Engineering and Grafting

Tissue culture, or cell culture, is the growth and maintenance of live cells outside the body. The techniques for doing this were invented in the early years of the 20th century and tissue culture became relatively easy from the 1950s, with the commercial availability of complex media, of sterile disposable containers, and of antibiotics to suppress microbial contamination.

Tissue culture is important for stem cell biology for several reasons. First, many stem cells exist only in culture and not in the intact organism. In particular, embryonic stem cells themselves are a tissue culture phenomenon as their equivalents in the early embryo are very short-lived and soon turn into something else. Second, the generation of useful differentiated cell types, such as dopaminergic neurons, hepatocytes or pancreatic beta cells, from pluripotent stem cells, requires very sophisticated tissue culture protocols in which the cells are taken through a sequence of different media containing various active components. Third, the delivery of cell therapy products by simple injection of a cell suspension is known to be a poor method involving a lot of cell death. Superior methods involve culture of cells on scaffolds of biomaterials and the grafting of implants rather than the injection of cells. Lastly, the holy grail of development of whole organs for transplantation means growing several different cell populations in close proximity on complex three dimensional scaffolds, and this requires very sophisticated tissue culture indeed.

In tissue culture the environment of the cells is under precise control (Figure 4.1a). The cells are growing on a specific substrate, they are bathed in a specific medium which provides nutrients, oxygen and regulatory factors. They also often interact with each other either through secreted substances or contact mediated interactions. Cells are usually viewed through the base of the culture vessel using an inverted microscope and phase contrast optics, which makes the almost transparent live cells stand out very clearly. Some different cell types growing in culture are shown in Figures 4.1b and 4.C.1.

It is very important to understand that cells in culture do not necessarily behave in the same way as their precursors in vivo. For example neurospheres, containing neural stem cells, can be cultivated from regions of the central nervous system known to contain no stem cells. Mesenchymal stem cells from the bone marrow are the precursors for bone in vivo, but in vitro they may also be caused to differentiate into adipocytes or smooth muscle. Embryonic stem cells can be maintained indefinitely in a growing, pluripotent state, but in vivo they rapidly become cells of more restricted potency. There are many reasons for these changes of behavior. The media used for culture are very rich and are designed to promote cell division, whereas the in vivo environment is usually appropriate for quiescence. Culture media contain bioactive growth factors and hormones that may differ from the repertoire available to the cells in vivo. Also,

The Science of Stem Cells, First Edition. Jonathan M. W. Slack.
© 2018 John Wiley & Sons, Inc. Published 2018 by John Wiley & Sons, Inc.
Companion website: www.wiley.com/go/slack/thescienceofstemcells

(a)

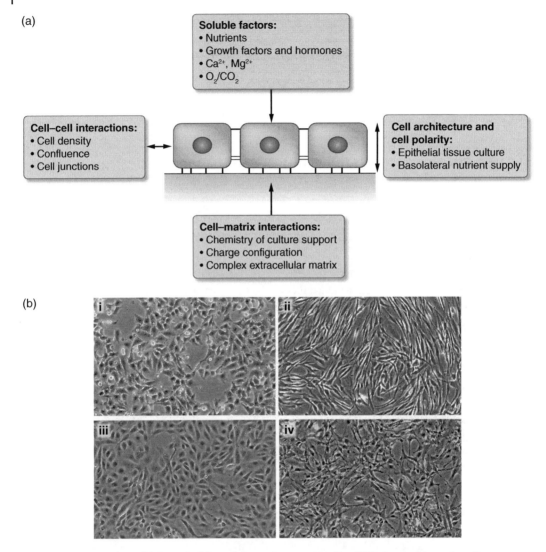

(b)

Figure 4.1 Tissue culture. (a) Control of the cellular environment in vitro. (b) Various cell types in culture. (i) Epithelial (HeLa); (ii) Fibroblastic (human mammary); (iii) Endothelial (CPAE); (iv) Astrocytes (human). (Sources: (a) From: Daniel Brunner et al. (2010) Serum-free Cell Culture: The Serum-free Media Interactive Online Database. Altex 27, 1/10 (2010). (b): ALTEX, ATCC.)

because of the high growth rates, tissue culture is a very selective environment so minority cell types with a growth advantage rapidly come to predominate over the others and hence change the composition of the culture. Some cell variants subject to selection also arise from epigenetic shifts or from somatic mutation and represent permanent changes from the original cell type. Regardless of these issues, tissue culture is a very important technology, especially for stem cell research. For many applied goals, such as generating a large enough population of cells for transplantation, the artificial nature of the tissue culture environment does not matter. However, it does need to be borne in mind in relation to those experiments which utilize tissue culture to gain insight into the actual situation within the organism.

Tissue engineering has grown up as a distinct discipline since the 1990s, and consists largely of the development of tissue culture technology by engineers. It is characterized by a more quantitative approach than is usual among biologists, the development of a number of novel chemical scaffolds for cell culture, and the adoption of various novel fabrication methods, such as lithofabrication or 3D printing, for making tissue and organ parts. The original aim of tissue engineering was to generate whole organs for implantation. This has turned out more difficult than originally envisaged, but will doubtless be achieved in due course.

Simple Tissue Culture

On a laboratory scale cells are usually grown in small plastic containers. These are made of polystyrene, which is optically clear and easily sterilized by irradiation. Polystyrene is a hydrophobic aromatic compound to which cells do not adhere. Untreated plates are used for bacterial culture or for animal cells in situations where cell adherence is not wanted, such as the formation of embryoid bodies from ES cells. To enable adherence the plastic is partly oxidized to generate some hydroxyl and carboxyl groups which make it "wettable". Proteins such as vitronectin and fibronectin from the medium will then bind to the plastic and provide a suitable surface for cell attachment. For specific application the plastic may be treated by adding a layer of fibronectin or collagen or polylysine to further improve cell attachment.

Some tissue culture vessels are shown in Figure 4.C.2. The most familiar container is the tissue culture flask ("T flask"). This has a flat bottom providing a good area for cell growth, and good visibility through the inverted microscope. It has a narrow mouth which helps to avoid infection by microorganisms, and is stackable, enabling many to be kept in an incubator. Cells are also often grown in simple petri dishes made of treated polystyrene, or in multiwell plates

that have different numbers of petri dish-like impressions made in a flat rectangular plate. These range from 6 to 96 well in number, the 96-well plate being the standard format for a lot of automatic machinery. Where large amounts of cells are required, roller bottles or cell factories may be used. Roller bottles are large cylindrical bottles which are rotated continuously so as to bathe the whole inner surface using a small amount of medium. The cells adhere and grow on the cylindrical surface. Cell factories are rectangular multi-storey plates which have a very large area for cell adherence. For both of these large scale methods visual observation of the cells is harder than with the small scale methods.

Some cell types can be grown in suspension culture. This may just be a simple dish or flask, or larger scale culture is performed in bottles incorporating a rotating stirrer arm. Since such bottles may be nearly filled with medium they allow for large amounts of cell production relative to similar sized surface culture methods, but for visual observation a sample must be taken out of the vessel for examination.

For really large scale cell production some kind of bioreactor must be used. This is a temperature controlled vessel with controlled flows of medium and gases and continuous monitoring of conditions (Figure 4.2). Massive bioreactors are used in the pharmaceutical industry for the production of antibodies or recombinant proteins from mammalian cell cultures. Bioengineering principles can be used to adapt these methods for the production of stem cells and many small bioreactor types have been developed for specific purposes.

Media

Tissue culture media (singular: medium) have a rational foundation in terms of the substances known to be important for cell nutrition and growth, but they have also evolved empirically and the choice of medium for a particular purpose often

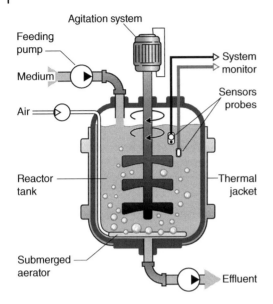

Figure 4.2 A typical design of bioreactor. (Wikimedia Commons.)

depends on usage and tradition. There is a wide range available ranging from simple "minimal" mixtures with about 30 components to very rich mixtures with over a hundred components.

Because of the importance of oxidative metabolism for ATP generation, cells need oxygen to support themselves. Tissue culture cells are usually grown in atmospheric oxygen concentration (about 20% by volume), although the optimum concentration is generally believed to be lower than this since the oxygen level in most parts of an animal's body is lower than the external atmosphere. Too much oxygen can be deleterious because it leads to the formation of free radicals that cause damage to cells. So, especially for stem cell culture, tissue culture incubators may be run at lower oxygen levels such as 5% by volume.

Mammalian cells will only remain in good condition very close to the normal body temperature, so good temperature control is essential. Most tissue culture incubators run at the human body temperature of 37 °C.

The salts in the medium often approximate to the levels found in blood plasma. Sodium

is the most abundant cation, with low levels of potassium, magnesium and calcium. The most abundant anion is chloride. Because water can pass across the plasma membranes of animal cells, the medium must match the osmolarity of the cell interior, otherwise cells will swell or shrink due to osmotic pressure difference. Media for mammalian cells generally have a total osmolarity about 350 mOsm. Apart from their importance in maintaining the overall osmolarity, the salts have other functions. Sodium and potassium are involved in the establishment of membrane potential which ranges from −10 mV internal in erythrocytes to −90 mV in excitable muscle fibers and neurons. Magnesium is a component of many enzymes and maintains the secondary structure of nucleic acids. Calcium is a key intracellular signaling molecule. Normally cytoplasmic levels of calcium ion are very low and admission of small amounts from the exterior medium has major biological effects.

The pH needs to be tightly controlled, 7.4 being normal. The pH control is usually achieved with bicarbonate-CO_2 buffers. Incubators are normally run at 5% CO_2 and media contain between 14–26 mM bicarbonate, depending on what other components are present. These give better results with most animal cells than other buffers, perhaps because bicarbonate is also a type of nutrient. Also often used is Hepes buffer, which has its pK at 7.4, although it can be slightly toxic to some cell types. There is one commonly used medium, Leibovitz L15, which relies on its amino acids for buffering and does not require a high CO_2 environment. Hepes buffered media or L15 medium equilibrate with normal atmospheric CO_2 so should not be placed in a 5% CO_2 incubator where they will become too acidic. Most media contain Phenol Red indicator. This has a red color at neutral pH and goes purple at alkaline pH and yellow at acid pH.

Glucose is usually present as an energy source at 5.5 mM, which is the normal resting level in human plasma. Some media have higher levels, up to as much as 25 mM, although this would correspond to serious diabetes in vivo. Amino acids are needed for protein synthesis. Media must contain the essential amino acids, which are those mammalian cells cannot synthesize. Richer media also contain the non-essential amino acids. Although these can be synthesized by the cells, if they are supplied ready formed the cells can put all their energy into new growth. Glutamine is an essential amino acid which is rather unstable and tends to become depleted from media over a few days, so is often added separately.

Most tissue culture media also contain animal serum, usually 10% fetal calf serum. This serves a wide variety of functions. It contains platelet derived growth factor (PDGF) which is released from platelets when the serum coagulates from whole blood. It also contains other growth factors including insulin, together with many vitamins, essential fatty acids, and trace elements. The high protein content of serum provides mechanical protection for the cells from shearing forces. Serum also provides additional pH-buffering capacity and neutralizes various toxins. The use of serum for tissue culture is a long standing practice but has two substantial disadvantages. Serum can never be completely characterized and there are often differences between batches that can be critical for experimental results. Ideally several batches should be screened for the purpose required, and then a sufficient supply of the best batch acquired to complete the program of work. There is also a small, but finite, risk of transmitting animal diseases via serum. This includes viruses, such as that causing bovine viral diarrhea, which may evade routine filter sterilization. It might also include prions, such as the causative agent of bovine spongiform encephalopathy (BSE), which was widespread in British cattle in the 1980s and 1990s. Serum for tissue culture is sourced from BSE-free cattle, but the lingering doubts over safety have led to a requirement for clinical grade cells to be grown without serum or other animal-derived materials.

Because of the various problems associated with the use of serum, there has been a progressive adoption of serum-free media for many purposes. The most important components are albumin, lipids (triglycerides, essential fatty acids, phospholipids, cholesterol), insulin, transferrin (an iron carrier protein), selenium (found in the amino acid selenocysteine), and an antioxidant such as 2-mercaptoethanol. But the optimal medium for each application is of course different in terms of specific nutrients and signaling molecules. One benefit of avoiding serum is that it becomes possible to give more attention to optimization of media for each specific purpose.

Contamination

Because tissue culture media are very rich in all sorts of nutrient they are an ideal situation for the growth of bacteria and fungi. Most media contain the pH indicator Phenol Red, which turns from red at neutral pH to yellow at acidic pH. Microorganisms generally grow fast and their metabolism generate a lot of CO_2 and organic acids so a contaminated medium rapidly turns yellow. Contaminated cultures should be discarded wherever possible as attempting to "cure" them with antibiotics risks selecting for antibiotic resistant bacteria. The usual precautions against contamination are as follows. First the use of a class 2 microbiological safety cabinet. A class 2 cabinet supplies a downwards flow of sterile, filtered air to the work and also draws in air downwards from the window at the front. Class 2 cabinets are found in most tissue culture labs and provide good protection for the cells against contamination and for the worker against any pathogens being handled in the cabinet. Another line of defense is the use of careful sterile procedure by the worker, in particular avoiding touching or breathing upon any surface which may contact the cultures. Often forgotten is the importance of sterilizing the waste medium which is removed from flasks during routine subculturing. This is often sucked out and accumulates in a non-sterile container all ready to grow microorganisms. In a well-managed lab frequent sterilization of waste lines is essential. Finally, many tissue culture media incorporate antibiotics which make the prospect of microbial growth very small. The standard ones are penicillin and streptomycin for bacteria and amphotericin B (fungizone) for fungi. Gentamycin is also often used as it has a wider antibacterial spectrum than penicillin and streptomycin, and can be used effectively to culture cells from contaminated sources. Nowadays it is generally considered bad practice to rely on the continuous presence of antibiotics to avoid contamination because it leads to less good aseptic technique and the antibiotics themselves may be slightly toxic to the cells. If antibiotics are used then careful disposal of any cultures that do become contaminated is all the more important as the contaminating organisms will now be antibiotic resistant.

Mycoplasmas pose a particular contamination problem in most tissue culture laboratories. They are very tiny bacteria which lack a cell wall and so are resistant to penicillin. Unlike normal bacteria or fungi, their effects on cell cultures can be quite subtle and they can persist unobserved for some time. They can be detected by staining the cultures with fluorescent DNA dyes and observing small positive bodies distinct from the cell nuclei. Alternatively there are now available PCR-based kits for mycoplasma detection. If a culture does develop mycoplasma it is good practice to discard it.

Quite apart from microbial contamination, it is also important to be alert for the possibility of contamination of one cell line for another. Many tissue culture lines look the same so contamination is often not obvious. If more than one line is handled in the same cabinet there is a small but real chance of cross-contamination and this has sometimes led to serious errors in high profile publications. Because of the super-selective environment of tissue culture, it does not take long for one line to be completely overgrown by a slightly faster growing one which was introduced by accident. Human cell lines can be verified by DNA-based tests similar to those used for forensic identification. These are based on PCR of a set of short tandem repeat loci where the number of repeats is very variable between individuals. Methods for non-human cells are as yet not so well standardized, but there is a variety of DNA-based and other tests available.

Growth in Culture

Primary cell cultures are grown directly from tissue explants and often retain some differentiated character from the cells in vivo. However tissue culture is a highly selective environment and unless care is taken to prevent it most primary cultures become overgrown by fibroblasts. Even if they can be passaged (i.e. subcultured), primary cultures have a finite lifespan. This is due to various causes including the shortening of the telomeres at the ends of chromosomes at each cell division, which eventually leads to chromosome damage; and to the gradual accumulation of Cdk inhibitors such as p16 and p21, which inhibit the cell cycle. In order to make available large numbers of cells from one primary culture a master bank should be set up and from this a number of working banks created which are of low passage number. Most primary cells lines sold by suppliers are low passage number cultures. The age of cultures can be expressed either as passage number, which can be known precisely, or as cell doublings which is always an approximation as cells in a dish do not all divide at the same rate.

Permanent cell lines have no limit to their propagation. They have all undergone mutation and selection to establish tumor-like properties which usually include restoration of telomeres on division, lack of accumulation of Cdk inhibitors, and often constitutive overactivity of one or more intracellular signaling pathways favorable to cell division. Permanent cell lines are always far removed in phenotype from their cells of origin although they may retain a few differentiated characters and because of their convenience are often used for experimentation.

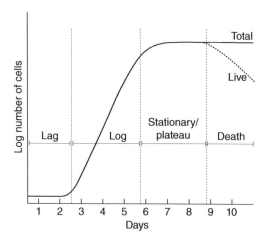

Figure 4.3 Growth curve of cells in culture. (From ATCC Primary Cell Culture Guide p.2.)

A typical growth curve for a cell line in culture is shown in Figure 4.3. Initially, after subculturing, there is a lag phase during which the cells attach and adapt themselves to the new environment. Then there is a phase of logarithmic growth when the medium and space are present in excess and most or all of the cells are dividing. Eventually the cells are all contacting one another (confluent) and growth is slowed by contact inhibition. Some tumor lines do not show contact inhibition and the cells continue to pile on top of each other to achieve an even higher density. Cells are at their healthiest during the log phase and cell culture procedures usually aim to keep them in log phase for as much time as possible by replacing the medium regularly and subculturing the cells before they reach confluence. This practice underpins one of the biggest differences between cells in tissue culture and those in vivo. In vivo cells are almost never undergoing exponential growth. Instead they are usually quiescent or undergoing slow growth.

Subculturing is usually carried out by treatment with the enzyme trypsin, which degrades much of the extracellular and cell surface protein and makes the cells drop off the substrate and become roughly spherical bodies in suspension. The trypsin is added in the absence of serum, often with EDTA to remove Ca^{2+}, which disrupts the cadherin-based cell contacts. After the cells have dropped off, new serum-containing medium is added. Among other things, serum contains a trypsin inhibitor which terminates the action of the enzyme. The cells are transferred at lower density into new flasks. They take a few hours to resynthesize their surface molecules and can then adhere to the new substrate. Other, milder, protease mixtures are now available instead of the traditional trypsin. When cells are subcultured the degree of dilution depends on the characteristics of the cells. Usually something between 1:2 and 1:10 is used. Too much dilution leads to a lot of cell death as the cells depend on survival factors secreted by their own number. For this reason, cloning of cells, i.e. growth of a colony from a single isolated cell, is often difficult, requiring specialized media and lots of care and attention. It is sometimes thought that clonogenic cells in culture are the same as stem cells but, as indicated elsewhere in this book, this is not necessarily the case.

Cryopreservation and Banking

It is bad practice to cultivate cells for too long by continual subculturing. To do so inevitably brings about genetic changes due to somatic mutation and selection in the rapidly growing culture situation and this will eventually alter the properties of the cells quite markedly. Prolonged culture also increases the risk of cross contamination with other cell lines, or of microbial contamination. Moreover, as mentioned above, primary cell lines have a finite lifespan and will senesce as this limit is approached.

To store live cells, cryopreservation is standard. Normal freezing will kill cells because of the formation of ice crystals and the consequent mechanical damage. But this can be avoided by slow cooling (1 °C/minute) in the presence of a cryoprotectivate agent, usually 10% dimethyl sulfoxide or glycerol. Serum also has some cryoprotective effect, so extra serum may be added, or for serum free cultures some alternative such as methyl cellulose will be added. Once frozen and reduced to below −70°C, the vials are stored in liquid nitrogen which has a boiling point of -196°C. Storage may be actually within the liquid nitrogen itself, or more usually in the vapor phase above it which has a temperature of around −130°C. For recovery, vials are warmed rapidly to 37°C and diluted into complete growth medium. There is always some cell death but preservation of the line is generally reliable.

Especially for a newly isolated primary cell line, it is essential to bank it before it is lost or becomes contaminated. Commercial suppliers and national cell repositories have elaborate cell banks to ensure consistency and reliability in their supplies. To make a cell bank, a small number of vials are frozen at low passage number to make a master bank. One vial will be thawed and expanded for a few more passages to make a large working bank. At both stages there is careful control for authenticity and absence of microbial contamination. Then for a group of experiments one working vial will be thawed and used. After a limited number of passages the cells will be discarded and replaced by those from another working vial. This procedure ensures consistent performance from the cells and also enables them to be donated to a repository in good condition so that others may also use them. Eventually the bank will run out, but if it is well planned this may not be for many years and by then it should be possible to generate similar cells from a new primary isolation.

GMP Cultivation

As mammalian cells have become the source of various pharmaceutical products, and as live cells derived from cultured stem cells have started to enter the clinic for transplantation therapy, the regulatory requirements for cell culture have been considerably stiffened. It is normally required that cells be grown using "Good Manufacturing Practice" (GMP). The precise standards for this differ somewhat in the USA and Europe, but the essence is very precise regulation of condi-

tions and very detailed record keeping. All the substances used for cultivation must have their origin traced and all the manipulations must be carried out using approved standard operating procedures (SOPs). For GMP cell culture it is generally felt that media should avoid all animal-derived components to avoid any risk of infection with animal-derived viruses. Ideally serum should be replaced by serum free culture formulae, and animal derived proteins by recombinant equivalents. The actual operations need to be undertaken to a very high standard, often requiring a "clean room" to work in, which is a self-contained area with a sterile filtered air supply. The costs of GMP are considerable compared to those of normal laboratory practice. But if application in the clinic is contemplated in the foreseeable future then it is wise to ensure that as many as possible of the less costly elements underpinning GMP production of the necessary cells are already in place.

Complex Tissue Culture

Induced Differentiation

Most tissue culture is conducted with a small set of media incorporating 10% fetal bovine serum. For simple expansion, this usually works, but it is obvious that to elicit the full range of cell behaviors much more specificity of media and culture conditions is necessary.

From a morphological point of view most tissue culture cells can be regarded as epithelial or mesenchymal. These terms relate to cell shape and behavior rather than to embryonic origin. An epithelium is a sheet of cells, arranged on a basement membrane, each cell joined to its neighbors by specialized junctions, and showing a distinct apical–basal polarity. Mesenchyme is a descriptive term for scattered stellate-shaped cells embedded in loose extracellular matrix. The origin of the ubiquitous fibroblasts that grow out of most tissue explants in primary culture is not entirely clear but they are generally thought to be derived either from the fibroblasts of the dermis or from the pericytes of the vascular system.

The majority of differentiated cell types do not divide in culture, so after differentiation they will be held in a static culture while they are used for experiments. The methods of tissue culture include various procedures for causing cells to differentiate, often evolved empirically. For example the mouse C2C12 cell line is composed of myoblasts that will differentiate into myotubes if the fetal bovine serum is replaced by horse serum. The Friend erythroleukemic cell line will differentiate into erythrocytes if treated with dimethyl sulfoxide. In the case of pluripotent stem cells (embryonic stem cells or induced pluripotent stem cells) enormous efforts have been devoted to devising methods to control their differentiation. In general the protocols have been designed based on the understanding of the normal course of events in the embryo, established by developmental biologists. The protocols therefore are multistep, involving treatment of the cells for appropriate periods with as many as eight different media containing specific hormones, growth factors or small molecule agonists and antagonists. This topic will be further discussed in Chapter 9.

Three Dimensional Cell Culture

As indicated above, most tissue culture involves growing cells on flat, two dimensional surfaces, or in suspension. But it has always been recognized that this is a very artificial situation and various methods for growing cultures in three dimensions have been developed (Figure 4.4a). There is a long tradition of in vitro culture of organ rudiments such as those of kidney, lung or salivary gland from mammalian embryos (Figure 4.4b). These cultures are of short

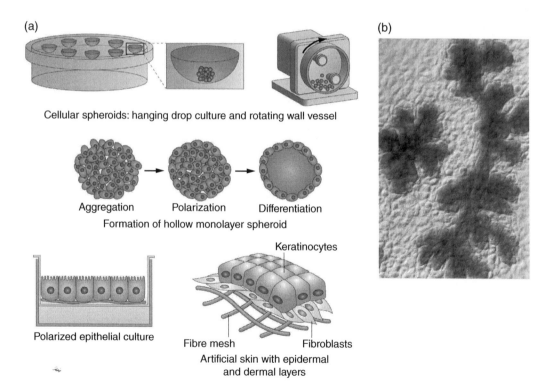

Figure 4.4 (a) Various procedures for growing cells in three dimensional configurations. (From: Pampaloni, F., Reynaud, E.G. and Stelzer, E.H.K. (2007) The third dimension bridges the gap between cell culture and live tissue. Nature Reviews Molecular Cell Biology 8, 839–845. Reproduced with the permission of Nature Publishing Group.) (b) Organ culture of a mouse embryo pancreas. The epithelium is stained for a β-galactosidase reporter, the mesenchyme is unstained. (Author's photo.)

duration and are used for developmental biology research. They consist of multiple cell types in an intimate relationship with each other. Although such cultures do expand they do not grow nearly as much as normal tissue culture cells and are terminated once the experiment is over. The experience of embryonic organ culture has resulted in the production of two new techniques: the use of three dimensional (3D) substrates such as collagen gels, and the culture of explants at an air-medium interface on a porous filter.

The key requirement for 3D culture is a suitable substrate. In vivo, cells are surrounded by extracellular matrices. These contain collagen, fibronectin, laminin, glycosaminoglycans and many other components. A commercial matrix preparation called Matrigel has been extensively used for 3D culture. This consists of matrix secreted by Engelbreth–Holm–Swarm (EHS) mouse sarcoma cells. It is liquid at low temperature but solidifies to a gel at 37 °C and is a very compatible environment for cell culture. Some remarkable differences in cell behavior occur in Matrigel compared to cultures on plastic. For example, Madin–Derby canine kidney (MDCK) epithelial cells will form polarized cysts in Matrigel, and when treated with hepatocyte growth factor (HGF) will form branched tubules. Much publicity has been accorded to the ability of human pluripotent stem cells to generate well organized organ rudiments in culture including correctly layered portions of retina or cerebral cortex (Figure 4.C.3). Useful though Matrigel is, it will clearly be necessary to devise more discriminating and specific substrates for different purposes.

One approach to this is to start from the hydrogels introduced by tissue engineers. These are polymers such as polylactic acid or polyethylene glycol which can absorb large amounts of water to produce a loose matrix allowing good diffusion of nutrients and dissolved gases. The chemical composition and polymerization characteristics of the hydrogel determine its porosity and stiffness. The polymers used tend not to be very adhesive

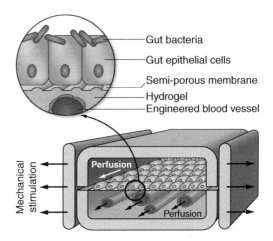

Figure 4.5 "Gut on a chip". (From: Gjorevski, N., Ranga, A. and Lutolf, M.P. (2014) Bioengineering approaches to guide stem cell-based organogenesis. Development 141, 1794–1804. Reproduced with the permission of The Company of Biologists Ltd.)

to cells, but can be modified in various ways. For example the cell binding domain from fibronectin can be added to promote cell adhesion. Growth factors can be covalently bound, or sequestered on heparin chains incorporated into the matrix. For products intended for implantation in to animals or patients, hydrolyzable cross links can be incorporated to promote gradual degradation of the matrix. Sometimes the continuous application of mechanical forces is required for optimum development of the cells, for example periodic stretching for development and maturation of cardiac muscle. In principle, synthetic scaffolds are superior to natural ones because of the ability to tailor them for precise requirements. Moreover the existing technology of photolithography can be used to shape the scaffold to a desired form ready for use.

An example of a complex tissue engineered model is the "gut on a chip" shown in Figure 4.5. This consists of a layer of gut epithelial cells (actually a permanent cell line) on a porous membrane. Beneath this is a hydrogel containing channels lined with endothelial cells to mimic blood vessels. Medium is perfused continuously through these vessels and the whole is subjected to peristaltic movements by periodically evacuating vacuum chambers on either side. It is even possible to add appropriate gut bacteria to the luminal side of the cell sheet to achieve a close approximation of the in vivo situation.

Artificial Organs and Organoids

The holy grail of tissue engineering is the creation of entire artificial organs, or at least pieces of organized tissue, that can be grafted into patients. At present these are relatively simple, such as artificial skin consisting of an epidermal and a fibroblastic layer, or small cartilage implants which have been grown in vitro. For more demanding applications there are various major problems which need to be overcome.

First, there is the supply of cells. Many of the cell types desired, such as neurons or cardiomyocytes, are post-mitotic and do not grow either in vivo or in vitro. Human differentiated cells are especially hard to obtain, but in future we can expect them to be manufactured from pluripotent stem cells, as discussed in Chapters 5 and 9, and methods for doing this on a large scale are continually being improved. Second, there is the maintenance of a structure containing more than one cell type in a stable form. Where there are multiple cell types in the same environment, one can easily overgrow another. It seems likely that stability will require some mutual support between the cell types such that they each secrete factors required for the other's survival. Third, there is the issue of nutrition. The limit for supply of nutrients and removal of waste products by diffusion is about 100 μm, or 4–7 cells linear dimension. Any structure thicker than this needs to incorporate a vascular system through which medium can be perfused before transplantation such that host blood can enter after transplantation. Finally the whole should maintain itself in a stable condition. If there is no cell growth at all then there can also be no cell death or the whole implant would die. However, cell growth needs to feed cell renewal in a stable

manner and not be uncontrolled as this would lead to formation of a tumor. These are demanding requirements and the problems have not all been solved. In recent years two types of approach have been developed for making 3D structures that solve at least some of the problems: 3D printing and recellularization of decellularized organs.

3D printing of cells follows the same principle as 3D printing of physical objects. The structure required is built up in layers using a device similar to an inkjet printer, controlled by a program containing the digital information for the required structure. A prototype device for printing tissue implants is shown in Figure 4.6. This consists of two parts: an xyz stage controller to move the stage while different components are extruded from the nozzle; and a cartridge for each component in liquid form, with a pressure regulator for dispensing them. The stage is enclosed in a temperature controlled, humidified container and the cartridges are kept warm enough to maintain the contents liquid. For those containing cells this obviously cannot be more than 37 °C. The structure required is digitized from a CT or MRI scan. The digital information goes into a computer-aided design program which controls the pressure regulator and the xyz movements of the nozzle. The cells are mixed into a medium containing gelatin, fibrinogen and hyaluronic acid. This is viscous enough to keep shape after extrusion and can later have the fibrinogen polymerized to fibrin by addition of thrombin, which causes gelation and makes the whole structure reasonably self-supporting. A supporting polymer, poly ε-caprolactone, is also woven through the cell pattern to provide extra structural rigidity. The cells and matrix are deposited in a weave that leaves many channels for diffusion of nutrients and waste materials. This method has been used to create pieces of bone and of skeletal muscle which have been successfully grafted into animals.

If 3D printing is a bottom up approach then decellularization is the top down approach to solving the same problems.

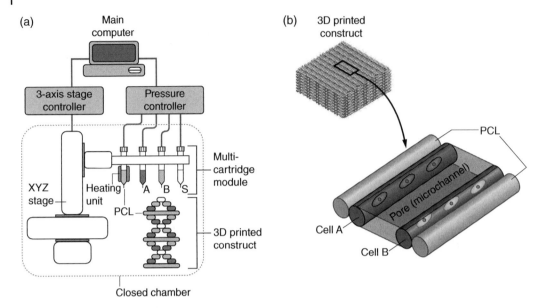

Figure 4.6 3D printing of cells to generate an artificial tissue implant. (a) Design of the apparatus. (b) A tissue construct containing two cells types with structural support and diffusion channels for nutrients. PCL = polycaprolactone. (From: Kang, H.W., Lee, S.J., Ko, I.K., Kengla, C., Yoo, J.J. and Atala, A. (2016) A 3D bioprinting system to produce human-scale tissue constructs with structural integrity. Nature Biotechnology 34, 312–319. Reproduced with the permission of Nature Publishing Group.)

The idea is that nature has done a very good job in evolving organs so why not start from these? If an organ is perfused over a few days with a 1% solution of sodium dodecyl sulfate (SDS) most of the cellular components are dissolved and washed away, leaving just the extracellular material. This is surprisingly substantial and decellularized organs are the same shape as when they started but are somewhat translucent or have a deathly white color (Figure 4.C.4). The idea is that cells of the appropriate types can then be reintroduced into the decellularized structure. These might, for example, be human cells into a decellularized pig heart. Decellularization removes all the components involved in the immune rejection of grafts so this does raise the prospect of creating an unlimited supply of human organs for transplantation. However there is real difficulty reintroducing cells. The blood vessels can be lined with endothelial cells, and some cells will migrate from the blood vessel channels into the matrix around them, but it can be difficult introducing large numbers of cells by direct injection into the matrix, which tends to be rather rigid. Nonetheless, even if it is a while before whole organs can be assembled containing the correct cell populations, decellularization and recellularization is likely to be an important route for creating smaller tissue grafts.

Grafting

The immune system of vertebrate animals has evolved to combat infection, but as a by-product it also acts as a major barrier to the grafting of cells, tissues or organs from one individual to another. Because the aim of much stem cell research is the production of cells for transplantation, it is important to understand some of the issues related to grafting and graft rejection. This discussion relates to graft rejection in adult, or at least in postnatal, organisms, as there is no graft rejection in embryos before the immune system has developed.

Grafts are described as being auto-, iso-, allo- or xeno-grafts. Autografts are from another part of the same individual. They are genetically identical and so provoke no immune response. Isografts are from a different individual who is genetically identical or almost so, such as an identical twin or another mouse of the same inbred strain. These also usually provoke no rejection. Allografts are from another individual of the same species and are normally rejected. They differ in the gene variants (alleles) they carry at numerous genetic loci. A genetic locus is considered to be polymorphic at the population level if it has more than one allele present at an appreciable frequency. Particularly important for graft rejection is the polymorphism of the Major Histocompatibility Complex (MHC) which is found in all vertebrate animals. The MHC of humans is called the Human Leukocyte Antigen (HLA) complex, and that of mice is the H-2 complex. There are certain sites within the body where the immune reaction to grafts is less severe. In particular this is true of the eye and to some extent the brain. The reasons for this so-called "immune privilege" are complex and it is due to active factors as well as to a relative absence of some of the usual factors leading to rejection.

Around the world there are currently about 100,000 solid organ grafts carried out in human patients each year. Because the procedures are often life-saving, the number would be much higher if two critical problems could be solved: the shortage of donor organs, and the difficulties associated with graft rejection and immunosuppression. Stem cell biology has the potential to address both of these problems. If cell, tissue or organ grafts were made from induced pluripotent stem cells (iPS cells) derived from the patient him- or herself, then they should in theory be a perfect immunological match requiring no immunosuppression. Furthermore, in vitro stem cell technology now offers the possibility of growing sufficient cells to overcome the shortage of donors. At present there are still many practical and financial obstacles to both of these possibilities but they remain important goals for the future.

The Immune System

The immune system comprises the innate system and the adaptive system. The innate system is based on recognition of various molecular motifs characteristic of microorganisms and is rapid in action. It is relevant to graft rejection especially through the properties of natural killer (NK) lymphocytes, which recognize and kill cells lacking the self-MHC specificity. The adaptive system is based on the recognition of just about any foreign molecule and the subsequent amplification of the cells recognizing it to mount a specific immune response. The adaptive response is slower than the innate response but has enormous scope. It generates both cytotoxic cells and antibodies directed against the antigen, and also generates specific memory cells which can rapidly proliferate and react again if the same antigen is detected in the future.

T Cells

A key role in adaptive immunity is played by the T (= thymus-derived) lymphocytes which carry T cell receptors (TCR) for the recognition of antigens. The receptors vary considerably between different individual lymphocytes because they are formed after a process of DNA rearrangement. Each receptor is a heterodimer of α and β chains and each chain is transcribed from a gene assembled in that specific cell from many coding regions in the genome. The α chain consists of a Vα sequence joined to a Jα sequence, and the β chain consists of Vβ, Dβ and Jβ, each of which is selected from a large ensemble of germ line sequences. The TCR is present on the cell surface as a complex with a four chain coreceptor called CD3, which is essential for TCR function. The DNA cutting of the sequences making up the TCR is carried out by lymphocyte-specific nucleases encoded by Recombinase Activating Genes (*RAG1* and *2*) and the splicing together is carried out by a set of enzymes involved in normal DNA repair. The number of possible T cell receptors arising from this process is extremely large and enables the recognition of a huge range of possible antigens. T cells get their name because their maturation occurs in the thymus gland. The DNA rearrangements occur in the outer region and then there are two stages of selection in the inner thymus. First there is a selection for affinity to MHC proteins, and second there is a selection against recognition of self-proteins, most of which are produced at a low level within the thymic medulla itself. Because of the random nature of the DNA rearrangement, only a few percent of the new T cells survive the full selection process and exit to the circulation. There are many types of T cell, but the two main types are the cytotoxic T cell, which carries the CD8 glycoprotein on its surface, and the helper T cell which carries the CD4 glycoprotein.

Immunological tolerance can develop in two ways. One is central tolerance due to selection against T cells in the thymus. This requires introduction of the antigen to the thymus before the major phase of T cell maturation, and its maintenance throughout life. The second is peripheral tolerance due to inactivation of specific T cells clones because of insufficient co-stimulation or excessive co-inhibition. Peripheral tolerance can develop under various circumstances but in general tissue grafts require some life-long immunosuppression if they are to survive.

The Major Histocompatibility Complex

In most circumstances T cells cannot recognize antigens on their own using the TCR, but only when the antigen is presented as a complex with proteins of the Major Histocompatibility Complex (MHC). What follows is based on the human HLA system. HLA consists of a number of gene loci many of which are very highly polymorphic. There are two groups of loci, called class I and class II which have somewhat different functions.

The class I proteins, encoded by the genetic loci A, B and C, are found on the cell surface of all nucleated cells although most antigen presentation to T cells is carried out by dendritic cells and macrophages. The MHC class I proteins each consist of an α-chain combined with a molecule of β2-microglobulin. Foreign proteins within the presenting cell are processed to peptides and the peptide is loaded onto the class I molecule within the endoplasmic reticulum, then transported to the cell surface where it is displayed in conjunction with costimulatory molecules such as CD80 and 86 (also called B7-1 and 7-2). This complex can be recognized by a cytotoxic T cell having a TCR complementary to the peptide-MHC combination, combined with the CD3 co-receptor and the CD8 glycoprotein.

The class II HLA genes are called DR, DQ and DP. They are expressed on B lymphocytes (B for "bone marrow-derived)", dendritic cells, monocytes and macrophages. They can also be expressed on endothelial cells, especially when stimulated by interferon γ, and also on some epithelia. The proteins consist of α and β chains. Like the class I proteins these present antigenic peptides on the cell surface. The difference is that the peptides derive from proteins taken up by the cells and processed in endosomes, before being introduced into the endoplasmic reticulum to combine with the class II molecules. The class II presentation is to helper T cells, which recognize the peptide/class II/co-stimulator complex by means of their own TCR/CD3/CD4 complex (Figure 4.7).

This system was evolved to eliminate microorganisms, especially viruses, that were present within cells, and, as we shall see, it leads to cell killing by several methods. The reason that this is important for graft rejection is that T cells perceive the peptides from graft antigens, especially those from the allogeneic MHC molecules on the graft cells, as being foreign. The situation has arisen because of the extensive polymorphism of the MHC complex although this polymorphism presumably did not evolve simply in

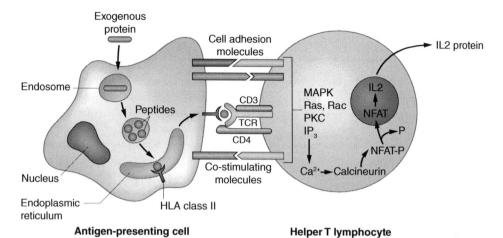

Figure 4.7 T cell activation. This shows activation of a T helper cell by an antigen presenting cell which has absorbed an exogenous protein, processed it to peptide and is presenting it with class II HLA to the T cell. One consequence of the activation is the secretion of IL2. MAPK = MAP kinase, PKC = protein kinase C, NFAT = nuclear factor of activated T cells, IL2 = interleukin 2.

order to resist grafting by human surgeons. It may be that the high polymorphism increases the diversity of T cell recognition capacity at the population level and hence enables at least some members of a population to survive an assault by a novel pathogen. But this problem lies in the realm of evolutionary biology and continues to be debated.

T and B Cell Responses

The stimulation of the T cells leads to phosphorylation of the cytoplasmic region of the CD3 complex and this activates several internal signal transduction pathways, especially the inositol trisphosphate (IP_3)/ protein kinase C (PKC) pathway, the mitogen activated protein (MAP) kinase pathway, and the GTP exchange proteins Ras and Rac (these pathways are depicted in Figure 7.4). An important limb of the signal transduction is the elevation of intracellular calcium ion provoked by the action of IP_3. Elevated Ca^{2+} combined with calmodulin activates the protein phosphatase calcineurin. This dephosphorylates a group of transcription factors, the Nuclear Factors of Activated T cells (NFATs), and thereby enables them to enter the nucleus. A key target of NFATs is the gene encoding the cytokine

interleukin 2 (IL2), so activated T cells produce IL2, which strongly stimulates their own proliferation. The process of stimulation of cytotoxic and of helper T cells is similar. In both cases the clones of cells showing specific recognition of the antigen become expanded over several days, increasing the graft rejection capacity. In addition a population of cells with the same TCR specificity is expanded as a reserve population of memory T cells which can quickly be activated in the event of a second stimulus of the same sort. This is why a repeat graft from the same individual is more rapidly rejected than the first graft.

Cytotoxic T cells kill their target cells directly. They secrete perforins, which punch holes in the target cell membrane, and granzymes which are proteases activating caspases in the target cell and thus bringing about apoptosis. Helper T cells function mostly by production of a range of cytokines which attract and stimulate other immune cells including various types of phagocyte. A subset of helper T cells become regulatory T cells ("T-regs") which damp down or inhibit specific immune responses. Helper T cells are also critical in enabling antibody production by B lymphocytes.

The B (bone marrow-derived) lymphocytes produce antibodies, also called immunoglobulins. Immunoglobulins, like T cell receptors, are generated following DNA rearrangements. Each molecule contains two heavy and two light chains and the N-terminal halves of these are the variable (V) regions assembled by DNA splicing of many germ line sequences. As for T cell receptors, the DNA cutting is carried out by the RAG1 and 2 nucleases. When B cells mature they carry antibody molecules of the IgM and D class, each with the same V regions, on their surfaces as part of a B cell receptor complex. When this recognizes an antigen it can present it to a helper T cell, along with class II MHC and costimulatory molecules. If a helper T cell recognizes this with its own TCR then it signals back to the B cell via the CD40 molecule as well as via the secretion of various cytokines. This provokes a further DNA rearrangement enabling production of secreted immunoglobulin of a different class, especially IgG, as well as proliferation of the B cell. Activated B lymphocytes either become antibody secreting plasma cells or memory B cells, which, like the memory T cells, are available to mount a rapid assault if the same antigen appears again. Antibodies kill target cells by binding to them and activating the complement cascade of serum proteins, which culminates in cell lysis or phagocytosis.

Reactions to a Graft

Most accounts of graft rejection deal with human organ transplantation. Because of the complexity of human organs processes of rejection involve events that would not occur following a graft of purified cells grown in vitro. For example, rejection of solid organ grafts depends a lot on the migration of dendritic cells out of the graft, which cannot occur if dendritic cells are not present. However, the more complex are the tissue engineered structures that are grafted, the more the situation will resemble that of a human organ graft.

The first problem is the potential for hyperacute rejection due to the presence of reactive antibodies in the host. These cause complement activation in the graft leading to thrombosis and death. There can also be rapid reactions, based on the innate immune system, from dendritic cells, macrophages, neutrophils and natural killer (NK) lymphocytes of the host. Starting from the moment of grafting, conditions are created for immune rejection arising from the injury associated with the surgery and the consequent inflammation. This involves secretion of numerous cytokines, such as IL2 and interferon γ (IFNγ), by inflammatory cells, and the recruitment of more immune cells into the graft. The processes of acute cellular rejection may occur over the first 6 months. An important component is due to dendritic cells from the graft stimulating T cells from the host with activation of cytotoxic and helper responses (direct response). Antibody presenting cells from the host will also pick up cellular debris, including donor MHC molecules, from the graft and present this to host T cells (indirect response). In addition to the T cell responses there may also be a host antibody response against antigens from the graft. If the graft survives all this then there is still the potential for chronic rejection leading to progressive tissue damage and fibrosis. This may involve some additional mechanisms as it is less susceptible to treatment with immunosuppressive drugs than acute rejection. Bone marrow transplants involve their own particular problems because the graft itself contains so many immune cells and new ones are being continuously generated from the hematopoietic stem cells of the graft.

The level of immune reaction against a graft depends in large measure on the degree of mismatch of the MHC alleles between donor and host. Various methods of HLA typing based on antibodies or on DNA sequencing are used to determine this. Because the HLA loci are closely linked to one another they are usually inherited together as a "haplotype" and a single individual will have one haplotype from the mother and another from the father. This means that there is a 1/4 chance that any two siblings will have identical haplotypes and such siblings are often favored as graft donors if they are available. The availability of good immunosuppressive drugs means that HLA matching is now considered less important for kidney grafts, although it remains very important for bone marrow grafts. It needs to be remembered that even if there is a perfect match of the principal MHC loci, there are also many minor histocompatibility loci that can still lead to recognition and rejection by the immune system.

Because there are never enough organs for human transplantation, much attention has been given to the possibility of using animal organs, for example those from pigs, which are a similar size and physiology to humans. This is called xenografting. However it involves substantially greater problems of graft rejection than found for allografts. Hyperacute rejection occurs based on the presence of antibody in humans directed against the cell surface carbohydrate group: Gal-α1-3 Gal (the α-Gal epitope) which is present on endothelial cells of animals other than primates. Genetically modified pigs have been bred lacking this epitope but the strength of graft rejection is still considerable.

Immunosuppressive Drugs

Apart from the special case of identical twins, the feasibility of human organ or cell grafting depends very largely on the use of a set of highly effective immunosuppressive drugs that have been discovered in recent decades which interrupt various steps in the immune response (Figure 4.8). Usually an aggressive regime is applied at the time of the graft and dosage is then reduced to the minimum required for graft maintenance. Because the immune system was evolved to fight infection, immunosuppression which depresses immune activity inevitably increases the risk of infection. Also, the drugs have many side effects because they target some generic

biochemical processes that are operative in many situations apart from the immune system. In general human recipients of solid organ grafts have to receive immunosuppression for life and the consequent morbidity is one of the limitations of transplantation.

A few of the main drugs important in human transplantation, and also used in animal experiments, are mentioned here. One class of agent is made up of antibodies against T cells which are used to disable T cells during the grafting procedure itself. These include OKT3, a monoclonal antibody directed against CD3 which is part of the T cell receptor complex. The blockage of CD3 by OKT3, or similar antibodies, prevents T cell activation. Next there are inhibitors of calcineurin which block the production of IL2 by activated T cells. Calcineurin is a protein phosphatase, activated by Ca^{2+}, which removes a phosphate group from the NFAT transcription factors, and thereby allows them to enter the nucleus and activate the transcription of the *IL2* gene. Cyclosporine is a cyclic peptide from a fungus and was discovered in 1976. It binds to an intracellular protein cyclophilin, and this complex inhibits calcineurin. A similar biochemical activity is shared by Tacrolimus (= FK506, Fujimycin). This is a macrolide compound from a streptomycete, discovered in 1987. It inhibits calcineurin as a complex with the cytoplasmic protein FKBP12, and is somewhat more potent than cyclosporine. Sirolimus (= rapamycin) is another macrolide from another streptomycete, discovered in 1972. It is an inhibitor of Mammalian Target of Rapamycin (mTOR), a central component of cellular signal transduction which needed for the cell division of lymphocytes stimulated by IL2. Mycophenolic acid (or its prodrug mycophenolate mofetil, MMF) was discovered as long ago as the nineteenth century but its immunosuppressive action has only been exploited recently. It also inhibits cell division, but through an inhibition of purine synthesis, and thereby suppresses B and T cell multiplication. There are various other drugs used in clinical practice, and, in particular, new

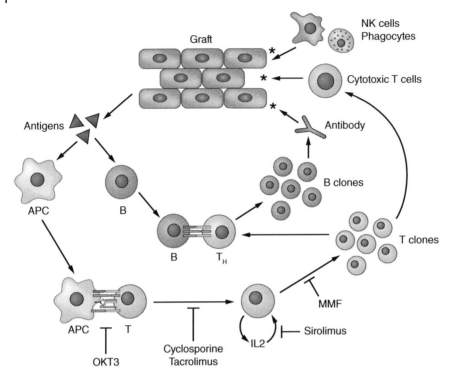

Figure 4.8 Processes of graft rejection. A cellular graft is shown, without its own dendritic cells. Debris are picked up by antigen-presenting cells which activate T cells. B cells are also stimulated to produce antibodies. The action point of various immunosuppressive drugs is shown: OKT3 is anti CD3, cyclosporine and tacrolimus antagonize calcineurin, sirolimus antagonizes IL2 action, mycophenolate mofetil (MMF) is anti-proliferative. APC = antigen-presenting cell; T = T lymphocyte; T_H = T helper lymphocyte; B = B lymphocyte.

monoclonal antibodies are constantly being introduced to suppress various essential components of antigen presentation and other steps in the immune response.

Animal Experiments Involving Grafting

The same principles apply to animal experiments as to human organ grafts. However, the situations differ in many ways. Laboratory mice have been in use for a very long time and a number of strains have been inbred to homozygosity (Figure 4.C.5). These strains differ from each other in many ways, not least in their immunological properties. But grafts may be carried out between members of the same inbred mouse strain because they are virtually genetically identical.

For experiments for which it is not possible to make the grafts within an inbred strain, there are various mouse strains available that are to a greater or lesser degree immuno-compromised. The nude mouse lacks a thymus gland and is seriously deficient in T cells. It arises from a loss of function mutation in the gene for the transcription factor FOXN1, which is needed for the later stages of thymus development. This strain also lacks body hair, hence the name "nude". Nude mice can often accept xenografts, but they do retain some T cell function and so their use has been overtaken by other, more severely immunocompromised, strains. The Severe Combined Immunodeficiency (SCID) mouse has a loss of function mutation of the gene encoding a DNA repair enzyme (Prkdc) that is necessary for the DNA rearrangements involved in T and B cell maturation. It is therefore severely deficient in B and T cells. It is usually used in the form of the NOD-SCID mouse, on a background of the Non-Obese Diabetic (NOD) strain which has multiple mutations affecting, among other things, the MHC complex and IL2 production. NOD mice have a number of immunological defects including a susceptibility to spontaneous development of autoimmune diabetes. The NOD-SCID mouse is very popular for a whole range of experiments involving allo- or xenografting and has become well established in the laboratory. Very similar in their properties are mice with knockouts of the *RAG1* or *-2* genes, which also required for DNA rearrangement. These mice also lack most T and B cells.

However, even these strains are not completely free of immune activity and some xenografting experiments require even more immunocompromised hosts. The beige mouse has a mutation of *Lyst*, a gene encoding an endosomal trafficking component, and is severely deficient in NK cells. So SCID-Beige mice lack most T, B and NK

function. γc (= CD132) is a common component of the IL2 and some other cytokine receptors, and it is necessary for signal transduction following binding of the cytokine to its receptor. Hence the loss of function mutant of the gene for γc has another form of severe combined immunodeficiency. Mouse strains that combine SCID and γc, or RAG and γc, are the most immunocompromised currently available and are used for the most demanding applications. Of course the more immunocompromised the mouse strain the more delicate it is, particularly in terms of susceptibility to infection, and this needs to be taken into account when planning experiments.

As far as non-mouse hosts are concerned, rats are quite often used and there is a nude rat, also a loss of function of *Foxn1*, which has similar properties to the nude mouse. For large animals such as pigs, immunodeficient models are under development, but most experiments are conducted using similar regimes of drug-induced immunosuppression as is used for human organ transplantation.

Further Reading

Chaplin, D.D. (2010) Overview of the immune response. Journal of Allergy and Clinical Immunology 125, S3–S23.

Chinen, J. and Buckley, R.H. (2010) Transplantation immunology: Solid organ and bone marrow. Journal of Allergy and Clinical Immunology 125, S324–S335.

Freshney, R.I. (2010) Culture of Animal Cells. A Manual of Basic Technique and Specialized Applications, 6th edn. John Wiley and Son, Hoboken NJ.

Geraghty, R.J., Capes-Davis, A., Davis, J.M., Downward, J., et al. (2014) Guidelines for the use of cell lines in biomedical research. British Journal of Cancer 111, 1021–1046.

Gjorevski, N., Ranga, A. and Lutolf, M.P. (2014) Bioengineering approaches to guide stem cell-based organogenesis. Development 141, 1794–1804.

Kang, H.-W., Lee, S.J., Ko, I.K., Kengla, C., Yoo, J.J. and Atala, A. (2016) A 3D bioprinting system to produce human-scale tissue constructs with structural integrity. Nature Biotechnology 34, 312–319.

King, J.A. and Miller, W.M. (2007) Bioreactor development for stem cell expansion and controlled differentiation. Current Opinion in Chemical Biology 11, 394–398.

Langer, R. and Tirrell, D.A. (2004) Designing materials for biology and medicine. Nature 428, 487–492.

Liu, N., Zang, R., Yang, S.-T. and Li, Y. (2014) Stem cell engineering in bioreactors for large-scale bioprocessing. Engineering in Life Sciences 14, 4–15.

McKay, D.B., Park, K. and Perkins, D. (2010) What is transplant immunology and why are allografts rejected? Chapter 3 in D.B. McKay and Steinberg, S.M. (eds). Kidney Transplantation: A Guide to the Care of Kidney Transplant Recipients. Springer Science + Business Media, NY, pp. 25–39.

Pampaloni, F., Reynaud, E.G. and Stelzer, E.H.K. (2007) The third dimension bridges the gap between cell culture and live tissue. Nature Reviews Molecular Cell Biology 8, 839–845.

Sasai, Y. (2013) Next-generation regenerative medicine: organogenesis from stem cells in 3D culture. Cell Stem Cell 12, 520–530.

Seliktar, D. (2012) Designing cell-compatible hydrogels for biomedical applications. Science 336, 1124–1128.

Zia, S., Mozafari, M., Natasha, G., Tan, A., Cui, Z. and Seifalian, A.M. (2016) Hearts beating through decellularized scaffolds: whole-organ engineering for cardiac regeneration and transplantation. Critical Reviews in Biotechnology 36, 705–715.

5

Early Mouse and Human Development

Although many important discoveries in developmental biology have been made using other organisms, including *Drosophila, Caenorhabditis elegans* and *Xenopus*, the mouse has served as the principal model for the study of developmental mechanisms in mammals. Numerous aspects of mammalian developmental biology are relevant to stem cell research, not least the fact that the study of mouse development led to the original discovery of embryonic stem cells. This chapter deals with the preimplantation and early postimplantation stages of development. The story continues in Chapter 7 which deals with formation of the main body parts of the embryo and Chapter 8 which deals with organogenesis.

Developmental mechanisms are highly conserved in evolution and most of the inducing factors and transcription factors discovered using *Xenopus* and zebrafish have turned out to have similar functions in the mouse. This means that they are almost certain also to play a similar role in human development. The direct study of human development has been quite limited due to the obvious ethical limitations on what research is acceptable, and to some quite stringent legal restrictions in some countries. Although it is safe to assume that much of the molecular genetics of human development resembles that of the mouse, there are also some significant differences in the course of early development and placentation and these will be noted throughout this chapter.

Much of the work on mouse development has depended on the advanced genetic methods that are available, such as knockouts and knock-ins, and the use of the Cre systems for conditional knockouts and for lineage labeling. Some microsurgical methods have also been used. For the first few days after fertilization, mouse embryos lie free in the reproductive tract. This is called the preimplantation phase and during this period embryos can be removed and manipulated in vitro. After about 4.5 days the embryo implants in the uterus and the placenta starts to develop. But it is still possible to culture whole embryos from early postimplantation stages for about 2 days in vitro. After this, whole embryos can no longer be maintained in vitro but individual organ rudiments can, and they are able to survive for considerable periods in suitable media. If modified embryos are to be raised to term they must be reimplanted into the reproductive tract of a "foster mother", as described in Chapter 3. The gestation of mouse embryos takes about 20 days from fertilization to birth. Because mice mate in the night, the stage of embryonic development is often expressed as days and a half, e.g. a 7.5 day embryo, designated E7.5, is recovered on the 8th day after the parental mice were put together.

It is important to note that an early mammalian development, in contrast to that of free living embryos like *Xenopus* or zebrafish, is largely devoted to the formation of a set of extraembryonic structures associated with

The Science of Stem Cells, First Edition. Jonathan M. W. Slack.
© 2018 John Wiley & Sons, Inc. Published 2018 by John Wiley & Sons, Inc.
Companion website: www.wiley.com/go/slack/thescienceofstemcells

vivaparity. The totality of what develops from the zygote is called the conceptus, and comprises the embryo itself plus all the extraembryonic structures, most of which form parts of the placenta. Until about the primitive streak stage of development there is no clear distinction between the future embryo and the extraembryonic structures as some cells from both the inner cell mass and the epiblast become parts of the placenta. Perhaps unfortunately, early mammalian conceptuses are often referred to colloquially as embryos, but it should be remembered that they are just as much precursors to the placenta as to the embryo itself.

In this book, human developmental times are given from fertilization. This is about two weeks later than the conventional gestational age, measured from commencement of the last menstrual period. It means that the average human developmental time from fertilization to birth is 38 weeks, not 40 weeks.

Gametogenesis

Germ Cells

Sexual reproduction requires the fusion of male and female gametes (sperm and egg) to form a fertilized egg (zygote). So the first phase of development concerns the formation of these gametes, a process which begins when the mouse itself is still an early embryo. The cells that will become the gametes are called germ cells or the germ line. All other cells in the body are called somatic cells, or the soma. This distinction is important because it is only genetic changes to the germ cells that can find their way into the next generation; genetic changes to the soma may affect the individual organism but not the offspring. This is relevant to some of the legal framework for stem cell and reproductive biology research: in most countries genetic modification of the human germ line is prohibited while that of the soma is permitted, and is normally referred to as gene therapy.

The germ line of mammals is protected by various mechanisms from the senescence that affects the soma. One aspect of this is the high level of selection among germ cells and gametes, which is much greater than found for somatic cells. The selective mechanisms include a requirement for correct migration behavior of the primordial germ cells, the massive death of primary oocytes before reproductive maturity, the ovulation of just one (in humans) out of several mature ovarian follicles, a requirement for appropriate sperm motility to traverse the female reproductive tract, and selection of one out of hundreds of sperm at the time of fertilization. By selecting repeatedly against low viability cells, these processes all reduce the number of mutations, or other deleterious nongenetic changes, passing to the zygote. In addition, in the female germ line, the number of cell divisions is quite small. The mitosis of oogonia ceases in late gestation and so a female mouse, or a girl, is born with all the oocytes she will ever have.

Mitosis and Meiosis

Normal somatic cells, as well as germ cells before meiosis, are diploid. That is they contain a complete set of chromosomes from each parent. The pairs of similar chromosomes from father and mother are called homologous chromosomes. Each chromosome can be thought of as being a single long molecule of double-stranded DNA. In the normal mitotic cell cycle (Figures 2.5; 5.1a) this is the situation during the G1 phase. During the S phase the DNA becomes replicated so that in the G2 phase each chromosome now consists of two long, identical, double-stranded DNA molecules. These are called chromatids. It is possible for DNA strands to break and rejoin and when this occurs between identical chromatids it is called sister chromatid exchange. This is normally of no consequence because the strands are identical, but it is relevant when considering the Cairns hypothesis of stem cell character (the "immortal strand", Chapter 2), which is incompatible with sister chromatid exchange. At cell division (the M phase) the identical chromatids separate and one set becomes carried to each of the two daughter cells by the mitotic spindle apparatus. During the normal mitotic cell cycle the pairs of homologous chromosomes derived from the father and mother behave independently.

During gamete formation, meiosis generates gametes with half the chromosome complement of the normal number and is a much more complex process (Figure 5.1b). Meiosis commences following the S phase of the last mitotic cycle, so each chromosome consists of two identical chromatids. During the prophase of meiosis, homologous chromosome pairs come together so that the four chromatids

from the two homologous chromosomes are associated. These are called bivalent chromosomes. They too may undergo DNA breakage and rejoining, and if this occurs between chromatids from different parents it will lead to separation of alleles such that the gametes later formed each have a different combinations of alleles drawn from the two parents. This is called crossing over and it leads to genetic recombination. Meiosis involves two cell divisions. The first division leads to separation of the bivalents into chromosomes, each with two chromatids. These are not exactly the same as the parental chromatids because of the occurrence of crossing over and genetic recombination. The second meiotic division leads to the formation of the

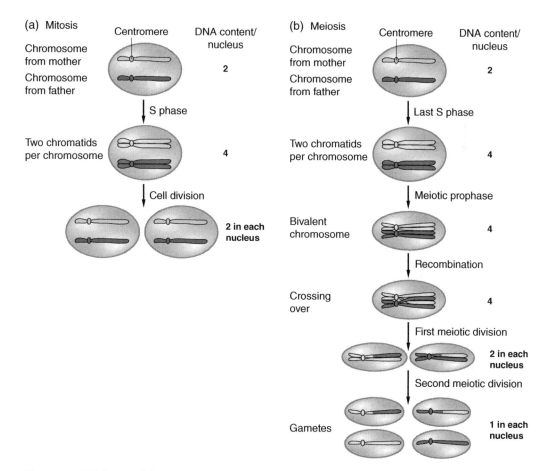

Figure 5.1 (a) Behavior of chromosomes during mitosis. (b) Behavior of chromosomes during meiosis. (Slack, J.M.W. (2013) *Essential Developmental Biology*, 3rd edn. Reproduced with the permission of John Wiley and Sons.)

actual gametes, each with a single set of chromatids. Note that the terms "haploid" and "diploid" refer to the number of distinct chromatids present, not the total amount of DNA, which always doubles at S phase.

Primordial Germ Cells (PGCs)

In all animals, germ cells are formed early in development. Particularly in invertebrate animals this can be as early as the fertilized egg itself, in the form of a cytoplasmic determinant called germ plasm which programs those cells that contain it to become germ cells. However in mammals the primordial germ cells (PGCs) are formed by induction. In mice, they arise around E6.25 from the posterior part of the epiblast (see below) in response to a BMP signal from the adjacent extraembryonic ectoderm (Figure 5.2). The evidence for this is that embryos lacking the *Bmp4* gene form no germ cells, and that treatment of competent epiblast with BMP4

will generate germ cells. In cynomolgus monkeys, and probably also in humans, PGCs arise not from the posterior epiblast but from the amnion. This is doubtless associated with the different origin of the amnion in primates as compared to rodents (see below). One of the earliest specific markers of PGCs is the zinc finger transcription factor BLIMP1, which controls many aspects of germ cell development. PGCs can be visualized at E7.5 by staining for activity of the enzyme alkaline phosphatase. They maintain expression of *Oct4* and *Nanog* and reactivate expression of *Sox2*, three genes encoding key pluripotency factors in the early embryo. Shortly after their formation, PGCs undergo a degree of global DNA demethylation, and the pattern of histone methylation shifts towards that of pluripotent stem cells. These characteristics indicate that PGCs have many of the properties of pluripotent stem cells, and it is in fact possible to culture pluripotent cells called EG cells from mouse PGCs.

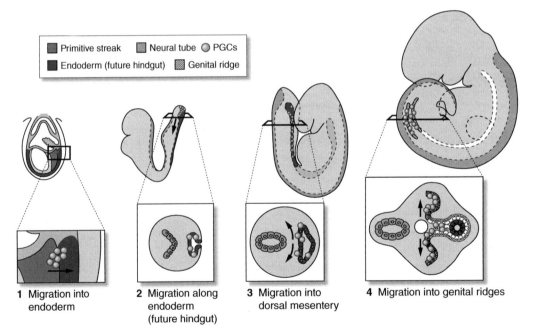

Legend:
- ■ Primitive streak
- ■ Neural tube
- ○ PGCs
- ■ Endoderm (future hindgut)
- ▨ Genital ridge

1 Migration into endoderm

2 Migration along endoderm (future hindgut)

3 Migration into dorsal mesentery

4 Migration into genital ridges

Figure 5.2 Migration route of primordial germ cells in the mouse embryo from the site of formation to the gonads. (From: Richardson, B.E. and Lehmann, R. (2010) Mechanisms guiding primordial germ cell migration: strategies from different organisms. Nature Reviews. Molecular Cell Biology 11, 37–49. Reproduced with the permission of Nature Publishing Group.)

During development of all animals, PGCs undergo some sort of migration from their position of formation into the developing gonads. In mice, they first enter the hindgut endoderm, then they migrate along the hindgut, then into the dorsal mesentery and into the lateral mesoderm that forms the gonads (Figure 5.2). This migration occurs mostly between E9.5–11.5. The migration pathway is controlled by a system frequently involved in developmental cell migration: the binding of SDF1 (stromal cell derived factor 1 or CXCL12), secreted by the lateral mesoderm, by CXCR4, a G-protein coupled receptor on the PGCs. In embryos lacking the *Cxcr4* gene, the PGCs fail to reach the gonads. From this point on, the future of the germ cells depends on whether the embryo is male or female. The mechanism of sex determination is discussed below, and affects both the structure of the gonads and many secondary sexual characters, as well as the germ cells themselves. In human embryos, the migration events are similar to those in the mouse but with slower timing: the PGCs become visible in the proximal yolk sac, using alkaline phosphatase staining, by about the 4th week, and migration to the prospective gonads occurs over the following 2 weeks.

Spermatogenesis

In mouse the germ cells become incorporated into testis cords at about E12.5 and mitosis is suspended 1–2 days later (Figure 5.C.1). In the first postnatal week the testis become organized into a set of seminiferous tubules, and the germ cells become spermatogonia which remain continuously mitotic and produce sperm throughout life (Figure 5.3). The seminiferous tubules also contain Sertoli cells, large cells providing nutrient support to the spermatogonia, and Leydig cells, which produce testosterone. Sertoli and Leydig cells both arise from the mesodermal tissue of the gonad. In mice, the PGC migration is completely finished by E13.5. At this time the germ cells are called gonocytes and they proliferate until E16.

They then become quiescent until about postnatal day 4 (P4) when they resume proliferation as spermatogonia. The expression of *Oct4* falls off after the stage of PGC migration, but persists at a low level in a few spermatogonia well into adult life. The organization of the spermatogonial stem cells will be described later in Chapter 10. In the process of spermatogenesis the spermatogonia divide to form paired or "aligned" groups joined by cytoplasmic bridges. At this stage they are called A type spermatogonia. They then enter a differentiation pathway through A1, A2, A3, A4, Intermediate and B type spermatogonia. These become primary spermatocytes which undergo the first meiotic division to form two equal secondary spermatocytes, and the second meiotic division to form four spermatids, each of which matures to become a single haploid sperm. During maturation of the spermatids, histones in the nucleus are mostly replaced by more basic proteins called protamines, which promote a condensed chromatin state in the sperm head.

In human males, the situation is similar with spermatogenesis commencing at the time of puberty. Earlier suppression of meiosis in spermatogonia is due to the upregulation of Cyp26b1 in the embryonic testis. This degrades retinoic acid, which normally promotes entry of female germ cells into meiosis. Removal of the *Cyp26b1* gene causes male germ cells to enter meiosis in the embryonic testis and subsequently to die.

Oogenesis

The mitotic female germ cells of the ovaries are called oogonia. In lower vertebrates oogonia continue to produce oocytes during adult life, but this is not the case in mammals. In mouse all of the oogonia have finished dividing by about 13.5 days, and in humans by 3 months of gestation. A primary oocyte is a cell that has completed its last mitosis and it can remain dormant for a considerable time (up to 50 years in humans). Oocytes are packaged into follicles, each consisting of

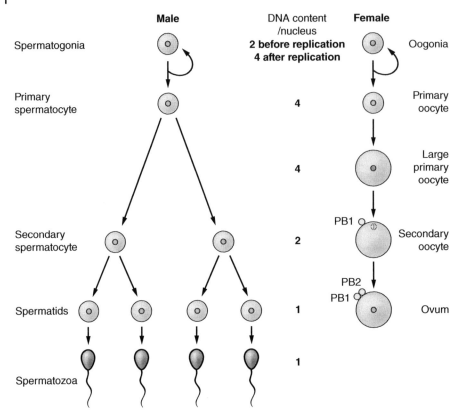

Figure 5.3 Outline of gamete maturation. (Slack, J.M.W. (2013) Essential Developmental Biology, 3rd edn. Reproduced with the permission of John Wiley and Sons.)

one oocyte surrounded by many granulosa cells, and a layer of thecal cells which produce estrogens (Figure 5.4a). Granulosa and thecal cells are somatic cells derived from the mesoderm of the gonads. The oocyte itself is surrounded by a clear layer of extracellular material called the zona pellucida, or just zona, which is secreted both by the oocyte itself and by the granulosa cells.

In mice the oogonia enter meiotic prophase and become primary oocytes after completion of their mitotic divisions around E13.5. Reproductive maturity is reached at about 6 weeks after birth, by which time many follicles have regressed, and each ovary contains about 10,000 viable follicles at various stages of development. These primordial follicles are stimulated to grow by follicle stimulating hormone (FSH) from the pituitary gland. The granulosa cells proliferate

and the follicle enlarges considerably with a fluid filled cavity, the antrum, appearing within (Figure 5.C.2). Under optimal conditions ovulation of 8–12 oocytes occurs every 4–6 days, stimulated by a surge of luteinizing hormone (LH) from the pituitary (Figure 5.4b). Female mice are only receptive to mating during the ovulation phase of their estrus cycle. In the course of ovulation the primary oocytes undergo the first meiotic division, to generate a secondary oocyte suspended in the metaphase of the second meiotic division, and a polar body which is a small plasma membrane sac containing the unused chromosome set. The secondary oocyte secretes a protease, tissue plasminogen activator, which helps it to exit the ovary. The oocytes then find their way into the open ends of the oviducts where, if mating has occurred, they may become fertilized by the

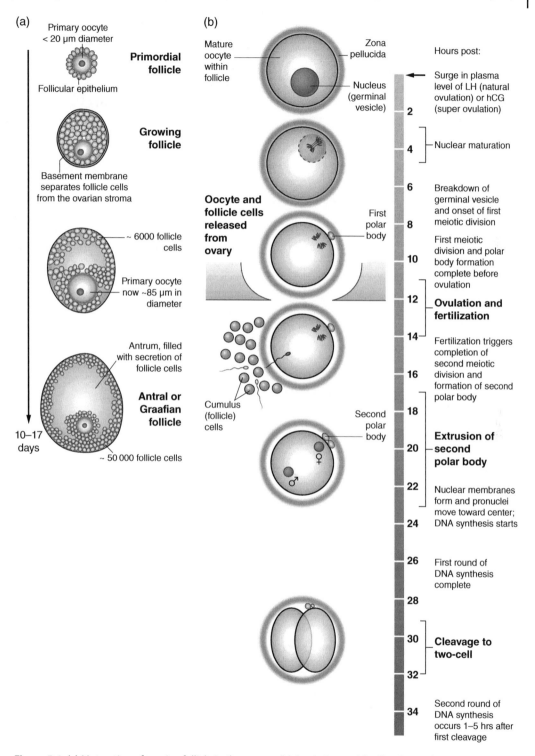

Figure 5.4 (a) Maturation of ovarian follicle in the mouse. (b) Ovulation and fertilization in the mouse. (From: Hogan, B., Beddington, R., Costantini, F. and Lacy, E. (1994) Manipulating the Mouse Embryo. A Laboratory Manual, 2nd edn. Cold Spring Harbor Laboratory Press, NY: Plainview.)

previously deposited sperm. Each secondary oocyte, colloquially called an egg, is surrounded by its zona and a cluster of cumulus cells, which are granulosa cells from the follicle. Following fertilization, the second meiotic division is completed with ejection of a second polar body containing the unused chromosome set.

In humans about 7 million primary oocytes are formed but by birth the number has reduced to about 1 million, and by puberty to about 40,000. As in primates generally, there is an approximately monthly reproductive (menstrual) cycle controlled by the pituitary gland. This secretes follicle stimulating hormone (FSH) which promotes growth of up to 20 follicles, only one of which is likely to be ovulated. Over a whole reproductive lifetime about 400 oocytes are actually released and become available for fertilization. At day 13–14 of the cycle, a surge of FSH and luteinizing hormone (LH) from the pituitary brings about ovulation. This involves onset of the first meiotic division, generating a secondary oocyte and polar body, and release of the oocyte, together with a mass of follicle (cumulus) cells, from the ovary. In humans the oviduct is called the Fallopian tube, or uterine tube, and the opening is called the ampulla. If sperm are present then fertilization may occur in the ampulla and the second meiotic division is completed with formation of the second polar body. The first polar body may also divide at this stage, giving a maximum of three polar bodies in all.

Fertilization

Although the basic fusion of sperm and egg to produce a diploid zygote is a feature of all sexually reproducing organisms, the details of fertilization mechanisms differ considerably between invertebrate models such as the sea urchin, and mammals. Mammalian fertilization is best known through study of the mouse. Human fertilization has been studied in relation to the large industry of in vitro fertilization used for fertility treatment

and although there are differences of detail, the main events are similar to those in the mouse.

The mouse sperm has a head containing the nucleus, a structure called the acrosome, a centriole and some mitochondria. The tail is a reinforced flagellum responsible for motility. Human sperm have the same components with a slightly different morphology (Figure 5.5). Sperm need to be activated in a process called capacitation that normally occurs in the female reproductive tract and is due to entry of Ca^{2+} and HCO_3^{2-} ions. Capacitation involves a rise in intracellular pH, an increase in the magnitude of membrane potential from -30 to $-50\,mV$, and some cell surface changes needed for interaction with the zona pellucida of the egg.

The sperm carries a hyaluronidase enzyme which helps penetrate the mass of cumulus

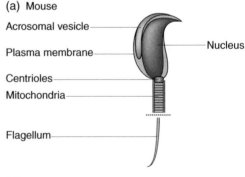

(a) Mouse

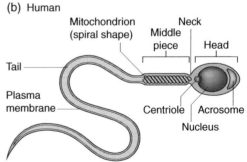

(b) Human

Figure 5.5 (a) Mouse sperm. (b) Human sperm. ((a) from: Darszon, A. et al. (1999) Ion channels in sperm physiology. Physiological Reviews 79 (2). The American Physiological Society. (b) from: http://imagebasket.net/789-structure-of-human-sperm-images.php?pics=true)

cells around the oocyte. Once it reaches the zona there is a sequence of specific recognition events required for successful fertilization (Figure 5.6):

1) A cell surface galactosyl transferase on the sperm binds to species-specific carbohydrate chains carried by ZP3, a glycoprotein component of the zona. This provokes the acrosome reaction via an elevation of intracellular calcium. The acrosome of the sperm undergoes exocytosis releasing, among other things, a protease that digests the zona allowing the sperm to reach the oocyte surface.
2) ADAM (disintegrin) proteins on the sperm bind to integrins on the oocyte membrane. This initiates fusion of sperm and egg membranes in a process requiring the oocyte protein tetraspanin (= CD9).
3) Fusion introduces phospholipase Cζ from the sperm; this activates the inositol trisphosphate pathway and provokes a rise of Ca^{2+} in the oocyte cytoplasm.
4) The Ca^{2+} increase causes the exocytosis of cortical granules from the oocyte, the completion of the second meiotic division, the resumption of DNA synthesis, the recruitment of maternal mRNA into

polysomes, and a general metabolic activation. The cortical granule contents include glycosidases and proteases which modify the zona so that it cannot bind further sperm, thus reducing the risk of fertilization by additional sperm (polyspermy).

The sperm nucleus decondenses and the protamines are replaced by histones. The DNA of both nuclei is replicated and they come together to form the mitotic spindle for the first cell division of embryonic development.

Early Development

Preimplantation Phase

The first few days of mammalian development are known as the preimplantation phase because during this time the embryos, or more precisely the conceptuses, are free living, initially in the lumen of the oviduct (in humans this is called the Fallopian tube or uterine tube) and later in the lumen of the uterus. During this time they undergo cell divisions and generate extraembryonic

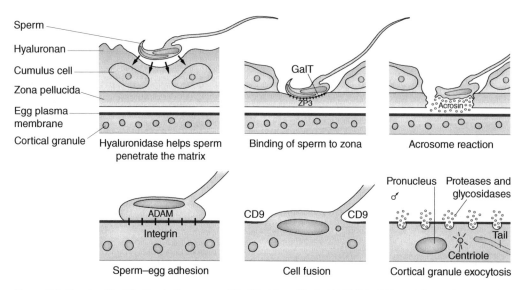

Figure 5.6 Events of fertilization in the mouse. (Modified from Slack, J.M.W. (2013) Essential Developmental Biology, 3rd edn. Reproduced with the permission of John Wiley and Sons.)

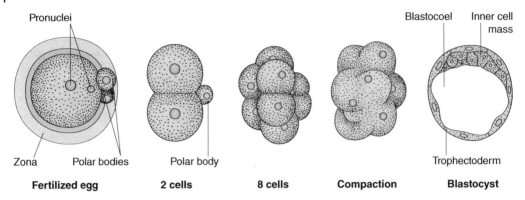

Figure 5.7 Preimplantation development of the mouse. The zona remains present during these stages. (Modified from Slack, J.M.W. (2013) Essential Developmental Biology, 3rd edn. Reproduced with the permission of John Wiley and Sons.)

structures (Figure 5.7). However they do not grow in mass because very little nutritional support is available from the surrounding fluid. The cell divisions are cleavage divisions in which the cells halve in volume with every division, the cells that result from cleavage divisions being called blastomeres. At the stage of implantation in the uterus (mouse 4 days, human 7 days) the blastocyst has a mass lower than the zygote due to metabolism, and the modest size increase that has occurred is due to the uptake of fluid. Implantation can only occur over a short time period during which the uterine lining (endometrium) is hormonally primed to be receptive. Once implantation has occurred the embryo is supported by a placenta and can grow rapidly in size. In fact the growth of the early postimplantation embryo is one of the few occasions in vivo when cell division approximates to the exponential rate that is characteristic of cell culture in vitro.

The first few cleavages of the preimplantation conceptus are rather slow and, unlike most non-mammalian early embryos, not synchronized. Zygotic gene expression is activated at quite an early stage (about 2 cells in mouse and 8 cells in human). If the first two blastomeres should separate, by accident or design, each is able to become a complete embryo, leading to the formation of identical twins. The first eight blastomeres are uniform

and similar to each other in appearance. At this stage the blastomeres become more adhesive, due to increased E-cadherin on the cell surface, and the conceptus becomes more compact in appearance. The first visible differentiation event is the formation of the trophectoderm (TE; called trophoblast in human conceptuses), as an epithelial layer around the outside. The cells on the inside then become known as the inner cell mass (ICM). At the 8-cell stage of the mouse, all the blastomeres become polarized (Figure 5.8). This probably occurs by a segregation of PAR proteins which are often involved in asymmetrical cell division (see Chapter 9), as it is possible experimentally to alter the proportions of ICM and TE by manipulating the levels of PAR proteins at this stage. Microvilli appear on the outer surfaces of the cells and during the next cell cycle some tangential divisions occur, generating cells lacking any of the outer membrane region. These are known as apolar cells and become the inner cell mass; the polar cells, retaining some outer membrane material, contribute mostly to the trophectoderm. For about two cell cycles the polar and apolar cells both have a labile specification, meaning that they will form the other cell type if moved in position between inside and outside. However, from the 64-cell stage they are determined, meaning that their fate is then resistant to moving position. The mechanism appears to depend on the Hippo signaling

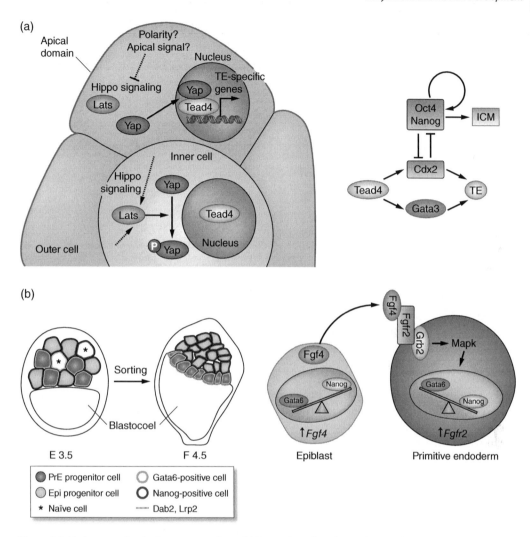

Figure 5.8 Early patterning in the mouse embryo. (a) Formation of trophectoderm and inner cell mass as a result of cell polarization. In the outer cells the suppression of the Hippo pathway causes YAP to enter the nucleus and to activate trophectoderm gene expression. (b) Formation of the primitive endoderm layer by sorting out. FGF signaling from the forming inner cell mass stabilizes the primitive endoderm. (Modified from Takaoka, K., Hamada, H., (2012) Cell fate decisions and axis determination in the early mouse embryo. Development. 139, 3–14.)

pathway, often involved in contact-mediated cell behaviors. The apolar cells, with contacts all round, activate the kinase STK3/MST; this phosphorylates the transcription factor YAP1 retaining it in the cytoplasm. In the polar cells, the pathway is less active so YAP1 can enter the nucleus and, together with TEAD4, upregulates expression of the transcription factor CDX2. CDX2 suppresses expression of genes for the pluripotency factors OCT4 and NANOG, and activates expression of various

components required for trophectoderm differentiation.

About this stage the conceptus cavitates such that a fluid-filled space called the blastocoel appears inside, and it is then called a blastocyst (Figure 5.C.3). The blastocyst hatches from the zona and enlarges somewhat by fluid uptake, and by 4.5 (mouse) or 6 (human) days, it is a fully expanded blastocyst ready for implantation. Because embryonic stem (ES) cells are grown from the inner cell mass there has been much discussion about the nature of inner cell mass cells and their relation to ES cells. Gene expression studies indicate that they are not the same, although they do share the biological property of pluripotency. One notable gene product that is important for characterization of ES cells is the enzyme alkaline phosphatase. In normal mouse development this is first expressed at the onset of zygotic gene expression. It is strongly expressed in the inner cell mass but lost from the trophectoderm. It is then lost from the epiblast after implantation but is re-expressed in primordial germ cells (see above).

The inner cell mass cells of the late blastocyst are of two types: respectively expressing the transcription factors NANOG and GATA6. These seem to be generated spontaneously and appear to sort themselves out because of different cell surface adhesivity, such that the GATA6 cells come to lie adjacent to the blastocoel (Figure 5.8b). These are known as primitive endoderm, while those remaining internally, which continue to express NANOG, are known as the epiblast. The difference between them is sustained by fibroblast growth factor (FGF) signaling, with FGF4 expressed in the epiblast and FGF receptor (FGFR2) in the primitive endoderm. Note that an inhibitor of FGF signaling is an important component of the 2i medium used for ES cell culture. ES cells have a tendency spontaneously to form primitive endoderm unless FGF signaling is actively suppressed. However they cannot form trophectoderm unless CDX2 is experimentally overexpressed.

Events in the human preimplantation conceptus are generally similar to those in the mouse (Figure 5.C.4), although on a somewhat slower time scale. Restricted expression of the genes CDX2, OCT4, NANOG, and GATA6 is somewhat later than in the mouse, not occurring until the late blastocyst stage (day 6–7), but following implantation there is more considerable divergence of events between human and mouse.

Implantation Period – Mouse

The uterus of the mouse is a Y-shaped structure with two long horns connected to the oviducts. It becomes primed for implantation by a high level of estrogen and progesterone. It also produces the cytokine leukemia inhibitory factor (LIF) well-known for its use in the cultivation of murine ES cells. The conceptuses move from the oviducts into the uterus during the preimplantation phase. At about 4.5 days, they attach to the uterine wall and provoke a decidual reaction, which is a proliferation of the stromal (mesenchymal) tissue of the uterus. The decidual tissue is loose and, with the help of secreted proteases from the trophectoderm, the blastocyst is able to burrow into it. During implantation the trophectoderm becomes regionalized. The part in contact with the ICM or epiblast continues to proliferate and is known as the polar trophectoderm. The remainder, known as the mural trophectoderm, continues DNA synthesis but ceases cell division and becomes transformed into polytene giant cells. These giant cells invade the decidual tissue, accompanied by polar trophectoderm and extraembryonic mesoderm, to form the placenta. The giant cell trophectoderm connects with maternal vessels and develops gaps allowing the formation of maternal blood sinusoids in close contact with the embryonic circulation. Also contributing to the placenta are the crypts of Duval lined with extraembryonic endoderm. Because the uterus in the mouse consists of two narrow horns, the decidua quickly fill up the lumen and in the normal multiple pregnancy form a line of swellings along each horn.

The epiblast now begins to project into the cavity of the blastocyst, accompanied by extraembryonic ectoderm derived from the polar trophectoderm. This projection is known as the egg cylinder, and a cavity, called the proamniotic cavity, develops in the interior (Figure 5.9). The cells of the primitive endoderm spread out from the epiblast over the entire internal surface of the mural trophectoderm. Cells in contact with the

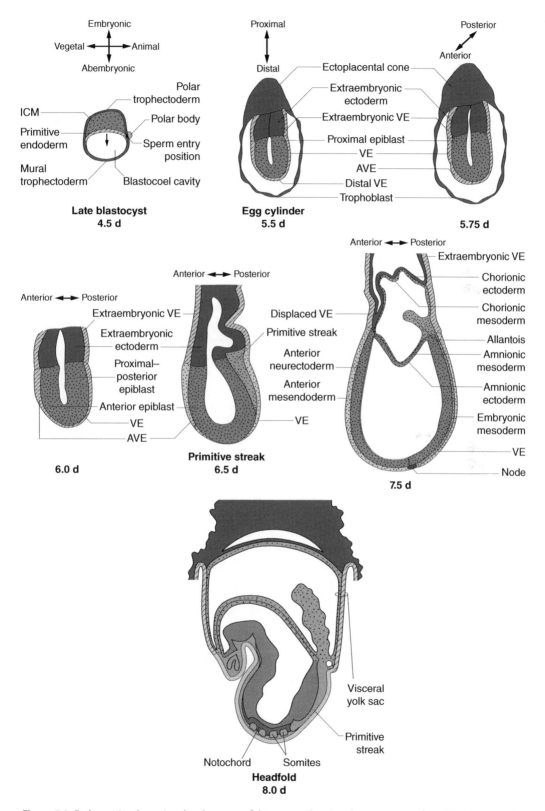

Figure 5.9 Early postimplantation development of the mouse showing the progression from late blastocyst to headfold stage. Over this period the embryo increases in size by a factor of about 8. VE = visceral endoderm, AVE = anterior VE. (Modified from Slack, J.M.W. (2013) Essential Developmental Biology, 3rd edn. Reproduced with the permission of John Wiley and Sons.)

trophectoderm become parietal endoderm, consisting of squamous cells secreting a basement membrane called Reichert's membrane. Those that remain in contact with the egg cylinder become the visceral endoderm, whose cells are cuboidal and have some similarities to the fetal liver, for example producing α-fetoprotein and transferrin.

The determination of the body plan depends on interactions between the epiblast, the visceral endoderm and the proximal tissues of the egg cylinder and will be described in Chapter 7, but in terms of descriptive embryology the first visible event is the formation of the primitive streak on one side of the epiblast. This is, by definition, the posterior side. The streak is a region both of cell proliferation and of cell movement. Cells are drawn from lateral parts of the epiblast and migrate through the streak to form layers of mesoderm and definitive endoderm. The streak elongates to reach the distal end of the egg cylinder and a cell condensation called the node forms at the distal end. A rod called the head process then forms anterior to the node. Over the next two days the node moves posteriorly and the general body plan appears in its wake.

Understanding the disposition of extraembryonic membranes in the mouse causes considerable confusion (Figures 5.9 and 5.10). A later stage mouse embryo will be found to be wrapped in three distinct layers. The outer is the parietal yolk sac, consisting of the original mural trophectoderm + parietal endoderm. The middle one is the visceral yolk sac, composed of visceral endoderm + extraembryonic mesoderm. This was originally the proximal part of the egg cylinder, joining the polar trophectoderm to the embryo. It becomes wrapped around the embryo in the course of a complex morphogenetic movement called "turning", which occurs between 7.5 and 9.5 days of gestation (Figure 5.10). The inner layer is the amnion, formed from the extraembryonic mesoderm nearest the embryo, and also wrapped around the embryo by the process of turning. The chorion arises from the polar trophectoderm + extraembryonic mesoderm. The

allantois, in the mouse, is a purely mesodermal structure arising from the posterior of the primitive streak. It extends towards the chorion and fuses with it, growing into the placenta and contributing blood vessels connected to the embryonic circulation.

Implantation Period – Human

Major divergence between the morphology of human and mouse development starts at the time of implantation. While mouse blastocysts implant with the embryonic side outward, human blastocysts implant with the embryonic side inward, usually at 8–10 days from fertilization. The trophectoderm, usually called trophoblast in humans, becomes divided into an inner cytotrophoblast which resembles the murine polar trophectoderm, but completely surrounds the embryo, and a syncytiotrophoblast, which comprises an acellular syncytium. Unlike the mouse there is no localized proliferation of a polar trophectoderm to generate extraembryonic ectoderm and an ectoplacental cone. The trophoblast begins to secrete human chorionic gonadotrophin β (HCGβ), the detection of which in urine is the basis for human pregnancy tests. In the first half of the menstrual cycle, estrogen from the growing follicles stimulates growth of the uterine lining. Following ovulation this process continues supported by progesterone from the residual follicle, called a corpus luteum, and the endometrium becomes more glandular and vascularized, forming a decidual tissue ready for implantation. During implantation the syncytiotrophoblast burrows into the decidual endometrium, assisted by the secretion of metalloproteinases (Figure 5.11a–e). Gaps open in the syncytiotrophoblast and become connected to the maternal vessels to form sinusoids filled with maternal blood.

Meanwhile the inner cell mass generates a layer of primitive endoderm, or hypoblast (Figure 5.11a). This resembles the primitive endoderm of the mouse, forming on the blastocelic surface of the inner cell mass and spreading around the inner surface of the trophoblast. The part remaining apposed to

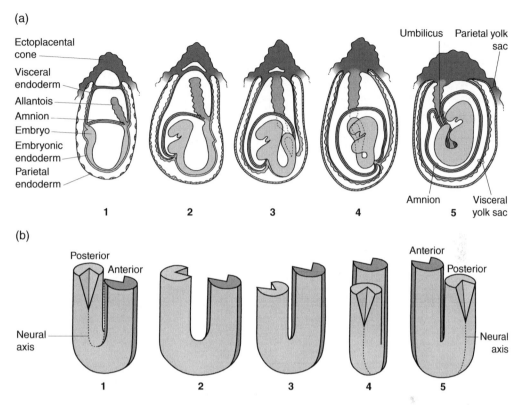

Figure 5.10 The turning process in mouse development. (a) From E7.5–9.5 the embryo becomes rotated around its own long axis leading to its envelopment by extraembryonic membranes and the ventral closure of the gut. (b) Diagram to show the rotation movement. (In: Copp & Cockroft, eds (1990) *Postimplantation Mammalian Embryos*. IRL Press, pp. 88–89). Modified from Kaufman, In Postimplantation Mammalian Embryos, Copp and Cockroft (eds) 1990. IRL Press, with permission from Oxford University Press.

the epiblast becomes the visceral endoderm while that lining the trophoblast becomes the parietal endoderm. The remainder of the inner cell mass becomes a flat plate of cells called the epiblast. A large amount of mesenchyme, called extraembryonic mesoderm, appears between the parietal endoderm and the trophoblast (Figure 5.11b). Its origin has been controversial, and variously ascribed to the trophoblast, the epiblast and the primitive endoderm. Recent RNAseq study of cynomolgus monkey embryos, which are very similar to human, indicate that a primitive endoderm origin is probably correct. Meanwhile the epiblast itself develops a cavity to form a flat epithelial blastodisc, apposed to the hypoblast, which forms the embryo proper, and a thin amnion apposed to the trophoblast. This configuration of the epiblast is very different to the rodent egg

cylinder, but is similar to that of many other types of mammal.

Projections called chorionic villi then grow into the developing placenta (Figure 5.11c–e). They consist of cytotrophoblast containing blood vessels formed from the extraembryonic mesoderm. They enter the blood sinusoids and provide for metabolic interchange between maternal and embryonic circulations. The chorionic villi are formed all around the conceptus, which eventually becomes enclosed by a decidual capsule on the side of the uterine lumen, although development of the placenta is most pronounced on the side of the original contact.

The arrangement of extraembryonic membranes in the human differs somewhat from the mouse although they are given the same names (Figure 5.12). Because stem cell lines are often isolated from extraembryonic

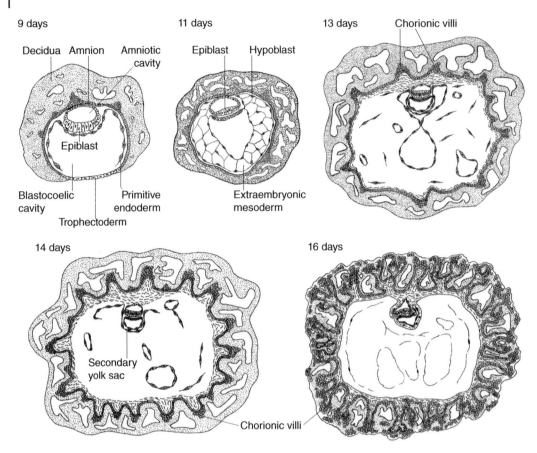

Figure 5.11 Early postimplantation development of the human conceptus. Unlike the mouse, the human epiblast is flat. The drawings cover the period of formation of the amnion, the extraembryonic mesoderm, the secondary yolk sac, and the chorionic villi. The diameter of the whole conceptus is about 0.6 mm at 9 days, 0.8 mm at 12 days and 2.6 mm at 16 days from fertilization. (Modified from Luckett, W.P. (1978) Origin and differentiation of the yolk sac and extraembryonic mesoderm in presomite human and rhesus monkey embryos. American Journal of Anatomy 152, 59–97. Reproduced with the permission of John Wiley & Sons.)

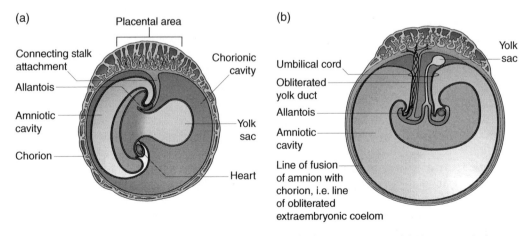

Figure 5.12 Formation of the extraembryonic membranes in the human conceptus. (a) About 3 weeks from fertilization. (b) About 4 weeks from fertilization. (Redrawn from Gray's Anatomy 41st edn. 2015. Reproduced with the permission of Elsevier.)

tissues, it is of some interest to understand their developmental provenance. In mouse the chorion derives from the small area of polar trophectoderm, lined with extraembryonic mesoderm, at the proximal end of the egg cylinder. In humans the name chorion is given to the whole of the trophoblast together with its underlying extraembryonic "mesoderm" which is the early-appearing tissue that may come from the primitive endoderm. The chorionic cavity is thus derived from the original blastocoelic cavity. In mouse the amnion arises from extraembryonic mesoderm caudal to the primitive streak, which becomes lined with epithelium from the epiblast. In humans the amnion forms by cavitation of the epiblast at about day 9. In later gestation the amniotic cavity expands relative to the chorionic cavity, and the amnion becomes fused with the chorion. Because human identical twins occasionally have a common amnion, such twins are presumed to have arisen by fragmentation of the epiblast after implantation. The allantois in the mouse is a mesodermal outgrowth from the extraembryonic mesoderm caudal to the epiblast which becomes incorporated into the umbilical cord and contributes blood vessels to the placenta. In humans the allantois is an endodermal outgrowth of the hindgut. It also projects into the umbilical cord, but is a somewhat vestigial structure and its blood vessels are formed from the surrounding extraembryonic mesoderm. The term "yolk sac" is used for two structures in the mouse: the parietal yolk sac consisting of the trophectoderm lined with parietal endoderm, and the visceral yolk sac consisting of visceral endoderm lined with extraembryonic mesoderm. In humans the yolk sac is a cavity formed from the primitive endoderm (hypoblast) and abutting the extraembryonic mesoderm of the chorionic cavity. This later fragments and reforms as a secondary yolk sac connected to the embryo (Figure 5.11c–e). Once the umbilical cord is formed, the secondary yolk sac becomes an extended structure projecting through the cord into the residual chorionic cavity (Figure 5.12b).

The selection for optimal viability which is such a prominent feature of germ cell development and function continues in the period after fertilization. Examination of normal and defective human conceptuses in hysterectomy specimens which were taken shortly after ovulation indicates that more than 50% of conceptuses never do implant or are lost shortly after implantation.

Ethical and Legal Issues Concerning the Early Human Conceptus

All issues to do with early human conceptuses, usually referred to as embryos, have been fraught with controversy. This is because the human zygote, and the conceptus arising from it, has the potential to become a complete new human being if it is implanted into a receptive uterus. The chief issues have been:

- Research on human preimplantation conceptuses.
- Use of human preimplantation conceptuses to generate embryonic stem cell lines.
- Genetic modification of human embryos.

Some national jurisdictions ban all of these activities. Others allow some limited activity under license. The USA is unusual in that it imposes a blanket ban if federal funds are used, but imposes no restrictions on the same activities supported by private funds.

The opponents of human embryo research and human ES cell generation argue that early conceptuses are human beings and should have full human rights. If they are not accorded human rights, then a slippery slope has been entered that will eventually lead to more permissive attitudes towards abortion and euthanasia. Supporters argue that preimplantation embryos are more akin to human tissue culture cells, which do not have human rights, and furthermore that great scientific and health benefits are likely to arise from such research.

Some of the biological issues relevant to the debate are as follows. As we have seen, the early "embryo" is really a conceptus which generates both an embryo and parts of the placenta. It is also possible in various ways to split the conceptus into two parts such that two complete embryos later develop: the well-known phenomenon of identical twinning. The first time in development that it is really possible to point to a group of cells which are going to become one embryo rather than two, and which is no longer contributing to extraembryonic structures, is the primitive streak stage, reached at about two weeks of development. It is also known that a large proportion of human conceptuses do not implant successfully so never develop further in the normal course of events.

A more practical consideration concerns what happens to human preimplantation conceptuses that are frozen in in vitro fertilization clinics. There are huge numbers of these world-wide which for one reason or another are no longer required by their parents. They are the normal source of material for both early embryo research and generation of human ES cell lines. In the USA federal funds may be used to work with human ES cell lines generated from such embryos, so long as a prescribed set of ethical guidelines have been followed. However the actual ES cell derivation work must be carried out using private funds.

The issue of germ line genetic modification is another difficult area. At present all jurisdictions prohibit this, although in the UK mitochondrial replacement therapy is permitted under license. The main demand for germ line modification comes from patient groups with single gene genetic disorders who would like to have unaffected children which are biologically their own. The advent of the CRISPR-Cas9 method means that genetic modification is now much easier to carry out than it was before. For reasons both of political pressure and technological possibility it is probable that it will one day be permitted. However there are doubts both about the safety aspects in terms of avoiding any damage to the embryo, and longer term to the problem of opening the door to the adoption of unsavory eugenic practices in the future.

Sex Determination

Mammalian sex determination is controlled by the sex chromosomes. Females have two copies of the X chromosome and males have one X and one Y chromosome. A small region of the two sex chromosomes is homologous, carrying the same loci, and is known as the pseudoautosomal region. However most parts are different, carrying different loci. The switch that controls whether an embryo becomes male or female is initiated by the presence of the Y chromosome. A gene called *Sry*, present only on the Y chromosome, becomes expressed in the gonads around 12.5 days gestation in mice. The SRY protein, which is a transcription factor belonging to the HMG (high mobility group) family, upregulates *Sox9*, encoding another transcription factor of the same family. *Sry* is only expressed transiently, but *Sox9* expression continues through development and, directly or indirectly, turns on a set of male-specific functions including the synthesis of testosterone by Leydig cells; and the production of Anti-Mullerian hormone, which causes degeneration of the Mullerian duct that would otherwise become the oviduct and uterus in females.

Evidence for the role of *Sry* is that, if it is deleted, a chromosomally male mouse will develop as a female. If it is overexpressed in a female mouse, then it causes development as a male, although without producing sperm. The situation in humans is very similar. This explains why an XXY individual (Kleinfelter's syndrome) is male, while an XO individual (Turner's syndrome) is female.

The development of the germ cells as spermatogonia or oogonia depends on the sex of the surrounding gonad. It is therefore possible in experimental conditions to generate oocytes of XY constitution and spermatogonia of XX constitution. These do not, however, form viable gametes.

X-Inactivation

Most of the genes on the X chromosome are not related to sexual differences and have the same functions in males and females. Because there are two X chromosomes in female cells but only one in male cells, all genes on the X chromosome are present at double the number in females compared to males. Gene dosage anomalies like this can cause serious problems so it is necessary to correct the situation to make the gene dosage from the X chromosome the same in both sexes. This occurs through inactivation of one of the X chromosomes in all cells of female mammals.

In the mouse there is an early inactivation of the paternal X. It remains inactive in the extraembryonic tissues, but is reactivated in the ICM. Later, in the egg cylinder epiblast, there is a random inactivation of the maternal or paternal X chromosome in each cell. X-inactivation depends on a gene *Xist* (pronounced "ex-ist") which encodes a nontranslated RNA. This coats the chromosome from which it is produced but not the others in the cell, and initiates a permanent condensation and inactivation of the chromatin. The condensed X chromosome is known as a Barr body. This situation is permanent in all somatic tissues of the female mouse, but the oogonia reactivate the inactive X by about 12.5 d of embryonic development. The promoter of *Xist* is demethylated in sperm and methylated in oocytes. This leads to the early inactivation of the paternal X in extraembryonic tissues. In the epiblast this imprint (see below) is erased and replaced by a random de novo methylation of the promoter associated with the random selection of the X to be inactivated in each cell.

The situation in human embryos is somewhat different. *Xist* is expressed from the 8-cell stage in both male and female embryos, and declines in males by the blastocyst stage. It continues in females until at least the 7-day blastocyst, but there is no X-inactivation by this stage. Instead the transcription from both X chromosomes is reduced ("dampening").

The X inactivation status of ES cells derived from female embryos has also been examined. Mouse female ES and iPS cells normally have both X chromosomes active, similar to the situation in the inner cell mass, but human ES and iPS cells are more variable, sometimes showing X-inactivation and sometimes not. Human pluripotent cells converted to a naïve state (see Chapter 6) do have both X chromosomes active, despite expression of *Xist*, and one X-chromosome becomes randomly inactivated during differentiation in vitro. In normal embryonic development human X-inactivation is thought to occur early in the postimplantation period and to do so randomly with respect to the maternal and paternal chromosome.

Imprinting

Most genes are expressed identically from the paternal and maternal autosomes. However there are some genes that are preferentially expressed from one or the other parental chromosome (Figure 5.13). This is known as imprinting. It is due to de novo DNA methylations which are introduced during germ cell development in a sex-specific manner. Methylation of cytosine residues in GC sequences of DNA is an important mechanism for gene control which will be described in Chapter 9. Its key feature is that it persists through DNA replication unless actively removed. In general, demethylation is associated with gene activity and methylation with gene repression.

Many imprinted genes are concerned with growth and growth control. A well-known example is the pair of genes encoding insulin-like growth factor 2 (IGF2) and the IGF2 "receptor" (IGF2R), which is actually an inhibitor of IGF2 function. *Igf2* is preferentially expressed from the paternal gene, whereas the inhibitory *Igf2R* is preferentially expressed from the maternal gene. The practical effect of this is that if the active allele of an imprinted gene is inactivated by mutation then it will generate a complete loss of function mutation, more likely to have harmful

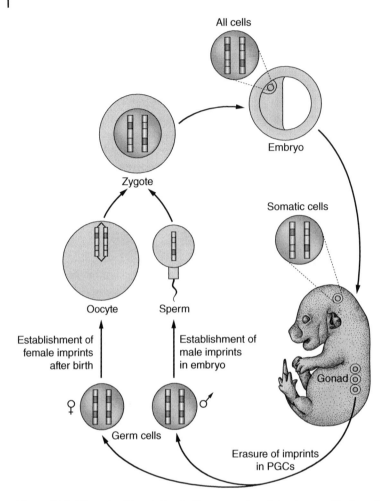

Figure 5.13 Life cycle of imprinting in the mouse. Imprints are erased in the primordial germ cells and subsequently reset in a sex-specific manner. (Modified from Slack, J.M.W. (2013) Essential Developmental Biology, 3rd edn. Reproduced with the permission of John Wiley and Sons.)

consequences than the loss of one out of two similar alleles. Conversely a mutation causing loss of imprinting can generate pathology due to overexpression of that gene. For example, loss of imprinting of *Igf2* with consequent overexpression is found in many cases of Wilms' tumor, a pediatric tumor of the kidney.

Following fertilization in the mouse, the embryo undergoes a demethylation of the whole genome, reaching a low point at the blastocyst stage, after which some remethylation occurs. However these global changes do not affect the imprinted genes. The imprints become erased in germ cells

and re-established in a sex-specific manner during gametogenesis. The time of erasure in PGCs is considered to be around E11.5 on the basis of removal of the specific methylation patterns of imprinted genes. In males, imprints are re-established in germ cells during mitotic arrest at E12.5–14.5. Female-specific imprints become re-established after birth, during growth of the primary oocytes. The pluripotent stem cells called EG cells are derived from PGCs and are very similar to ES cells except that their imprints have been erased.

Imprinting is the main reason why parthenogenesis works poorly in mammals.

It is possible to create parthenogenetic mouse embryos by activating the egg and suppressing the formation of the second polar body such that the two maternal chromosome sets come together to form a diploid zygote nucleus. However such embryos develop poorly because they cannot form the extraembryonic tissues in adequate quantity and quality. Conversely, androgenetic diploids, formed by introducing two sperm nuclei into an enucleated egg, tend to form too much extraembryonic tissue and the embryo itself arrests at an early stage. Imprinting may also be the reason for the tendency of cattle fetuses resulting from cloning, or even from in vitro fertilization, to become too large for easy delivery. This is ascribed to a disturbance of the normal imprints due to the abnormal environment experienced by the nucleus.

The term imprinting is also used in some quite different contexts which should not be confused with that under discussion here. It has been shown by epidemiologists that various deleterious consequences in adult life can arise from stresses during development, for example caused by maternal malnutrition. This may involve epigenetic mechanisms, but not the type of sex specificity described above. The term is also used in psychology to describe the fixation by a newborn individual on a specific object or parent.

Cloning by Nuclear Transplantation (SCNT)

The first successful cloning experiments involving replacement of an oocyte nucleus by a somatic nucleus were carried out in frogs. These indicated that quite high frequencies of clones could be generated from nuclei of early embryo blastomeres, but later stage cells, and particularly terminally differentiated cells gave poor results. The first example of successful cloning in mammals from a differentiated cell was the famous Dolly the sheep in 1997, using a mammary cell nucleus as donor. These experiments gave the impetus to overcome the technical problems in mice, and it is now possible to obtain cloned mice by somatic cell nuclear transfer (SCNT). It is necessary to use mature secondary oocytes, not fertilized eggs, for this purpose. Oocytes are obtained by superovulating females with chorionic gonadotrophin. The cumulus cells are removed and a special pipette is used to penetrate the zone and suck out the second meiotic metaphase plate containing the chromosomes. The donor nucleus, often from the readily available cumulus cells, is injected and allowed to become reprogrammed for 1–6 hours. Then the oocyte is activated by exposure to strontium ions, which induce the repetitive internal calcium pulses typical of natural egg activation. Following this the SCNT oocyte is allowed to develop to a blastocyst. This occurs with quite good frequency (50%). The blastocysts are implanted into the uterus of foster mothers, previously mated with a sterile male to render the uterus competent for implantation. Only a very small proportion of embryos survive to term: about 1% would be normal in a good lab. As with frogs, a much better success rate is found if early blastomere nuclei are used as donors, presumably because they need little or no reprogramming. Embryonic stem cell nuclei also give better results, of the order of 10%.

Although the success rate with differentiated cells is particularly low, it is not zero. Mice have been cloned from B or T lymphocytes in which the characteristic DNA rearrangements of antibody or TCR genes have occurred. The same rearrangement is subsequently found in all cells of the cloned mouse. Whatever the source of donor nuclei they must be in the G1 (or G0) phase of the cell cycle, not in G2. In clones made from female donors, the inactivated X chromosome becomes reactivated during the reprogramming process. Partly random re-inactivation then occurs, although there is a tendency for the same X chromosome to be inactivated as before, suggesting a persistent epigenetic memory of inactivation. The telomeres of the donor nucleus may be short, due to telomere erosion during its development, but telomere length becomes reestablished during development of the cloned individual. In cloning experiments, not only is the overall success rate in terms of live births very low, but also many clones of mice, as well as cattle, have abnormalities on birth, often due to problems with placental development. Furthermore, the gene expression patterns of both cloned mice and their placentas are often abnormal. Human reproductive cloning is banned in all jurisdictions that have regulations on such matters, but creation of human conceptuses by SCNT, followed by their use to make an ES cell line, is permitted in some countries.

Further Reading

Arnold, S.J. and Robertson, E.J. (2009) Making a commitment: cell lineage allocation and axis patterning in the early mouse embryo. Nature Reviews. Molecular Cell Biology 10, 91–103.

Augui, S., Nora, E.P. and Heard, E. (2011) Regulation of X-chromosome inactivation by the X-inactivation centre. Nature Reviews. Genetics 12, 429–442.

Durcova-Hills, G. and Capel, B. (2008) Development of germ cells in the mouse. In: Current Topics in Developmental Biology. Academic Press, Vol. 83, pp. 185–212.

Ferguson-Smith, A.C. (2011) Genomic imprinting: the emergence of an epigenetic paradigm. Nature Reviews. Genetics 12, 565–575.

Hayashi, K., de Sousa Lopes, S.M.C. and Surani, M.A. (2007) Germ cell specification in mice. Science 316, 394–396.

Hochedlinger, K. and Jaenisch, R. (2002) Nuclear transplantation: lessons from frogs and mice. Current Opinion in Cell Biology 14, 741–748.

Johnson, M.H. (2009) From mouse egg to mouse embryo: polarities, axes, and tissues. Annual Review of Cell and Developmental Biology 25, 483–512.

Kashimada, K. and Koopman, P. (2010) Sry: the master switch in mammalian sex determination. Development 137, 3921–3930.

Nomikos, M., Swann, K. and Lai, F.A. (2012) Starting a new life: Sperm PLC-zeta mobilizes the Ca2+ signal that induces egg activation and embryo development. BioEssays 34, 126–134.

Ord, T. (2008) The scourge: moral implications of natural embryo loss. American Journal of Bioethics 8, 12–19.

Richardson, B.E. and Lehmann, R. (2010) Mechanisms guiding primordial germ cell migration: strategies from different organisms. Nature Reviews. Molecular Cell Biology 11, 37–49.

Rossant, J. and Tam, P.P.L. (2009) Blastocyst lineage formation, early embryonic asymmetries and axis patterning in the mouse. Development 136, 701–713.

Rossant, J., Tam, P.P.L. (2017) New insights into early human development: lessons for stem cell derivation and differentiation. Cell Stem Cell 20, 18–28.

Saitou, M. and Yamaji, M. (2012) Primordial germ cells in mice. Cold Spring Harbor Perspectives in Biology 4.

Sasaki, K., Nakamura, T., Okamoto, I., Yabuta, Y., et al. (2016) The germ cell fate of cynomolgus monkeys is specified in the nascent amnion. Developmental Cell 39, 169–185.

Slack, J.M.W. (2013) Essential Developmental Biology. Wiley-Blackwell, Oxford.

Takaoka, K. and Hamada, H. (2012) Cell fate decisions and axis determination in the early mouse embryo. Development 139, 3–14.

Trounson, A. and Bongso, A. (1996) Fertilization and development in humans. In: Pedersen, R.A. and Schatten, G.P. (eds). Current Topics in Developmental Biology, Vol 32, pp. 59–101.

Wakayama, T. and Yanagimachi, R. (1999) Cloning the laboratory mouse. Seminars in Cell and Developmental Biology 10, 253–258.

Wassarman, P.M., Jovine, L. and Litscher, E.S. (2001) A profile of fertilization in mammals. Nature Cell Biology 3, E59–E64.

Wilkins, J.F. (2005) Genomic imprinting and methylation: epigenetic canalization and conflict. Trends in Genetics 21, 356–365.

6

Pluripotent Stem Cells

Pluripotent stem cells are those which resemble the epiblast of the early embryo. They can be grown without limit in vitro, and they are capable of differentiating into most or all cell types of the body. Conventionally the term totipotent is now reserved for cells that can form an entire conceptus: comprising the trophectoderm as well as the inner cell mass. According to this definition, the only totipotent cells in mammals are the zygote itself and the first few blastomeres formed by division of the zygote. Pluripotent cells are defined as those which can produce all cell types except the trophectoderm, including not just the fetal tissues but also various types of extraembryonic tissue such as the visceral endoderm. Multipotent is the term now usually used for cells that can form a particular subset of cell types under normal circumstances, and would cover most of the tissue-specific stem cell populations found in adult organisms.

There are three ways of producing pluripotent stem cells. First, they can be derived by in vitro culture of the inner cell mass from an early conceptus. These are known as embryonic stem cells (ES cells). Second, they can be produced from somatic cells by introducing a group of pluripotency-inducing genes. These are called induced pluripotent stem cells (iPS cells). Finally they can be produced by transplanting a somatic cell nucleus into an enucleated secondary oocyte. This is called somatic cell nuclear transfer, or SCNT. The environment of the oocyte reprograms the nucleus to a state similar to a zygote nucleus, and the resulting conceptus can be cultured in vitro to generate a cell line. Pluripotent stem cells made by these three different methods are similar to each other but there are also some differences.

Mouse Pluripotent Stem Cells

Mouse Embryonic Stem Cells

Mouse embryonic stem cells were first isolated in 1980 by culturing in vitro the inner cell masses from mouse blastocysts (Figure 6.1a). Since then methods for culture have improved considerably. Originally ES cells had to be grown on feeder cells, which are fibroblasts irradiated to destroy their own ability to divide. The feeder cells remain alive and provide a number of factors important for the growth of the pluripotent cells, of which the most important is Leukemia Inhibitory Factor (LIF), which belongs to the interleukin-6 group of cytokines and works through the JAK-STAT signal transduction pathway (Figure 6.1b). Feeder cells are still widely used, but now it is also possible to culture mouse ES cells in a serum-free medium containing LIF plus inhibitors of the MEK and GSK3 signal transduction pathways, which are the pathways typically stimulated by FGF and Wnt factors respectively (see Figure 7.4). This is called 2i medium, the two inhibitors being the MEK inhibitor PD0325901 and the GSK3 inhibitor CHIR99021. Initially

The Science of Stem Cells, First Edition. Jonathan M. W. Slack.
© 2018 John Wiley & Sons, Inc. Published 2018 by John Wiley & Sons, Inc.
Companion website: www.wiley.com/go/slack/thescienceofstemcells

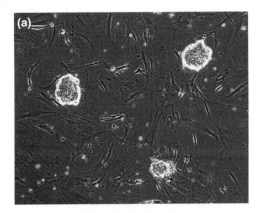

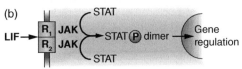

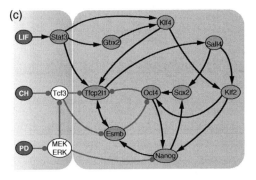

Figure 6.1 Mouse embryonic stem cells. (a) Three colonies of mouse ESC on a background of feeder cells. (b) The signal transduction pathway activated by LIF. Binding of ligand to receptors causes activation of Janus kinases (JAKs) which phosphorylate STAT transcription factors. These can then enter the nucleus and regulate target genes. (c) A model for the pluripotency gene network in mouse ESC. On the left are shown the extracellular factors LIF and 2i. On the right are shown the genes encoding the transcription factors associated with pluripotency, and the regulatory relationships between them. (b): from Slack, J.M.W. (2013) Essential Developmental Biology, 3rd edn. Reproduced with the permission of John Wiley and Sons. (c): from Dunn, S.-J., Martello, G., Yordanov, B., Emmott, S. and Smith, A.G. (2014) Defining an essential transcription factor program for naïve pluripotency. Science 344, 1156–1160. (Source: Austin Smith.)

it was only possible to produce ES cells from the 129 strain of mouse, but using 2i medium they can now be established from other strains as well, and also from rats.

In adherent cultures mouse ES cells grow as domed clumps of refractile cells. These clumps can be dissociated with suitable enzymes and be replated, after which single cells can reconstitute new colonies. The ability of a single cell to reconstitute a colony in vitro is referred to as clonogenic. Mouse ES cells can also be grown in suspension in the presence of FGF-2 where they form small clusters of cells.

The cells have large nuclei, clear nucleoli and little cytoplasm. Their pluripotent properties depend on a network of transcription factors of which OCT4 (=POU5F1), SOX2 and NANOG make up the core group (Figure 6.1c), OCT4 functioning as a dimer with SOX2. The three factors together promote their own transcription, thus generating a stable autocatalytic loop, which maintains the pluripotent state so long as the cell environment remains constant. They also bind to numerous sites throughout the genome and repress expression of a range of developmental control genes, namely those involved in the hierarchy of developmental decisions that characterizes early development. Many such genes have promoters lying in chromatin that carries both activating and inhibitory histone modifications, typically $H4K3Me_3$ (activating) and $H4K27Me_3$ (inhibitory). This state is called "poised", indicating that a relatively small stimulus can either activate or repress these developmental control genes.

In normal development global genome remethylation commences at the blastocyst stage, so the cells from which the ES cells are derived, and the ES cells themselves, have a near normal level of DNA methylation and preserve the imprints of their embryo of origin. But female mouse ES cells normally have both X-chromosomes active, indicating that they retain a developmental state prior to the random X-inactivation that normally occurs in the epiblast. Mouse ES cells exhibit other markers characteristic of early embryos, including the cell surface carbohydrate SSEA1 and the enzyme alkaline phosphatase, and also express various other pluripotency-specific transcription factors, including KLF4, KLF2, ESRRB, TFCP2L1, TBX3 and GBX2.

Although usually established from inner cell masses, mouse ES cells are believed to be derived from the early, preimplantation stage, epiblast. This is on the basis of the comparison of gene expression profiles and on the direct isolation of ES cells from epiblast cells of 4.5-day conceptuses. It is also possible to culture very similar cell lines from primordial germ cells, especially if they are cultured in 2i medium. These are called embryonic germ cells (EG cells). They can be grown either from the posterior ends of 8.5-day embryos or from the germinal ridges of 11.5 d embryos. Like the cells from which they are derived, EG cells from the later source do show imprint erasure.

Differentiation of Mouse ES cells

In order to establish the developmental potency of mouse ES cells, there are three different types of procedure in common use (Figure 6.2). The first is differentiation in vitro. This occurs simply on withdrawal of LIF and of the 2i inhibitors. If the cells are allowed to aggregate then the resulting structures are called embryoid bodies (Figure 6.C.1). The first tissue to develop in these is the primitive endoderm, which arises at random throughout the embryoid body. Older reports indicate the formation of a layer around the outside, but this only occurs if the endoderm is allowed to form before aggregation, so that sorting of endoderm cells to the surface is possible. As in the inner cell mass of the embryo, primitive endoderm formation involves the loss of NANOG and the upregulation of GATA6 expression. About 3 days after withdrawal of LIF, OCT4 expression is lost from many cells which upregulate markers of specific germ layers and other major tissue regions of the early embryo. These include for example BRACHYURY/T for the mesoderm and SOX17 for definitive

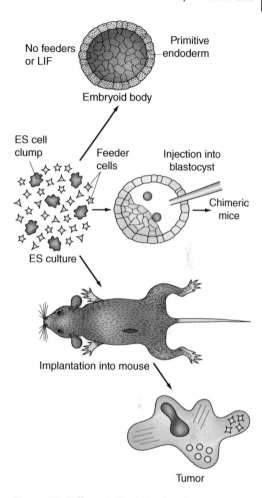

Figure 6.2 Differentiation behavior of mouse embryonic stem cells. They can form embryoid bodies in vitro, teratomas in vivo, and contribute to mouse embryos if introduced at an early stage. (Modified from Slack, J.M.W. (2013) Essential Developmental Biology, 3rd edn. Reproduced with the permission of John Wiley and Sons.)

endoderm. Subsequently markers of individual organs appear, for instance Troponin C for cardiac muscle. The time scale of development of embryoid bodies is similar to the normal time scale of embryonic development, but the spatial organization found in embryos is largely lacking. This is probably because embryoid bodies are usually much bigger than real inner cell masses or epiblasts, so they provide a very different environment in terms of cell contacts, extracellular matrix and inducing factors secreted by nearby cells.

The second type of differentiation-promoting procedure is the formation of teratomas (Figure 6.C.2). If mouse ES cells are injected into an immunologically compatible adult mouse, usually at a subcutaneous or intramuscular location, then a tumor is formed resembling the type of naturally occurring human tumor called a teratoma. This contains a jumble of structures including derivatives of all three embryonic germ layers, together with some persistent pluripotent cells which maintain expansion of the tumor. Teratomas require a minimum number of cells to be injected for their formation. Largely because of the variability in their structure, their biology is still poorly understood. For instance it is not known how many cells need to participate in the formation of a teratoma, nor whether there is an obligatory sequence of developmental decisions similar to that found in the embryo itself.

The third type of differentiation behavior is shown if mouse ES cells are reintroduced into an early conceptus by injection into the blastocoelic cavity, or by aggregation with early blastomeres (Figure 6.2). The ES cells will mingle with the cells of the inner cell mass and contribute descendants to all structures and tissue types of the embryo. In such experiments the donor cells need to be genetically distinct from the host to enable their subsequent identification, and the resulting embryo, being composed of two genetically distinct populations of cells, is known as a chimera. If the embryos are reimplanted into the uterus of a foster mother (see Chapter 3), they can be allowed to develop to term. This enables the resulting chimeric mice to be reared and mated. If it is possible to breed offspring carrying the genetic markers of the original ES cell line, then this proves that functional germ cells have arisen from the ES cells originally introduced into the embryo. This is called germ line chimerism. It is possible for a slightly defective ES cell line to generate chimeras but not germ line chimeras, so the formation of germ line chimeras is taken as the "gold standard" of pluripotent cell behavior. Even

more of a gold standard is tetraploid complementation, mentioned in Chapter 3 in the context of the analysis of knockout phenotypes. Tetraploid cells are those with a double complement of chromosomes. Mouse embryos can be made tetrapoid by electrofusion of the first two blastomeres, which effectively generates a new zygote with a double set of chromosomes. Tetraploid embryos cannot develop properly, but they can form extraembryonic structures sufficient to support gestation of a diploid fetus. So if normal ICM cells, or mouse ES cells, are implanted in the blastocoel of a tetraploid host, then the donor cells form the entire embryo while the tetraploid cells form the placenta. The ability to form an entire embryo is a more stringent criterion of pluripotent cell quality even than germ line chimerism, and only some mouse ES cells lines can do it successfully.

It is possible to generate pluripotent stem cells not only from mouse embryo ICMs, but also from epiblasts of early postimplantation stages. These are known as EpiSCs (see Figure 6.6f below). They satisfy the basic definition of pluripotent stem cells in terms of growth in culture without limit and the ability to form a wide range of cell types, either as embryoid bodies or as teratomas. But they differ in various ways from normal ES cells. First, the colony morphology is different: those of EpiSCs being flatter. Second, unlike normal mouse ES cells, EpiSC cannot be passaged clonally. Third, EpiSC do not grow in LIF but instead require activin and FGF in the medium. Fourth, they do not generate chimerism in mouse embryos. Furthermore the ancillary transcription factors found in mouse ES cells are absent, *Oct4* is controlled by a different enhancer, the overall level of genome methylation is higher, and, in female EpiSC, only one X-chromosome is active. It is now accepted that such cells are in a distinct pluripotent state from mouse ES cells. The ICM-derived cells are called naive or ground-state, and the epiblast-derived cells are called primed. It is possible to convert mouse ES cells to EpiSC simply by changing the culture medium to the activin/ FGF-containing medium used for EpiSC. The reverse transformation, of EpiSC to ES cells, can be achieved by overexpression of the *Klf4* gene, one of the group of ancillary pluripotency-associated transcription factors.

Mouse iPS Cells

In 2006 it was discovered that it was possible to reprogram ordinary somatic cells to pluripotent status by the introduction of a group of genes selected from the pluripotency gene network. The resulting cells are called induced pluripotent stem cells or iPS cells (Figures 6.3; 6.C.3 a,b). Since then reprogramming to pluripotency has been a very active field of research and considerable technical progress has been made. The original gene cocktail consisted of four genes: *Oct4, Sox2, Klf4* and *cMyc*. OCT4 and SOX2 are part of the core pluripotency network of transcription factors in ES cells. KLF4 is an ancillary transcription factor also associated with pluripotency. cMYC is a transcription factor that has many roles but in this context

serves to increase cell division rate which assists the reprogramming process, maybe simply by increasing accessibility of regulatory sites within the DNA. It has been found that various other combinations of genes may be used successfully but they almost always include *Oct4*. The original delivery was by means of retroviruses that integrated into the genome of the target cells, but subsequent work showed that integration is not essential and that reprogramming can be achieved using a variety of non-integrating delivery systems. What is essential is that the synthesis of the gene products occurs at a high level and that it can keep up with the division of the cells: for example using self-replicating episomes or Sendai virus (a virus with RNA–RNA replication), or repeated transfections with RNA encoding the required genes. Methods that do not achieve this continued high level expression may still work but they tend to have very low efficiency.

Typically only a few cells in the target population are reprogrammed and they are selected out by growth in mouse ES cell medium, where they appear as refractile

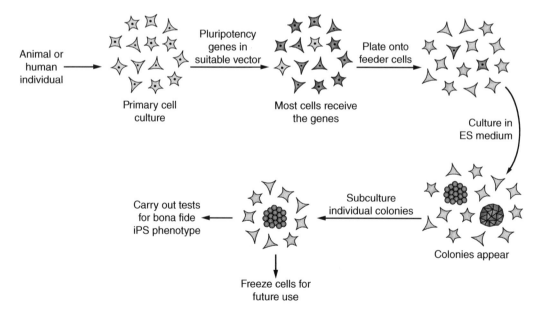

Figure 6.3 Procedure for making induced pluripotent stem cells (iPS cells). (Slack, J.M.W. (2013) Essential Developmental Biology, 3rd edn. Reproduced with the permission of John Wiley and Sons.)

colonies that can be isolated and further expanded. The target cells in mouse experiments are usually fibroblasts derived from embryos (murine embryonic fibroblasts, or MEFs). These are a poorly characterized and potentially heterogeneous source and there was early suspicion that the reprogrammed cells were already stem cells of some sort, although subsequent investigation has shown that this is not the case.

Investigation of the mechanism of iPS cell formation has depended a lot on the use of inducible gene constructs, especially the doxycycline inducible system (see Chapter 3). Here the gene of interest is placed under the control of a Tetracycline Response Element (TRE) based on the Tet operator from *E. coli.* The cells are also supplied with a gene

expressing the Tet activator (TA, usually the version called rtTA), under control of a constitutive promoter. When the drug doxycycline is supplied, it binds to the TA enabling the complex to activate transcription from the TRE. Initially this was used to regulate expression of the *Oct4, Sox2, Klf4* and *cMyc* (OKSM) genes in lentivirus vectors which can integrate into the DNA of the target cells. If iPS cells are induced by dosing the cells with doxycycline, and then the drug is withdrawn, the iPS cells still persist and continue to grow (Figure 6.4). This shows that persistent expression of the transgenes is not necessary for maintenance. Once they are established, the iPS cells are already expressing the pluripotency gene network including *Oct4, Sox2* and *Nanog*, from their own endogenous

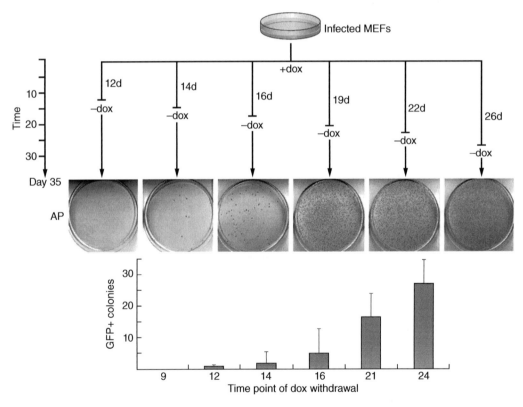

Figure 6.4 Experiment to show that the exogenous reprogramming factors are needed during the formation of iPS cells but not for their maintenance. Murine embryonic fibroblasts are infected with doxycycline inducible lentivirus encoding the OKSM factors. The doxycycline is withdrawn at different times and the resulting alkaline phosphatase or GFP reporter positive colonies are shown at day 35. (From: Brambrink, T., Foreman, R., Welstead, G.G., Lengner, C.J., et al. (2008) Sequential expression of pluripotency markers during direct reprogramming of mouse somatic cells. Cell Stem Cell 2, 151–159. Reproduced with the permission of Elsevier.)

genes, and doing so in a stable, self-sustaining manner. In fact persistent expression of the exogenous OSKM genes inhibits differentiation of the iPS cells and to achieve differentiation it is necessary to shut down their expression.

Next it was shown that iPS cells can arise from most or all somatic cells and not just from rare stem cells present in the starting material. Chimeric mice were prepared by making iPS cells containing OKSM under doxycycline control, and injecting these cells into mouse blastocysts. The resulting mice are used as a source of pre-B or mature B lymphocytes, isolated by FACS from the bone marrow and spleen respectively, and these were plated one per well in the presence of growth factors to enable them to grow. Addition of doxycycline induced the formation of iPS cell colonies in these clonal cultures, and all the iPS cell colonies are found to carry the same characteristic DNA rearrangement of antibody genes as their founding B lymphocyte (Figure 6.5). In the case of mature B cells, overexpression of an additional gene, C/EBPα, was required to obtain iPS cells. In a refinement of this experiment it was shown that all monoclonal cultures of pre-B cells exposed to doxycycline would generate some iPS cell colonies eventually. These experiments indicate that even a cell differentiating by means of DNA

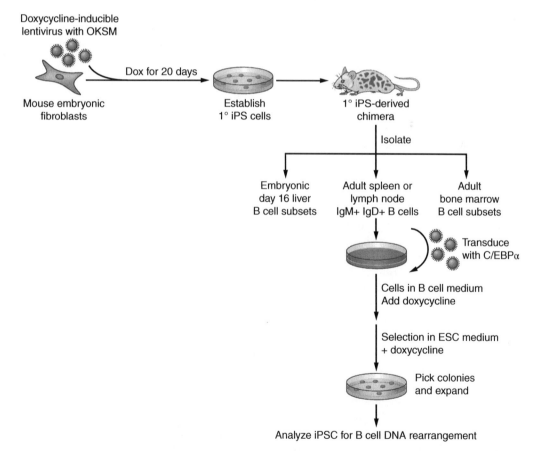

Figure 6.5 Formation of iPS cells from fully differentiated precursors. Here chimeric mice are generated with some cells containing doxycycline-inducible OKSM transgenes. B-lymphocytes are cultured and transformed into iPSC. (Modified from Hanna, J., Markoulaki, S., Schorderet, P., Carey, B.W., et al. (2008) Direct reprogramming of terminally differentiated mature B lymphocytes to pluripotency. Cell 133, 250–264. Reproduced with the permission of Elsevier.)

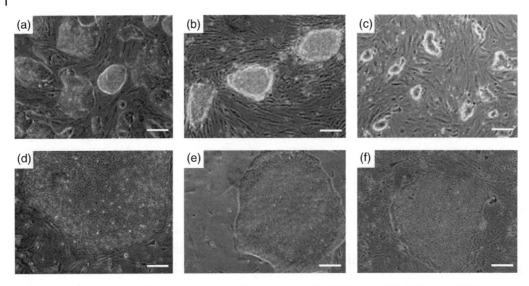

Figure 6.6 Different kinds of pluripotent stem cells. (a) Mouse ESC. (b) Mouse iPSC. (c) Human iPSC induced to a naive pluripotent state with chemical inhibitors. (d) Human ESC. (e) Human iPSC. (f) Mouse EpiSC. Note the flatter colony morphology for (d)–(f). (From: Robinton, D.A. and Daley, G.Q. (2012) The promise of induced pluripotent stem cells in research and therapy. Nature 481, 295–305. Reproduced with the permission of Nature Publishing Group.)

rearrangement can be converted into an iPS cell, and that every cell, not just a rare sub-population, could yield iPS cells among its descendants.

The properties of mouse iPS cells are very similar to those of mouse embryonic stem cells (Figure 6.6 a,b). They grow as refractile colonies in LIF/2i medium or in other mouse ES cell media. The cells have large nuclei with prominent nucleoli and little cytoplasm. They show self-sustaining expression of the pluripotency gene regulatory network, and examination of the endogenous pluripotency genes shows that inhibitory DNA methylations have been removed from their promoters. They also display other characteristic markers including the cell surface carbohydrate SSEA1 and the enzyme alkaline phosphatase. In female cells the inactive X-chromosome is reactivated. When LIF is withdrawn they will form embryoid bodies with typical patterns of differentiation. They will also form teratomas when injected into immunologically compatible adult animals and these produce derivatives of all three embryonic germ layers.

The best quality mouse iPS cells will generate germ line chimeras when injected into pre-implantation embryos and in some cases even tetraploid complementation has been achieved. Interestingly, during the process of iPS cell induction, the alkaline phosphatase appears early, followed by SSEA-1, and only later is endogenous Nanog upregulated and the autocatalytic pluripotency gene network established.

Mouse iPS cells are not however exactly the same as mouse ES cells. Careful study of the whole genome expression profiles using RNAseq indicates that the patterns are similar but not identical. Also there are often residual epigenetic features: DNA methylations and histone modifications, that resemble the parent cells and have not been completely erased. This means that iPS cells can sometimes show easier differentiation back into the cell type of origin than into other types. Investigations of this sort have shown that there is not just one single pluripotent state that is stable but a whole range of similar stable states differing in the degree to which they approach the normal ES cell state. These differences can affect the differentiation behavior of the cells so before commencing work with an iPS cell line it is important to characterize it thoroughly.

Human Pluripotent Stem Cells

Human embryonic stem cells were first produced in 1998, using similar methods to those used for mice (Figure 6.6d). Human preimplantation conceptuses were put into culture and pluripotent cells were grown from their inner cell masses. These cells grow on feeders as refractile colonies, and they express the core pluripotency genes and alkaline phosphatase. In terms of pluripotency they can differentiate in an embryo-like manner as embryoid bodies, and when injected into immunodeficient mice (see Chapter 4) they can form teratomas containing derivatives of all three germ layers.

Despite these similarities of the human ES cells to the mouse ES cells, it soon became clear that there were a number of differences. The human ES cells grow as flatter colonies. Instead of LIF, they require activin and FGF for growth, and they cannot be passaged clonally (i.e. single cells usually cannot re-establish new colonies). They do not express SSEA-1, although they do possess a set of antigens found in human embryonal carcinoma cells: SSEA-3 and -4, which are cell surface carbohydrates carried on glycolipids, and TRA 1-60 and 1-81, which are cell surface carbohydrates carried on the glycoprotein podocalyxin. Female lines have one X-chromosome inactivated. It is not known whether human ES cells are capable of forming chimeric embryos when injected into human blastocysts, as such an experiment is considered unethical. However similar experiments with primate ES cells, which are very similar to the human, indicate that chimerism is unlikely to occur. This list of characteristics actually matches very closely the "primed" mouse ES cells, derived from post-implantation epiblasts, rather than the "naive" mouse ES cells derived from the pre-implantation epiblast or inner cell mass. So it is now considered that human ES cells are the equivalent of the mouse EpiSC (= primed ES cells). Various methods including overexpression of genes and treatment with various factors and inhibitors have been described that can convert human ES cells into something similar to the mouse naive state. However it remains uncertain whether there is a real, stable, human naïve ES cell state.

Shortly after the production of the first mouse iPS cells, human iPS cells were also produced using similar methods (Figures 6.6e and 6.C.3c). These tend to resemble human ES cells rather than mouse ES cells, in other words to be in the primed state. Human iPS cells have been subjected to very detailed analysis because of their potential use for making patient-specific cells for transplantation which would potentially be immunologically compatible with the host. In terms of overall gene expression they are similar to human ES cells but show some divergence. Like mouse iPS cells, they tend to show epigenetic characteristics resembling the cells of origin, and to show a propensity to differentiate in this direction rather than others. Also, like all tissue culture cells, human iPS cell lines accumulate somatic mutations. On average human iPS cells carry about six more non-synonymous point mutations than the starting cells and may also carry copy number variations. The existence of such mutations is not surprising in view of the in vitro culture and selection procedures that are required to make an iPS cell line.

Human pluripotent stem cells (both ES cells and iPS cells) have a reputation for being hard to cultivate, largely because of their very poor cloning efficiency. So the most effective method of subculture has been manual dissection of colonies into smaller pieces which are re-plated and allowed to expand. However, better media are becoming available. For example culture on a surface of laminin 521 and E-cadherin in a defined medium free of animal products has been reported to enable derivation of human ES cells from single blastomeres and to enable clonal subculture thereafter.

SCNT-Derived Embryonic Stem Cells

Apart from ES cells and iPS cells, the third method of producing pluripotent cell lines is by somatic cell nuclear transfer (SCNT) into secondary oocytes, as described in Chapter 5. The somatic nucleus becomes reprogrammed in the oocyte environment and can support development into a blastocyst. At this stage the inner cell mass can be used as a source of cells in the same way as a naturally fertilized conceptus. Several ES cell lines have been established from mouse SNCT blastocysts in this way.

Since the method worked in mice it was an obvious step to attempt the same in humans. In the days before iPS cells, it was felt that this could be the route to the production of personalized ES cells, genetically the same as the donor of the somatic nucleus. If this person were a patient in need of a cell graft, the cells should be a perfect immunological match. Largely because of the small number of groups engaged in such work, and the difficulty of procuring human oocytes, human SCNT was not achieved until 2012. Comparison of human SCNT-derived ES cells with embryo-derived ES cells and with iPS cells indicates that the SCNT-derived ES cells are closer in overall gene expression pattern to the embryo-derived ES cells than they are to iPS cells. However given the practical and ethical difficulties in making such lines, it seems more likely that any future patient-specific cell therapy will be based on iPS cell-derived grafts.

Ethical Issues Concerning Human ES Cells

As with direct research work on human conceptuses, the derivation of human ES cells has caused significant controversy. The issue is very similar: to opponents of this work, preimplantation human conceptuses are deemed to be human beings which should be accorded full human rights. To set up a cell line from an inner cell mass is considered to represent destruction of the conceptus and therefore of a human being. Human iPS cells, although they are very similar to ES cells, are considered acceptable because they have not originated from human conceptuses. There is even more opposition to the practice of SCNT because this is a form of cloning. Although nobody is in favor of reimplanting a human SCNT conceptus into the womb to produce a live birth, some people consider that even using the conceptus to establish a cell line constitutes human cloning. Although this procedure does not involve the destruction of a naturally fertilized or in vitro fertilized egg, it is felt by some to violate human dignity.

Supporters of human ES cell work usually consider preimplantation conceptuses to be more akin to human tissue culture cells than to fully formed human beings, and hence not to have any special human rights. They further argue that the surplus embryos from IVF clinics that are used for this work would otherwise be discarded. These embryos are donated for research with the permission of the parents who desire no further children, and many feel that this is an honorable use. In relation to SCNT, although all agree that this is a form of cloning (called "therapeutic cloning" by some), because no actual cloned fetus is produced it is felt to be harmless. Moreover the advocates argue that significant medical benefits will accrue from the use of human ES cells, and that SCNT is a potential method of obtaining patient-matched ES cells whose differentiated derivatives would require no immunosuppression on implantation into the donor. The issue of human dignity is a complex one. There is at present no clear agreement among bioethicists about what human dignity actually is, and therefore it is hard to judge whether it is being violated by any given procedure.

Pluripotent Stem Cells from Postnatal Organisms

The types of pluripotent stem cell described up to now: mouse and human ES cells, made either from fertilized or SCNT blastocysts, together with iPS cells, all indubitably exist. They have been isolated in many labs and handled in many more, and their properties are substantially agreed upon. But this is not the case for pluripotent stem cells from postnatal animals or humans, or from extra-embryonic sources such as amniotic fluid or umbilical cord blood. Despite the fact that many labs work on such entities there is little reproducibility in this work and the properties of such cells, and even their existence, continues to be a source of controversy.

In principle the existence of such cells seems unlikely because after the epiblast has become divided into different body regions pluripotent cells have effectively been lost from the embryo. The hierarchical system of development, described in Chapters 7–9, is generally accepted as being correct and has no place for pluripotent cells after the first subdivision of the epiblast. Furthermore, when pluripotent cells are brought into existence within the adult body by induction of the iPSC-inducing gene cocktail from transgenes in mice, the effect is to generate teratomas, which are encountered only rarely in nature.

Nevertheless the number of reports of cells with wide potency being isolated indicates that there is a phenomenon to be explained. One possible explanation is that there are a few pluripotent cells remaining from the epiblast which are located too far away from the inducing signals and so do not participate in the normal process of epiblast subdivision and developmental commitment. Such cells might later become located almost anywhere but only those in positions not exposed to the usual inducing factors (Wnts, FGFs, Nodals, BMPs etc) at some later stage would be expected to retain the pluripotent ability until birth. One such position might, perhaps, be within the amniotic fluid. This explanation is compatible with the findings that pluripotent cells are hard to isolate, that they are more common in perinatal or young animals than in older animals, and that results differ between labs.

The second potential explanation is the role of in vitro culture itself. It is often found that the same cells show a wider range of behaviors in vitro then in vivo, because the culture conditions can be adjusted over a much wider range. For example, the mesenchymal stem cells from the bone marrow (see Chapter 11) probably function as a source for the renewal of bone in vivo. But in vitro they can be caused to become adipose tissue or smooth muscle by changing the culture conditions. It is likely that in vitro culture can at least widen the range of differentiation behavior in some cell populations.

Third there is the issue of selection in culture. Cells are constantly undergoing somatic mutations and spontaneous epigenetic changes and novel variants can easily be isolated if enough cells are subject to selection for a long enough period. So for example, after its loss from the epiblast the pluripotency gene *Oct4* is not normally expressed in any part of the mouse embryo or adult other than the germ cells. However if bone marrow cells are placed in culture in ES cell medium and cultured for some weeks, it is eventually possible to select out some clones of OCT4-positive cells.

Probably the truth is some combination of all these mechanisms: occasional cells left over from the epiblast, the effects of in vitro culture, and a buildup of somatic mutation. It is sometimes argued that it does not matter if a given type of stem cell that exists in vitro has no in vivo counterpart so long as it has some scientific or clinical utility in its own right. This may be true, after all, ES cells do not really exist in vivo, as their counterparts are progenitor cells rather than stem cells. However it is very important for everyone that scientific procedures can be made reliable and reproducible. This is the case for ES and iPS cells, but has not so far proved to be the case for pluripotent cells from peri- or postnatal sources.

Applications of Pluripotent Stem Cells

So far the chief application of pluripotent stem cells has been to make genetically modified mice, mostly knockouts of specific genes. This has been of great importance in developmental biology for helping to understand the developmental role of specific genes. It has also been of importance in generating mouse models for human diseases, to assist research on pathogenesis and new therapies.

Three main reasons are usually cited for the importance of human pluripotent stem cells. The first is to aid the study of human development, second to generate hard-to-obtain human cell types for drug testing, and third to generate cells, tissues or even organs for transplantation.

As we have seen in Chapter 5, early human development is similar to that of mouse, but there are a number of significant differences. Understanding these differences will be important for a whole variety of purposes including fertility research, causes of early abortion, causes of developmental defects, and understanding better how to control differentiation of pluripotent stem cells. There will always be severe limitations on what is practical or what is considered ethical in terms of utilization of actual human embryos for research, so the availability of in vitro cell-based model systems offers great potential. Progress in this area will depend on better methods for replicating a post-implantation environment for the cells to enable normal development to take place for a limited period.

In terms of drug development, animal testing remains necessary, but there are metabolic differences between animals and humans and certain tests benefit from the use of human cells. In particular the liver is the first destination of most drugs and the way they become metabolized by hepatocytes can be extremely important for their pharmacokinetics, activity and efficacy. Human hepatocytes are hard to obtain and so the availability of a limitless supply made from pluripotent stem cells will greatly assist the process of drug development. Likewise, many drug side effects are exerted on the heart and can be detected by the effect of the drug on cardiomyocytes in vitro, but human cardiomyocytes are even harder to obtain than hepatocytes. Therefore, a ready supply from pluripotent stem cells is very valuable.

The public face of pluripotent stem cells is their potential for making limitless amounts of differentiated cells for transplantation. This is now in early clinical trials in several areas including β-cell transplants for diabetes, retinal pigment cells for macular degeneration, and glial progenitors for spinal damage (see online supplement). Although iPS cells can in principle be made from the individual patient and the resulting grafts would be fully immunocompatible, the high cost of personalized therapy is likely to restrict applications in the near future to allografts requiring immunosuppression.

An exciting new possibility is that of making whole organs for transplantation by a procedure called blastocyst complementation. Here a host embryo is prepared lacking a gene essential for the formation of a specific organ. For example, lack of *Pdx1* prevents or at least greatly compromises, formation of the pancreas, and lack of *Nkx2.5* is similarly necessary for formation of the heart. It is now possible to ablate genes with good efficiency using CRISPR-Cas9 technology directly on the zygotes (see Chapter 3). Thus such embryos can be made in any species, not just in mice. Then pluripotent stem cells containing the gene in question are injected into the blastocyst and the chimeric embryo is returned to the female reproductive tract to develop to term. It is known from experiments with mice that the donor cells tend to form the organ that the host is unable to make because of its genetic defect. The long-term hope is to use this approach to grow human organs in animal hosts, by putting normal human pluripotent stem cells into an animal embryo with the relevant gene ablated. For example the pig has a size and physiology similar to the human and is often considered as a possible host. Successful experiments have been made generating rat organs in mice, and vice versa (Figure 6.C.4), but so far the integration of human pluripotent cells into large animal hosts has been poor, for reasons that are not understood. Even if human organs can be generated, a potential problem with this approach is that they will still have blood vessels derived from the host. Blood vessels are highly immunogenic and liable to provoke a strong xenograft rejection on transplantation. Overcoming these various technical problems is likely to take considerable time and effort, however the approach is an interesting example of the combined potential of stem cell biology and developmental biology.

Further Reading

Abad, M., Mosteiro, L., Pantoja, C., Canamero, M., et al. (2013) Reprogramming in vivo produces teratomas and iPS cells with totipotency features. Nature 502, 340–345.

Brons, I.G.M., Smithers, L.E., Trotter, M.W.B., Rugg-Gunn, P., et al. (2007) Derivation of pluripotent epiblast stem cells from mammalian embryos. Nature 448, 191–195.

Bulic-Jakus, F., Katusic Bojanac, A., Juric-Lekic, G., Vlahovic, M. and Sincic, N. (2016) Teratoma: from spontaneous tumors to the pluripotency/malignancy assay. Wiley Interdisciplinary Reviews: Developmental Biology 5, 186–209.

De Los Angeles, A., Ferrari, F., Xi, R., Fujiwara, Y., et al. (2015) Hallmarks of pluripotency. Nature 525, 469–478.

Duggal, G., Warrier, S., Ghimire, S., Broekaert, D., et al. (2015) Alternative routes to induce naïve pluripotency in human embryonic stem cells. Stem Cells 33, 2686–2698.

Dunn, S.-J., Martello, G., Yordanov, B., Emmott, S. and Smith, A.G. (2014) Defining an essential transcription factor program for naïve pluripotency. Science 344, 1156–1160.

Evans, M. (2011) Discovering pluripotency: 30 years of mouse embryonic stem cells. Nature Reviews. Molecular Cell Biology. 12, 680–686.

González, F. and Huangfu, D. (2016) Mechanisms underlying the formation of induced pluripotent stem cells. Wiley Interdisciplinary Reviews: Developmental Biology 5, 39–65.

Lensch, M.W., Schlaeger, T.M., Zon, L.I. and Daley, G.Q. (2007) Teratoma formation assays with human embryonic stem cells: a rationale for one type of human–animal chimera. Cell Stem Cell 1, 253–258.

Ma, H., Morey, R., O'Neil, R.C., He, Y., et al. (2014) Abnormalities in human pluripotent cells due to reprogramming mechanisms. Nature 511, 177–183.

Nichols, J. and Smith, A. (2009) Naive and primed pluripotent states. Cell Stem Cell 4, 487–492.

Nichols, J. and Smith, A. (2011) The origin and identity of embryonic stem cells. Development 138, 3–8.

Pera, M.F. and Trounson, A.O. (2004) Human embryonic stem cells – prospects for development. Development 131, 5515–5525.

President's-Council-on-Bioethics (2008) Human Dignity and Bioethics. Washington D.C.

Robinton, D.A. and Daley, G.Q. (2012) The promise of induced pluripotent stem cells in research and therapy. Nature 481, 295–305.

Silva, M., Daheron, L., Hurley, H., Bure, K., et al. (2015) Generating iPSCs: Translating cell reprogramming science into scalable and robust biomanufacturing strategies. Cell Stem Cell 16, 13–17.

Sohni, A. and Verfaillie, C.M. (2011) Multipotent adult progenitor cells. Best Practice and Research Clinical Haematology 24, 3–11.

Solter, D. (2006) From teratocarcinomas to embryonic stem cells and beyond: a history of embryonic stem cell research. Nature Reviews. Genetics 319.

Stadtfeld, M. and Hochedlinger, K. (2010) Induced pluripotency: history, mechanisms, and applications. Genes & Development 24, 2239–2263.

Tachibana, M., Amato, P., Sparman, M., Gutierrez, N.M., et al. (2013) human embryonic stem cells derived by somatic cell nuclear transfer. Cell 153, 1228–1238.

Tamm, C., Pijuan Galitó, S. and Annerén, C. (2013) A comparative study of protocols for mouse embryonic stem cell culturing. Plos One 8, e81156.

Tesar, P.J., Chenoweth, J.G., Brook, F.A., Davies, T.J., et al. (2007) New cell lines from mouse epiblast share defining features with human embryonic stem cells. Nature 448, 196–199.

Thomson, J.A., Itskovitz-Eldor, J., Shapiro, S.S., Waknitz, M.A., et al. (1998) Embryonic stem cell lines derived from human blastocysts. Science 282, 1145–1147.

Yamaguchi, T., Sato, H., Kato-Itoh, M., Goto, T., et al. (2017) Interspecies organogenesis generates autologous functional islets. Nature 542, 191–196.

7

Body Plan Formation

Within the implanted mammalian conceptus lies a sheet of similar-looking pluripotent cells called the epiblast. How does this become transformed into a miniature mouse or human having a set of body parts arranged in the correct relative positions? Elucidating these mechanisms has been the task of the science of developmental biology and it was accomplished largely between the years of 1980 and 2000, although many details still remain to be established. The key to understanding early development is to appreciate that it does not proceed in a single step from pluripotent cells to terminally differentiated ones. Instead, development proceeds through a hierarchy of steps in each of which a particular group of cells acquires a new state of developmental commitment. These states were postulated by classical experimental embryologists on the basis of labeling, explantation and grafting experiments on early embryos. Since the 1980s it has become possible to observe states of developmental commitment directly by in situ hybridization and by other methods of visualizing expression of the genes responsible.

Embryological Concepts

Developmental Commitment

Embryologists often speak of specification, determination and fate, and these terms each have distinct meanings. A specified state is one which is stable in the absence of external signals. For example, most parts of the upper hemisphere of an early frog embryo will develop into balls of epidermis if isolated from the embryo into a neutral medium (a simple buffered salt solution). They are said to be specified to form epidermis (Figure 7.1). However, if grafted into the equatorial or lower parts of the embryo they develop into mesodermal or endodermal structures respectively. By contrast, tissues which retain their commitment regardless of where they are grafted are said to be determined. Once gastrulation is underway, the dorsal part of the upper hemisphere of the frog has become determined to form the brain, and this commitment is not affected by grafting to a new position. Specification and determination represent different degrees of stability of a committed state. These concepts were originally defined for multicellular explants of tissue. They can also be applied to single cells, although single cells usually show behavior more labile than the tissues from which they came because of disruption of extracellular materials and cell contacts which often have a role to play in the state of commitment.

The normal fate of a tissue region is a different concept again. This indicates what the region will become in the course of normal development, undisturbed by the experimenter. Using cell labeling techniques, which were described in Chapter 3, it is possible to establish the fate of all parts of an embryo and thereby to construct a fate map of the whole embryo for a particular stage. It is important to note that the fate is not necessarily

The Science of Stem Cells, First Edition. Jonathan M. W. Slack.
© 2018 John Wiley & Sons, Inc. Published 2018 by John Wiley & Sons, Inc.
Companion website: www.wiley.com/go/slack/thescienceofstemcells

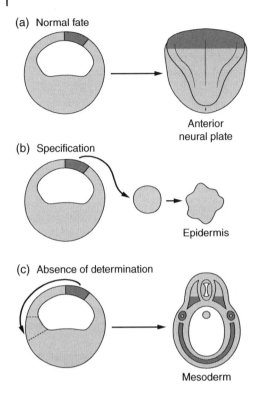

(a) Normal fate

Anterior
neural plate

(b) Specification

Epidermis

(c) Absence of determination

Mesoderm

Figure 7.1 Illustration of the meaning of fate, specification and determination. (a) The indicated region on the dorsal side of the upper hemisphere of a frog embryo will normally become parts of the brain. (b) If explanted into a neutral medium, the same tissue will become epidermis. (c) If grafted to an equatorial position, in this case ventrally, it will become parts of the mesoderm. At this stage the tissue is not determined and its fate can be altered by grafting to the new position.

the same as the specification or the determination of a region at a particular time. Again, consider the upper hemisphere of a frog blastula. In normal development the dorsal part becomes the brain, and the ventral part becomes epidermis. So the dorsal part of the animal hemisphere can behave in different ways: in isolation it becomes epidermis, in situ, in the environment of the embryo, it forms neural tissue, and if grafted to the lower part of the embryo it forms mesodermal tissue. These differences are indicated in Figure 7.1.

Classical embryologists realized that the reason for these differences was the presence

of short range signals within the embryo that controlled the behavior of cells, and specifically their states of developmental commitment. These signals were called inducing factors and the process referred to as embryonic induction. Furthermore, the classical embryologists realized that the states of commitment could not be for terminal cell types like neurons or muscle fibers, they must instead be for broad body regions, like heads and abdomens. This means that a succession of inducing signals, with successive subdivision of embryo regions, is necessary to generate the complex anatomy of the animal. Each of the successive states of commitment is defined by expression of one or more genes, now called selector genes or developmental control genes.

Before the relevant genes were identified there was often some difficulty in comprehension of embryological literature because the same words can be used to indicate both position in the embryo and the current state of developmental commitment. For example, the word "dorsal" can mean the upper part of the body, but it can also refer to a dorsal state of commitment of a tissue region that has been exposed to signals imparting a dorsal-type specification even though it may not be in a dorsal position. It is important to bear this in mind when reading older embryological papers.

Finally it was appreciated that if a region of the embryo was to become subdivided repeatedly in response to different inducing signals, then it had to have a specific competence to respond to these signals, and that this competence would change with each successive round of induction. This is now understood in terms of the presence of receptors for extracellular signals and the associated intracellular pathways of signal transduction which control gene expression.

One of the great successes of the era of molecular biology has been its explanation in molecular terms of the old embryological concepts of commitment, induction and competence. This understanding has also enabled very specific and targeted interventions to be made in embryonic development. Predictable changes to developmental events, and to the resulting anatomy of the organism, can now be brought about through overexpression or inhibition of specific genes or gene products at the appropriate times and places.

Embryonic Induction

Regulation of the key genes controlling the formation of early embryo body regions is done by means of extracellular signals emitted from one group of cells and affecting another. These signals, called inducing factors in the context of embryonic development, are a subset of those factors also known as growth factors or cytokines in cell biology. In development the signal often takes the form of a concentration gradient with the high end at the source region and the low end distant from it. If the competent cells can respond differently to several different concentrations then one graded signal can generate several different outcomes. This is the key to embryonic development because it means that each cycle of signaling and response can generate a pattern more complex than the initial one. Inducing signals that have multiple responses in this way are called morphogens.

The minimum starting complexity for an embryo is two regions: one emitting the signal and the other responding to it. If all the responding cells do the same thing then the pattern of the embryo would stay the same. However, if the responding cells show, say, three different responses at different concentration thresholds, then the interaction will produce four zones (the original signaling region plus three different induced zones) where there were two before. Since each zone can then become a signaling or responding zone for the next interaction, the number of regions in the embryo can increase rapidly.

Figure 7.2 shows one model for the operation of a morphogen. A source region produces the factor and maintains it at a constant concentration. The remaining cells of the embryo are competent to respond to the

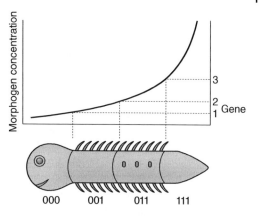

Figure 7.2 How a gradient of an inducing factor, in other words a morphogen, can generate a complex pattern. In this model, the gradient is high at the future caudal end of the embryo and low at the future rostral end. Three developmental control genes are activated at different thresholds. The resulting animal has a head and three segments, with states of gene activity 000, 001, 011, 111. (Slack, J.M.W. (2013) *Essential Developmental Biology*, 3rd edn. Reproduced with the permission of John Wiley and Sons.)

factor by activating specific genes at particular threshold concentrations. The factor can diffuse across the responding region and is destroyed at a rate proportional to its own local concentration. This model generates an exponential concentration gradient with the high end at the source. It is stable in time because the production and diffusion of the factor is balanced by its destruction. In actual embryos such gradients may not ever reach the steady state, but their behavior is still qualitatively the same.

The competence of the responding cells involves the ability to turn specific genes on or off at specific concentration thresholds of the gradient. These are precisely the developmental control genes whose activity determines the commitment of regions of the embryo. A simple gradient will naturally produce nested sets of gene activity, for example if there are three genes with different concentration thresholds, the nearest territory to the source will have all of them on, the next territory has one off, the next two off and the last territory, where no thresholds were exceeded, still has them all off. These states of gene activity can be represented as "0" for off and "1" for on. Thus the initial state is 000 all over, and the final state is a row of four territories with gene activities 111, 011, 001, 000 (Figure 7.2). This example shows that the gradient model not only explains how a simple embryo can autonomously become more complex. It also shows that a specific pattern can become imparted with a specific polarity in terms of the arrangement of new territories in relation to the source of inducing factor.

Gradients have certain properties which it is important to understand when designing differentiation protocols for pluripotent stem cells. First, a particular factor may have different effects depending on its precise concentration. Second, application of a single concentration might be expected to generate just a single outcome. But in reality clumps of cells will impede access, generating local gradients of an applied factor and causing a heterogeneity of outcomes. Third, in order to achieve the desired result the time of dosing must correspond to the relevant period of competence of the cells. Fourth, the duration of treatment must not be excessive as the system will develop to a new state of commitment which may embody competence to respond to the same factor but with a different outcome. By assessing each of these issues carefully, a developmental biology perspective can provides valuable guidance to the design of differentiation protocols.

Lastly there is the issue of the bistability of threshold responses. In general cellular responses to inducing factors are irreversible, and in this respect differ from responses of adult cells to hormones, which are usually reversible. The reason for the irreversibility is a positive feedback as shown in Figure 7.3. The signal activates gene expression and then the product also activates its own gene, maintaining activity after the signal has disappeared. In the longer term, the state of gene activity will be further stabilized by changes to chromatin organization. Real

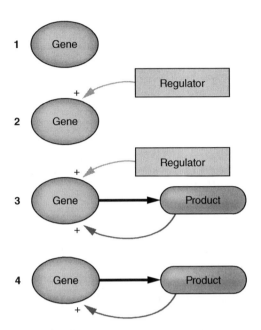

Figure 7.3 A very simple model for a bistable switch. This depicts a temporal sequence. At time 2 the gene is activated by a regulator, typically a signal transduction pathway activated by an inducing factor. At time 3 the gene product is being produced and also activates expression. At time 4, the signal has gone but expression is still retained because of the positive feedback. (Slack, J.M.W. (2013) Essential Developmental Biology, 3rd edn. Reproduced with the permission of John Wiley and Sons.)

systems are much more complex than this simple model, with feedback loops containing many more components, but the model does capture the essence of bistability: the developmental control gene is either on or off, and after transient exposure to a suprathreshold level of inducing factor, it will remain on permanently.

Symmetry Breaking

Embryonic development offers a number of examples of what are called symmetry-breaking processes. This means that the normal course of development generates an unstable situation which eventually breaks down and becomes resolved into a new stable state, but by doing so generates some spatial asymmetry. To give a simple mechanical example, imagine a thin coin being gradually raised to a vertical position and then left isolated. The process of raising it is completely deterministic. But once on its edge, the coin is susceptible to any minor perturbation such as a vibration or a stray air current, which will cause it to fall flat. The processes of falling is also deterministic, but the direction is not, so when it falls it could end up either way up, displaying either heads or tails. So the same starting situation can generate two different outcomes, and which actually occurs is a matter of chance. A population of such coins will generate a random mixture of heads and tails, in approximately equal numbers.

Biochemical symmetry-breaking systems generally start from a situation of homogeneous composition. This becomes unstable and resolves itself into zones of different chemical composition, the polarity and arrangement depending on random environmental perturbations at the outset which may even be as small as the thermal fluctuations of individual molecules. The systems of embryonic induction and gene regulation often embody elements of positive feedback which tend to generate symmetry breaking processes. Symmetry breaking is seen for example in the formation of primitive endoderm cells from the inner cell mass of the mouse embryo, and the acquisition of rostrocaudal polarity by the radially symmetrical egg cylinder. In real embryos there is often some slight pre-existing bias which causes symmetry breaking processes to occur with the same polarity each time, even though such systems can in principle break the symmetry starting from a completely homogeneous starting situation.

Key Molecules Controlling Development

Genes Encoding Developmental Commitment

There are several genes whose principal function is to encode early states of developmental commitment, namely the three germ layers: ectoderm, mesoderm and endoderm, and broad body regions such as head, trunk and tail. These genes encode transcription factors which regulate other genes required for functions such as specific morphogenetic movements and the competence to be subdivided in response to the next inducing signal. The germ layers have been known to descriptive embryologists for a long time because most animal embryos show three distinct tissue layers at an early stage. Endoderm is encoded by genes for transcription factors including GATA4, FOXA2 and SOX17. It occupies the inner part of the early embryo and later becomes the epithelial lining of the gut and respiratory system. The mesoderm is encoded by genes for transcription factors including BRACHYURY (often known as T in the mouse), TBX6, and TWIST. It occupies the middle part of the early embryo and later becomes the muscles, connective tissues, limbs, kidneys and gonads. The ectoderm occupies the outer surface of the early embryo and becomes the brain, spinal cord, epidermis and also the neural crest, which forms a variety of structures. It is not clear whether any transcription factor encodes the ectoderm as a whole, but certainly SOX1 and -2 are important for the formation of the early neuroepithelium, and at a somewhat later stage p63 is necessary for development of the epidermis.

When considering early body regions it is important to note that the terms "anterior" and "posterior" are used differently in zoology and in human anatomy. In animals such as the mouse, anterior means the head end and posterior means the tail end. Dorsal is the upper side and ventral is the lower side. In humans, walking as we do on two legs, anterior is the same as ventral, and posterior is the same as dorsal. The terms "rostral" and "caudal" can be used unambiguously to indicate the head and tail end for both humans and animal models, hence the usage in this book. The best known gene family concerned with rostro-caudal identity is the *HOX* family. This is a set of genes encoding transcription factors. In vertebrate animals there are four gene clusters and in invertebrates often just one. The *HOX* genes are typically expressed in embryos during the time that the structures along the rostro-caudal axis are being established, and their expression is nested such that all members of a cluster are active in the caudal region and expression is lost at a specific rostral level which different for each gene. Remarkably the order of the rostral expression limits of the *HOX* genes is the same as their physical order on the chromosome, with the 3′ genes having the most rostral limits and the 5′ genes having their expression confined to the caudal end. The name *HOX* is short for "homeobox" the characteristic DNA binding domain found in all members of this gene family. This relates to the fact that the earliest to be discovered, in the fruit fly *Drosophila*, displayed homeotic mutations in which one body segment was converted into another. Although the situation is more complex with four gene clusters, in general the rule is the same in vertebrate animals. Loss of function mutations of *HOX* genes generally transform vertebrae into more rostral types, and gain of function mutations (i.e. those leading to overexpression) generally transform vertebrae into more caudal types. The transformations are usually seen near the normal rostral expression limit of the gene in question. The occurrence of such homeotic mutations is very good evidence that the *HOX* genes, do in fact encode rostro-caudal body levels in the early embryo. There are also many other genes containing a homeobox which are not members of the *HOX* family. Most of these are also involved in some way with embryonic development but they do not necessarily show homeotic mutations.

The expression of the *HOX* genes reaches as far rostrally as the hindbrain, but beyond this other transcription factor genes are responsible for encoding subdivisions of the head. These include *Otx2* and *Lhx1*. At the caudal end of the body, a family of *Cdx* genes, which are non-*HOX* homeobox genes, are expressed, and they act as the first layer of downstream genes regulated by the caudally expressed *HOX* genes.

Inducing Factors

There are several signaling systems of particular importance in development. Accordingly the extracellular signaling molecules, the inducing factors themselves, have become very important in the control of the differentiation of pluripotent stem cells which will be dealt with in Chapter 9. All of them are encoded by multigene families, the products of each individual gene usually having a similar or overlapping biological activity with the others. Each factor binds to a specific cell surface receptor and the receptors are coupled to intracellular signal transduction pathways, usually leading to upregulation or repression of specific genes. Each pathway has a number of interacting proteins which increase or decrease its activity, but only the central components are mentioned here. Several of the signal transduction pathways contain negative steps, i.e. one component inhibits the next. This can lead to difficulties of presentation, but the key thing to remember is that two negative steps in a pathway are equivalent to one positive step. The essential features of some of the key pathways are shown in Figure 7.4.

Wnt System

The Wnt system is important for the formation of the rostrocaudal pattern of the early embryo, for the development of many organs, and is involved in the maintenance of nearly all types of stem cell. Wnts are single chain proteins with a covalently linked lipid chain, so are of low solubility in water. Wnt receptors are encoded by *Frizzled* genes. Wnt 1, 3A and 8 bind to Frizzleds which interact with an intracellular protein Dishevelled, which inhibits the enzyme glycogen synthase 3 (gsk3). When active, gsk3 phosphorylates β-catenin, thus keeping it in the cytoplasm. When it is inhibited, the β-catenin is dephosphorylated, and can carry the transcription factor Tcf1 into the nucleus where it regulates target genes (Figure 7.4a). This β-catenin pathway is sometimes called the canonical

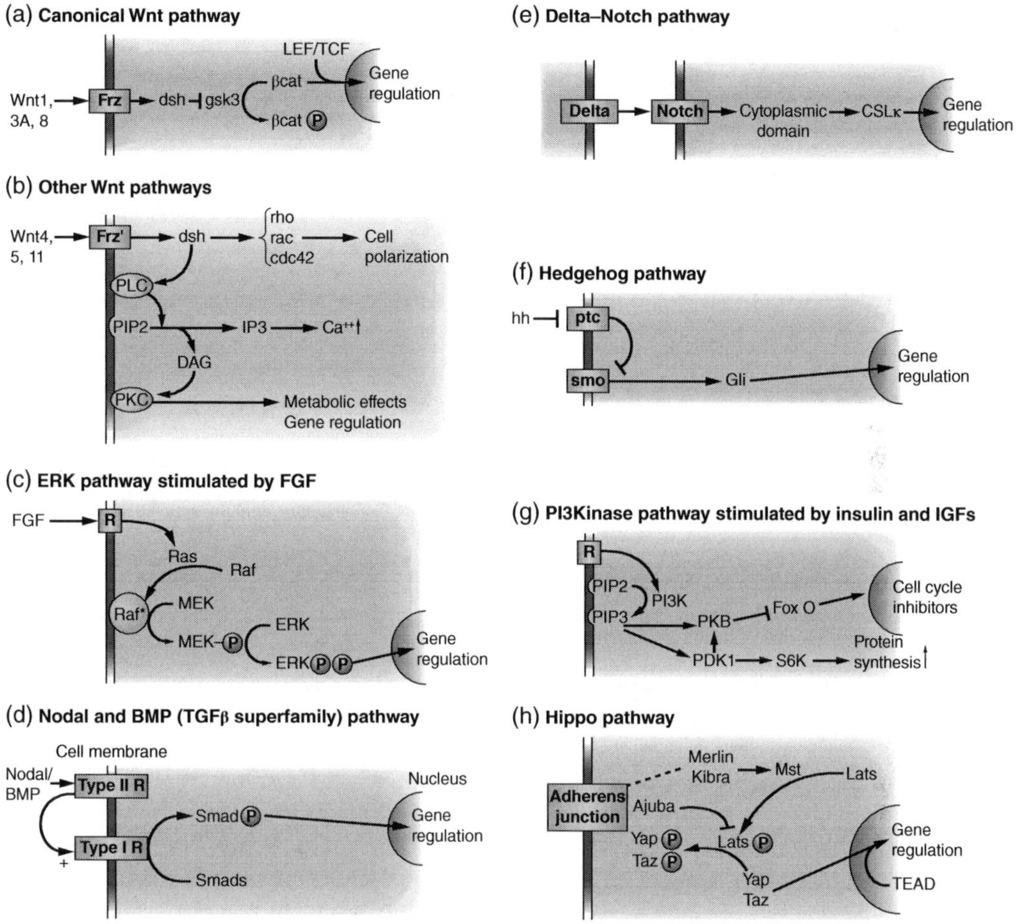

(a) **Canonical Wnt pathway**

(b) **Other Wnt pathways**

(c) **ERK pathway stimulated by FGF**

(d) **Nodal and BMP (TGFβ superfamily) pathway**

(e) **Delta–Notch pathway**

(f) **Hedgehog pathway**

(g) **PI3Kinase pathway stimulated by insulin and IGFs**

(h) **Hippo pathway**

Figure 7.4 Some signaling pathways important in early development. (a) The canonical Wnt pathway. (b) Other Wnt pathways. (c) The FGF pathway. (d) The Nodal or BMP pathway. (e) The Notch pathway. (f) The Hedgehog pathway. (g) The insulin/IGF pathway. (h) The Hippo pathway. Only key features are shown here, as all the pathways have many other interacting components. (Modified from Slack, J.M.W. (2013) Essential Developmental Biology, 3rd edn. Reproduced with the permission of John Wiley and Sons.)

Wnt pathway. β-catenin is also involved in cell adhesion by anchoring cadherin molecules to the cytoskeleton. This means that Wnt signaling can also reduce cell adhesion by removing β-catenin from the adhesion complex.

Other Wnts, including 4, 5 and 11, bind to other types of Frizzled which have different effects (Figure 7.4b). They can activate the small GTPases Rho, Rac and Cdc42, leading to changes in cell polarization. They can also activate phospholipase C, which generates inositol-1,4,5-trisphosphate and diacylglycerol from phosphatidyinositol-4,5-bisphosphate,

leading to a variety of metabolic and gene regulation events.

FGF System

FGF stands for fibroblast growth factor, but the FGFs have many functions in addition to that suggested by their name. They are single chain proteins binding to a family of tyrosine kinase receptors on the cell surface. Binding of FGF to the FGF receptors and consequent receptor dimerization is facilitated by cell surface heparan sulfate and the pattern of sulfation on this and other extracellular carbohydrates can affect the level of FGF signaling. Dimerization leads to activation of the receptor, and to autophosphorylation, leading to exchange of GDP for GTP on the associated Ras protein, and this causes translocation of the Raf protein to the cell surface. Raf is a kinase, activated by membrane lipids, which phosphorylates MEK (mitogen activated, ERK-activating kinase), which in turn phosphorylates ERK (extracellular signal regulated kinase). Activated ERK enters the nucleus and activates various transcription factors by phosphorylation, leading to gene regulation (Figure 7.4c).

This signal transduction pathway is often called the MAP kinase pathway, although it is better to call it the ERK pathway as there are other MAP kinases. Various other growth factors in addition to the FGFs, stimulate the ERK pathway using different tyrosine kinase receptors. For example epidermal growth factor (EGF), the neuregulins, platelet-derived growth factor (PDGF), hepatocyte growth factor (HGF) and ephrins can all activate the ERK pathway through their own specific receptors. Specificity between different types of FGF largely lies with the splice forms of the receptors: the R2b splice form being selective for FGF7 and 10, and the R2c form being selective for FGF4 and 8.

In early development, FGFs are required for formation of mesoderm and, along with Wnts, for specification of the rostrocaudal body pattern. Later they have key roles in development of many organ systems. In addition to being involved in the maintenance of various types of stem cell, FGFs are needed, along with activin, for the proliferation of human ES cells and the related mouse EpiSC.

Nodals and BMPs

Nodal gets its name from a mouse mutant defective in the node, the principal signaling center of the early embryo. BMPs (bone morphogenetic proteins) were given their name because of their promotion of bone formation in connective tissue, but they have many other functions. Both are members of the TGFβ superfamily of growth factors, which have a common molecular structure of two polypeptide chains joined by disulfide bridges. They are made as longer pro-forms which need proteolytic cleavage to become active. Nodals and BMPs have different biological activities based on binding to different receptors and utilizing different intracellular signaling. However the overall pattern is the same. The factor binds to a specific type 2 receptor on the cell surface, this forms a complex with, and thereby activates, a type 1 receptor. This is a serine-threonine kinase and it phosphoryates Smad proteins in the cytoplasm. Nodals cause phosphorylation of Smad 2 and 3, while BMPs cause phosphorylation of Smad 1, 5 and 8. The phosphorylated Smads enter the nucleus and function as transcription factors regulating target genes (Figure 7.4d).

Nodals are key factors in the initial induction of mesoderm and endoderm in the early embryo. They are biochemically very similar to activins, which, in the adult, causes release of follicle stimulating hormone from the pituitary. Because activins are generally more available than nodals, they are used instead of nodals in protocols to bring about differentiation of pluripotent stem cells, and in the cultivation of human ES cells.

BMPs are involved in numerous inductive interactions during development, including the induction of primordial germ cells and the repression of neural development, as well as in the formation and healing of bone in the adult.

Notch System

Notch is a cell surface receptor, originally named because loss of function mutants in the fruit fly *Drosophila* had notches in their wings. The ligands are called Deltas or Jaggeds and are also cell surface molecules. Because both the ligand and receptor are tethered to the cell surface, this system requires cell contact between ligand and receptor-bearing cells in order to operate. Contact causes activation of a membrane bound protease, γ-secretase, which cleaves Notch, liberating a portion called NICD

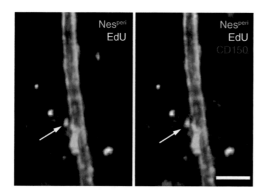

Figure 1.C.1 Hematopoietic stem cell, identified by staining with an antibody to CD150 (red), and also labeled with EdU (white) from a pulse given 30 days previously. The green color shows pericytes expressing nestin-GFP surrounding a small arteriole. (Kunisaki, Y., et al. (2013) Arteriolar niches maintain haematopoietic stem cell quiescence. Nature 502, 637–643. Reproduced with the permission of Nature Publishing Group.)

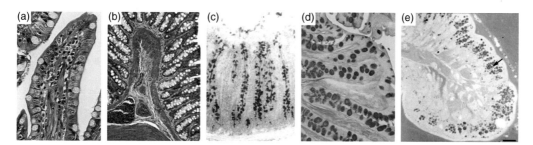

Figure 2.C.1 Histological sections of the gut stained with various different stains. (a) Hematoxylin and eosin. (http://histology-world.com/photoalbum/displayimage.php?album=35&pid=2838.) (b) Masson's Trichrome. (https://thumbs.dreamstime.com/z/histology-colon-trichrome-masson-s-pinkish-red-smooth-muscle-dark-granulated-red-red-blood-cells-42768592.jpg.) (c) Periodic acid–Schiff (PAS) (From: Iiboshi Y. et al. (1997). Developmental changes in the distribution of the mucous gel layer in rat small intestine. Asia Pacific Journal of Clinical Nutrition 6, 111–115.) (d) Alcian Blue (https://histologistics.files.wordpress.com/2015/08/rat-colon-alcian-blue-pas.jpg.) (e) Sudan Black. (arrow: lipid droplets) (From: Marza, E., Barthe, C., André, M., Villeneuve, L., Hélou, C. and Babin, P.J. (2005) Developmental expression and nutritional regulation of a zebrafish gene homologous to mammalian microsomal triglyceride transfer protein large subunit. Developmental Dynamics 232, 506–518.) (Courtesy of Anatech Ltd.; Marza et al., 2004. Reproduced with the permission of John Wiley and Sons; Courtesy of Hans Synder, Boshi et al., 1996. Reproduced with the permission of SAGE Publications.)

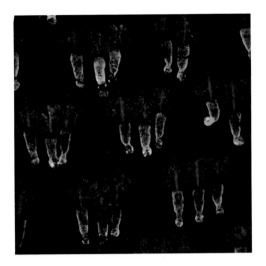

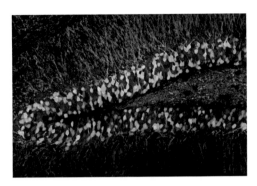

Figure 3.C.1 Multicolor labeling of clones in the mouse hippocampus using the Brainbow technique. (From: Cell Picture Show. Tamily Weissman http://www.cell.com/pictureshow/brainbow. Courtesy of Tamily Weissman, Harvard University.)

Figure 2.C.2 Wholemount of mouse tail hair follicles. This is immunostained for keratin 15 (green), indicating hair follicle stem cells, and Ki67 (red) to show proliferating cells. The blue color is DAPI stain for DNA. The image is a confocal z-stack. (From: http://www.smithsonianmag.com/science-nature/the-startling-beauty-of-the-microscopic-180949245/#YyePsBtJ5gx0Qxpu.99 *Yaron Fuchs, Howard Hughes Medical Institute/The Rockefeller University, New York, NY USA. Eighth Prize, 2013 Olympus BioScapes Digital Imaging Competition®.* Courtesy of Olympus BioScapes Digital Imaging Competition.)

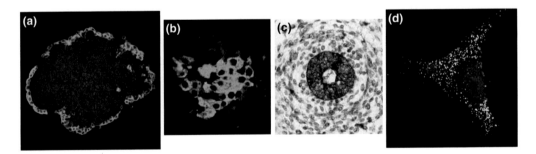

Figure 2.C.3 Immunostaining (a–c) and in situ hybridization (d). (a) An embryoid body made from mouse embryonic stem cells. The outer layer is positive for the primitive endoderm marker Disabled-2. The blue color is DAPI stain for DNA. (b) A pancreatic islet in the mouse. Endocrine cells are shown by immunostaining for glucagon (red), insulin (green) and somatostatin (blue). (c) Histochemical visualization using diaminobenzidine staining of keratin 8 in 13.5 d mouse embryo esophagus. (a–c author's photos.) (d) In situ hybridization on HeLa cell using the RNA Scope method. Four probes are used simultaneously, for β-actin (red), RPLP0 (60S acidic ribosomal protein P0, yellow), PPIB (peptidylprolyl isomerase B, light blue), and HPRT-1 (hypoxanthine phosphoribosyltransferase 1, green). Darker blue is DAPI stain for DNA. (Part D from Wang et al., 2012. RNAscope: A novel in situ RNA analysis platform for formalin-fixed, paraffin-embedded tissues. Journal of Molecular Diagnostics 14, 22–29. Reproduced with the permission of Elsevier.)

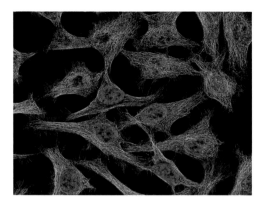

Figure 4.C.1 HeLa cells. Multiphoton fluorescence image of HeLa cells stained with the actin binding toxin phalloidin (red), microtubules (cyan) and cell nuclei (blue). (https://images.nigms.nih.gov/index.cfm?event=viewDetail&imageID=3521 **ID Number** 3521. National Center for Microscopy and Imaging Research.)

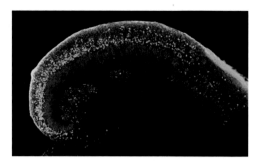

Figure 4.C.3 A human brain organoid made from pluripotent stem cells. It is immunostained for neurons (TUJ1, green) and neuronal precursors (SOX2, red). Cell nuclei are blue. (From: Bredenoord, A.L., Clevers, H. and Knoblich, J.A. (2017) Human tissues in a dish: The research and ethical implications of organoid technology. Science 355. eaaf9414. DOI: 10.1126/science.aaf9414)

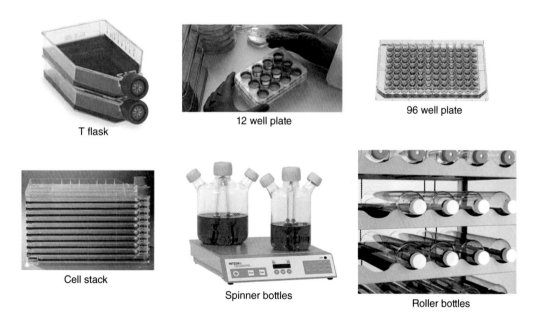

T flask

12 well plate

96 well plate

Cell stack

Spinner bottles

Roller bottles

Figure 4.C.2 Containers for tissue culture. **T flask:** https://www.thermofisher.com/us/en/home/life-science/cell-culture/cell-culture-plastics/cell-culture-flasks.html. **12 well plate:** https://www.thermofisher.com/us/en/home/life-science/cell-culture/cell-culture-plastics.html. **96 well plate:** http://www.coleparmer.com/Product/Thermo_Scientific_Nunc_Edge_96_Well_Plates_Sterile_No_Lid_Nunclon_Delta_160_Cs/EW-01930-63. **Cell stack:** http://csmedia2.corning.com/LifeSciences/Media/pdf/bp_cellstack_manual_1_27_03_cls_bp_007.pdf. **Spinner bottles:** http://www.laboratorynetwork.com/doc/cell-culture-devices-cellroll-and-cellspin-0001. **Roller bottles:** http://wheaton.com/lab/roller-culture.html.

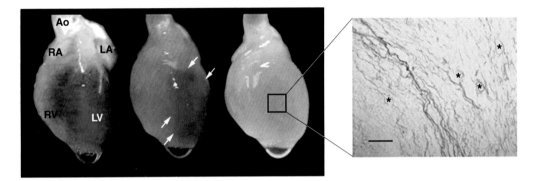

Figure 4.C.4 Decellularized heart. The three hearts are shown on a perfusion apparatus and show different degrees of decellularization. The one on the right, perfused with SDS, is fully decellularized. On the right is a histological section showing the absence of cells and the preservation of blood vessel spaces (*). Ao = aorta; RA = right atrium; LA = left atrium; RV = right ventricle; LV = left ventricle (From: Zia, S., Mozafari, M., Natasha, G., Tan, A., Cui, Z. and Seifalian, A.M. (2016) Hearts beating through decellularized scaffolds: whole-organ engineering for cardiac regeneration and transplantation. Critical Reviews in Biotechnology 36, 705–715. Reproduced with the permission of Taylor & Francis.)

Figure 4.C.5 Different inbred strains of mouse. (Stanton Short/Jennifer L. Torrance courtesy of the Jackson Laboratory.)

Figure 5.C.1 Germ cells in the mouse embryo testis at E13.5. They are green because of an Oct4 reporter. (Author's photo.)

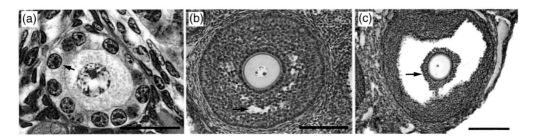

Figure 5.C.2 Maturation of mouse oocytes. (a) Primary follicle, arrow: granulosa cells, scale bar 20 μm. (b) Early antral follicle. Arrow: early antral cavity, scale bar 100 μm. (c) Preovulatory follicle. Arrowhead: granulosa cells, scale bar 200 μm. (From: Myers, M., Britt, K.L., Wreford, N.G.M., Ebling, F.J.P. and Kerr, J.B. (2004) Methods for quantifying follicular numbers within the mouse ovary. Reproduction 127, 569–580. Reproduced with the permission of BioScientifica Ltd.)

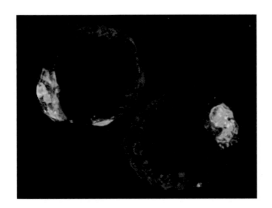

Figure 5.C.3 Mouse blastocysts. Cells expressing Oct4, mostly in the inner cell mass, are colored green by an Oct4 reporter. (Author's photo.)

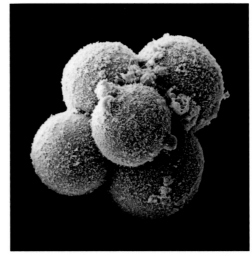

Figure 5.C.4 Early human conceptus, scanning electron micrograph. (From: B0003402 Early human embryo. Yorgos Nikas: Wellcome Images, images@ wellcome.ac.uk)

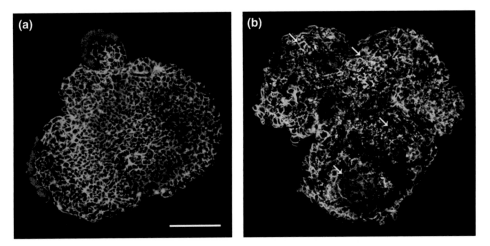

Figure 6.C.1 Embryoid bodies from mouse pluripotent stem cells. (a) Embryoid body made from mouse iPSC, with some time allowed to predifferentiate so that it has formed a partial external layer of primitive endoderm. Immunostained for the visceral endoderm marker HNF4 (lilac). Scale bar 100 μm. (b) Embryoid body made from mouse iPSC. After 8 days it contains many structures and clumps of differentiated cells. Arrows indicate clumps of cardiac muscle immunostained for cardiac troponin (lilac). (Author's photos.)

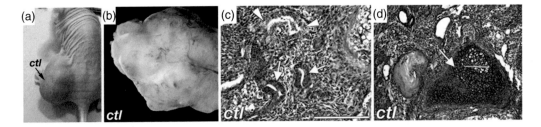

Figure 6.C.2 Teratoma arising from mouse embryonic stem cells. (a) and (b) 5×10^6 cells were injected into a SCID mouse and formed a tumor (ctrl). (c) and (d) The tumor contains various differentiated tissues normally arising from all three embryonic germ layers, including neuroepithelium (white arrows in c), mucous epithelium (white arrowheads in c) and cartilage (white arrows in d). (From: Fico, A., De Chevigny, A., Egea, J., Bösl, M.R., Cremer, H., Maina, F. and Dono, R. (2012) Modulating Glypican4 suppresses tumorigenicity of embryonic stem cells while preserving self-renewal and pluripotency. Stem Cells 30, 1863–1874. Reproduced with the permission of John Wiley and Sons.)

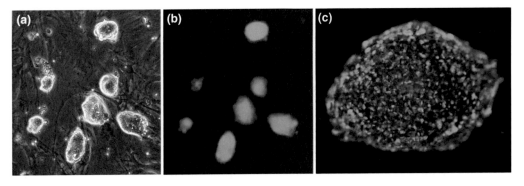

Figure 6.C.3 Induced pluripotential stem cells (iPSC). (a) Mouse iPSC colonies viewed by phase contrast. (b) The same colonies immunostained for NANOG protein (green). (c) A Human iPSC colony immunostained for TRA-118 (green) and DNA (blue). (Author's photos.)

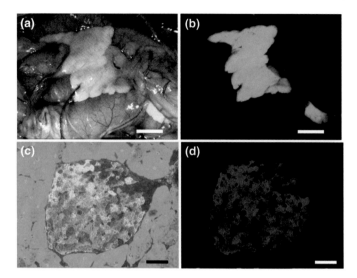

Figure 6.C.4 Generation of a mouse pancreas in a rat. The rat embryo had its *Pdx1* gene inactivated using the CRISPR-Cas9 technique, and GFP-positive mouse pluripotent stem cells were injected into the blastocyst. (a, b) In the resulting rats, the mouse pancreas fluoresces green while the rat gut is unlabeled. Scale bar 5 mm. (c, d) Immunostaining of insulin (red) in islet cells of the mouse pancreas. Scale bar 50 μm. (From: Yamaguchi, T., Sato, H., Kato-Itoh, M., Goto, T., et al. (2017) Interspecies organogenesis generates autologous functional islets. Nature 542, 191–196. Reproduced with the permission of Nature Publishing Group.)

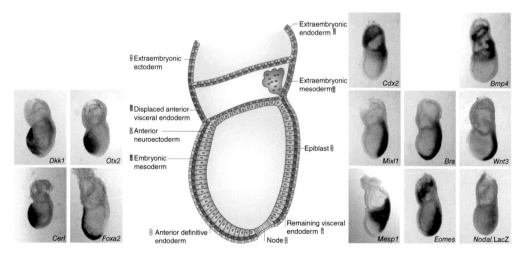

Figure 7.C.1 Expression patterns of a number of key developmental control genes in the mouse embryo at 6.5–7.5 days, visualized by in situ hybridization, or in the case of *Nodal*, by reporter expression. (From: Arnold, S.J. and Robertson, E.J. (2009) Making a commitment: cell lineage allocation and axis patterning in the early mouse embryo. Nature Reviews. Molecular Cell Biology 10, 91–103. Reproduced with the permission of Nature Publishing Group.)

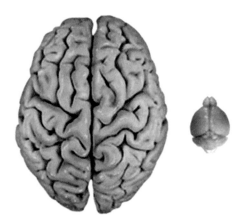

Figure 8.C.1 Relative size of the human and mouse brains. (From: https://corticalchauvinism. com/2013/01/20/of-mice-and-men-complications- of-animal-models-in-autism-research/WordPress. com.)

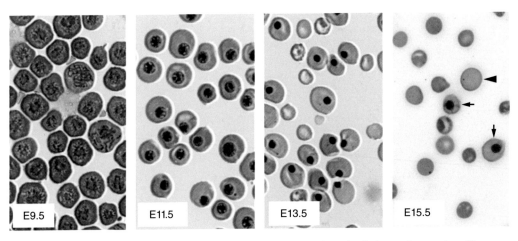

Figure 8.C.2 Blood cells found in mouse embryos. The primitive nucleated cells (arrows) are replaced by erythrocytes that lose their nuclei (arrowhead). (From: Palis, J., Malik, J., McGrath, K.E. and Kingsley, P.D. (2010) Primitive erythropoiesis in the mammalian embryo. International Journal of Developmental Biology 54, 1011–1018. (*Source:* Jim Palis, reproduced with permission from UPV/EHU Press.)

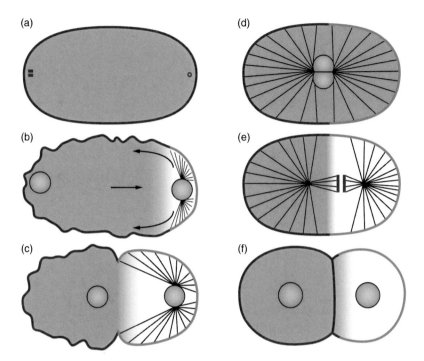

Figure 9.C.1 Sequence of events leading to the unequal first division of the *C. elegans* zygote, anterior to left. (a) Chromatin from the oocyte (left) and sperm (right) shown in blue, PAR-3 complex in red, MEX-5 in pink. (b) PAR-1/2 complex (in green) accumulates in posterior, pronuclei form, cytoplasmic flows commence. (c) PAR-1/2 zone expands, oocyte pronucleus moves posteriorly. (d) Pronuclear fusion. (e) First mitosis. (f) Two-cell stage: formation of AB and P₁ blastomeres. (From: Nance, J. (2005) PAR proteins and the establishment of cell polarity during *C. elegans* development. Bioessays 27, 126–135. Reproduced with the permission of John Wiley and Sons.)

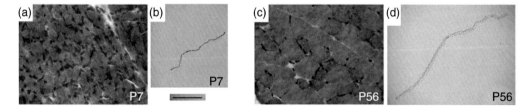

Figure 9.C.2 Postnatal growth of myofibers in the mouse. Sections (a, c) and isolated fibers (b, d) are shown at 7 and 56 days after birth. The scale bar represents 100 μm for the 7 day images, and 1 mm for the 56 day images. Nuclei are blue because the mice were transgenic for a nuclear-localized *lacZ* gene driven by the promoter for *Myf5*. (From: White, R., Bierinx, A.-S., Gnocchi, V. and Zammit, P. (2010) Dynamics of muscle fibre growth during postnatal mouse development. BMC Developmental Biology 10, 21.)

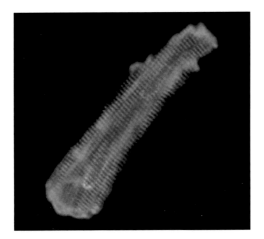

Figure 9.C.3 Rat cardiomyocyte. The green color is immunostaining for desmin, concentrated in Z bands, and the orange nuclear stain is propidium iodide. This cell is binucleate. (Novus Biologicals, LLC.)

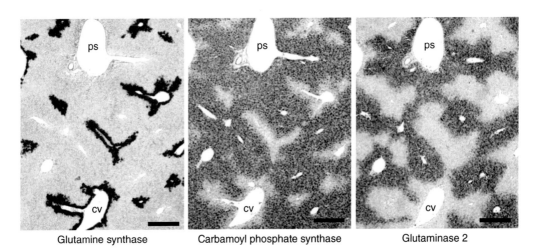

| Glutamine synthase | Carbamoyl phosphate synthase | Glutaminase 2 |

Figure 9.C.4 Metabolic zonation in mouse liver. In situ hybridizations are shown for mRNA encoding three liver enzymes. Glutamine synthase is perivenous, while carbamoyl phosphate synthase and glutaminase are periportal. PS = portal space; CV = central vein. (From: Benhamouche, S., Decaens, T., Godard, C., Chambrey, R., et al. (2006) Apc tumor suppressor gene is the "zonation-keeper" of mouse liver. Developmental Cell 10, 759–770. Reproduced with the permission of Elsevier.)

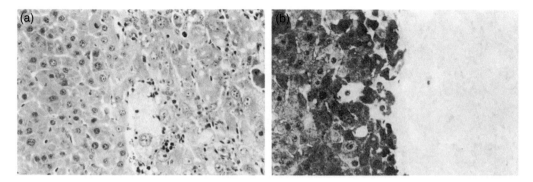

Figure 9.C.5 Population of a *FAH⁻* mouse liver by a graft of FAH⁺ hepatocytes. (a) shows an H&E stained section of the junction between graft and host. The degraded condition of the host cells on the right is apparent. (b) shows FAH histochemistry with the donor cells on the left. (From: Overturf, K., Aldhalimy, M., Ou, C.N., Finegold, M. and Grompe, M. (1997) Serial transplantation reveals the stem-cell-like regenerative potential of adult mouse hepatocytes. American Journal of Pathology 151, 1273–1280. Reproduced with the permission of Elsevier.)

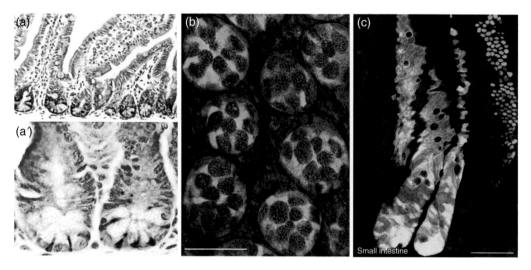

Figure 10.C.1 Intestinal stem cells. (a, a′) Intestinal stem cells are visualized using histochemistry for β-galactosidase (blue) in a reporter mouse in which the *lacZ* gene is driven by the *Lgr5* promoter. The positive cells lie in the crypt base interleaved with Paneth cells. (b) Transverse section of the crypt base. Here the reporter mouse is *Lgr5-GFP* so the stem cells are green and the Paneth cells, immunostained for lysozyme, are red. Scale bar 50 μm. (c) Confetti labeling of intestinal crypts. The reporter is a version of R26R in which four colors may be generated at random by Cre mediated recombination. Crypts are uniformly labeled by only one of the four colors indicating clonal selection of stem cells within each crypt. In the upper part of the figure, villi may contain more than one color as they are fed by more than one crypt. (*Sources:* (a), From Barker, N., van Es, J.H., Kuipers, J., Kujala, P., et al. (2007) Identification of stem cells in small intestine and colon by marker gene Lgr5. Nature 449, 1003–1007. Reproduced with the permission of Nature Publishing Group. (b&c), From Snippert, H.J., van der Flier, L.G., Sato, T., et al. (2010) Intestinal crypt homeostasis results from neutral competition between symmetrically dividing Lgr5 stem cells. Cell 143, 134–144. Reproduced with the permission of Elsevier.)

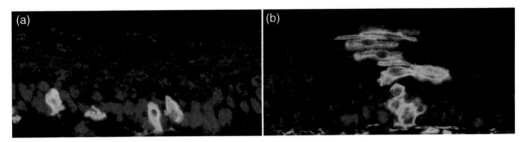

Figure 10.C.2 Lineage tracing of epidermal stem cells in mouse skin. The mice express *Axin2-CreER* with an *R26R* type reporter expressing GFP (green) following Cre mediated recombination. The blue color is DAPI stain for cell nuclei, and the red is immunostaining for Dickkopf3, a Wnt inhibitor present in the superficial layers. Tamoxifen was given on postnatal day 21 and the images show the situation at 1 day (a) and 2 months (b) thereafter. (a) 1 day, a few basal layer cells are labeled. (b) 2 months, clones of labeled cells are visible leading from the basal layer to the surface. (From: Lim, X., Tan, S.H., Koh, W.L.C., et al. (2013) Interfollicular epidermal stem cells self-renew via autocrine Wnt signaling. Science 342, 1226–1230. American Association for the Advancement of Science.)

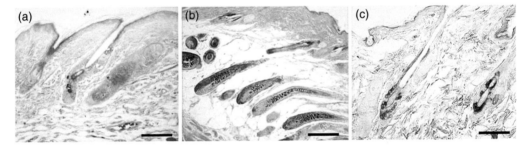

Figure 10.C.3 Lineage tracing of stem cells in the hair follicle bulge, using *Lgr5-CreER x R26R* mice. Mice were injected with tamoxifen at postnatal day 21, when the hair follicles are in telogen. (a) After 6 days, some bulge cells are labeled. Scale bar 50 μm. (b, c) After 16 and 25 days respectively, labeling extends to all parts of the hair follicle below the sebaceous glands. Scale bars 100 μm (b) and 50 μm (c). (From: Jaks, V., Barker, N., Kasper, M., van Es, J.H., et al. (2008) Lgr5 marks cycling, yet long-lived, hair follicle stem cells. Nature Genetics 40, 1291–1299. Reproduced with the permission of Nature Publishing Group.)

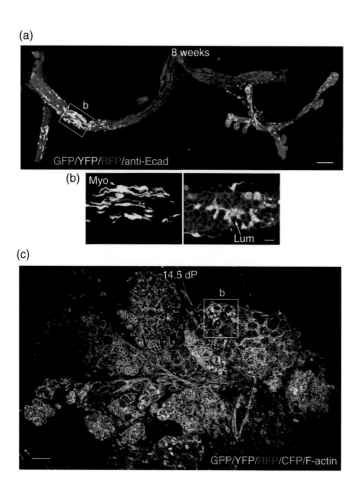

Figure 10.C.4 Lineage labeling of the mouse mammary gland. (a) The mouse is *keratin5-rtTA/TRE-Cre/R26R-Confetti*. Administration of doxycycline causes production of Cre and recombination of the reporter to generate clones of different color. Scale bar 100 μm. (b) Two optical sections through the yellow clone in box b, showing that both luminal and myoepithelial cells are labeled and are therefore derived from a single progenitor. Scale bar 10 μm. (c) Alveolar growth during pregnancy. The mouse is *Elf5-rtTA/TRE-cre/R26R-Confetti*. It was dosed with doxycycline before pregnancy and is now at 14.5 days. Individual alveoli have been generated from labeled stem cells, and may be of mixed color (box b), showing that more than one stem cell may contribute to one alveolus. Scale bar 150 μm. (From: Rios, A.C., Fu, N.Y., Lindeman, G.J. and Visvader, J.E. (2014) In situ identification of bipotent stem cells in the mammary gland. Nature 506, 322–327. Reproduced with the permission of Nature Publishing Group.)

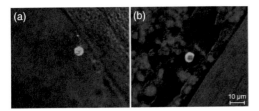

Figure 10.C.5 FACS purified HSC from a GFP expressing mouse were injected into an irradiated recipient and imaged in the marrow of femur slices after 4 hours. (a) Localization near endosteal surface. (b) Localization near blood vessels, visible as gaps, and bone. Blue color is DAPI stain for DNA. (From: Xie, Y., Yin, T., Wiegraebe, W., He, X.C., et al. (2009) Detection of functional haematopoietic stem cell niche using real-time imaging. Nature 457, 97–101. Reproduced with the permission of Nature Publishing Group.)

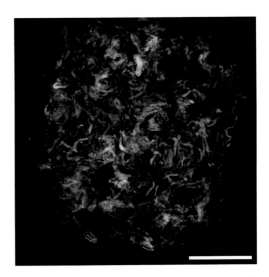

Figure 10.C.6 Labeling of spermatogonial stem cells using the mouse *Bmi-CreERxR26R-Brainbow*. 24 weeks after tamoxifen administration a large number of labeled clones persist in the testis, each filling a whole segment of a tubule. Scale bar 5 mm. (From: Komai, Y., Tanaka, T., Tokuyama, Y., Yanai, H., et al. (2014) Bmi1 expression in long-term germ stem cells. Scientific Reports 4. Reproduced with the permission of Nature Publishing Group.)

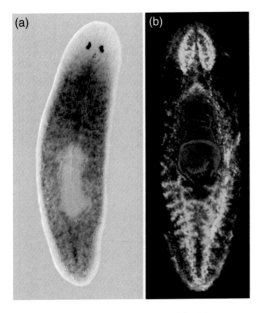

Figure 11.C.1 The planarian worm *Schmidtea mediterranea*. (a) External view. (b) Wholemount immunostaining with three different antibodies specific for neurons (yellow), pharynx (magenta) and gut (blue). These worms are about 1.4 mm long. (a), From: https://www.gesundheitsindustrie-bw.de/files/cache/db8dc3fb3338fac884250045ded70ba6_f11681.jpg. (b), from: Sánchez Alvarado, A. (2012) Q&A: What is regeneration, and why look to planarians for answers? BMC Biology 10, 88. *Source:* Dr. Siegfried Schloissnig.)

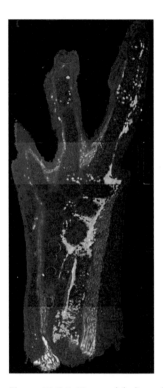

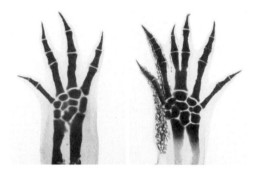

Figure 11.C.3 Example of the experiment depicted in Figure 11.4. On the left is a normal axolotl limb, stained to show the skeleton in blue. On the right is a limb which received a posterior skin graft, then was amputated through the graft. It has regenerated as a 6-digit double-posterior pattern. The contribution of graft and host tissue is approximately shown by the pigment cells, as the graft was from a pigmented donor. It may be seen that the forked central digit is almost entirely of host composition. (Author's photos, from Slack, J.M.W. (2013) Essential Developmental Biology, 3rd edn. Reproduced with the permission of John Wiley and Sons.)

Figure 11.C.2 Lineage labeling of connective tissue in urodele limb regeneration. The green color shows cells descended from a graft of GFP-positive cartilage into an unlabeled host limb. The limb was amputated through the graft and structures derived from the graft remain green. The muscle of the limb is immunostained red. It is clear that the graft cells have become cartilage, perichondrium, tendon, and a little dermis, but have not contributed to muscle or epidermis. (From: Kragl, M., Knapp, D., Nacu, E., et al. (2009) Cells keep a memory of their tissue origin during axolotl limb regeneration. Nature 460, 60–65. Reproduced with the permission of Nature Publishing Group.)

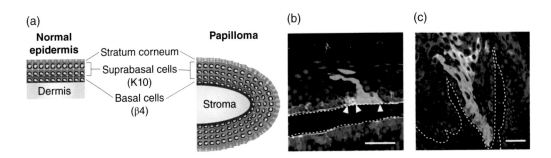

Figure 11.C.4 Visualization of stem cells in a mouse papilloma. The papillomas were induced in mouse skin by chemical carcinogenesis. The mice were *K14-CreER, Rosa-YFP* and lineage labeling of epidermal cells at clonal density was induced by a low dose of tamoxifen. (a) shows the similarity of organization of a papilloma to normal epidermis. (b) shows YFP label (green) in the papilloma after 6 days and (c) after 14 days. (b) and (c) are immunostained for β4 integrin (red) to mark the basal layer and keratin 10 (lilac) to mark upper layers of cells. Scale bars 50 μm. (From: Driessens, G., Beck, B., Caauwe, A., Simons, B.D. and Blanpain, C. (2012) Defining the mode of tumour growth by clonal analysis. Nature 488, 527–530. Reproduced with the permission of Nature Publishing Group.)

(Notch IntraCellular Domain), into the cytoplasm. NICD activates a transcription factor called CSLκ which then migrates to the nucleus and activates target genes (Figure 7.4e). The Notch system is involved in numerous developmental inductive interactions. In particular it is important for many cell differentiation processes, where one cell type emerges in close proximity to another.

Hedgehog System

The name Hedgehog arises from the bristly phenotype of the loss of function mutant of the gene in *Drosophila*. Hedgehogs are single chain secreted proteins. They are autoproteases, cleaving themselves to liberate the biologically active N-terminal half from the rest of the molecule. This biologically active fragment carries two lipid modifications, a fatty acyl chain, present in the original full length protein, and a cholesterol molecule added during cleavage. It binds to a receptor called Patched and inhibits its activity. Patched is a G-protein linked receptor which is constitutively active, normally repressing activity of another cell surface molecule called Smoothened. Smoothened normally represses proteolysis of Gli proteins. Gli proteins are transcription factors which are active in the full length form but inactive or even repressive in the truncated form. Hence in the absence of Hedgehog, Patched is active, Smoothened inactive and Gli inactive. In the presence of Hedgehog, Patched is inactive, Smoothened active and Gli active. This is a typical "double negative" pathway in which two inhibitions contribute to the overall activation of Gli by Hedgehog (Figure 7.4f).

Sonic hedgehog is very important for patterning of the central nervous system and the limb buds, while Indian hedgehog is important in skeletal development.

Growth Promoting Pathways

The proteins Insulin-like growth factor 1 and 2 (IGF1&2) are very important for stimulation of cell division during development. They bind to a tyrosine kinase type cell surface receptor, which activates phosphatidyl inositol-3-kinase. This phosphorylates membrane phosphatidyl inositols at the 3 position and these modified lipids stimulate protein kinase B (=Akt). This inhibits FOXO which normally activates various cyclin dependent kinase inhibitors. Since two inhibitions equals one activation, the net result is to stimulate cell division (Figure 7.4g). In postnatal life insulin acts through this pathway, although its effects are more largely metabolic, especially stimulating glucose uptake and glycogen synthesis.

The other important pathway controlling growth is the Hippo pathway (Figure 7.4h), whose name derives from the *hippo* gene in *Drosophila* whose loss of function mutants cause overgrowth. In mammals, there are two transcription factors, YAP and TAZ which promote growth by activating synthesis of cyclin E. These are inactivated by the kinases LATS-1 and 2, which are activated by phosphorylation by kinases MST1 and 2 (the Hippo homologs). This pathway does not appear to be controlled by an extracellular ligand, but the MSTs are connected to an atypical cadherin on the cell surface called FAT1-4, so the pathway may be regulated by cell contact events. In the preimplantation embryo, suppression of the pathway in the outer cells leads to their differentiation as trophectoderm.

Retinoic Acid

Unlike the other inducing factors mentioned here, retinoic acid is a small molecule, not a protein. It is made from dietary vitamin A (retinol) by oxidation. As a small hydrophobic molecule it can freely cross the plasma membrane and enter cells. Its receptor (RAR) is one of the class of nuclear hormone receptors and is normally found in the nucleus bound to its target promoters. In the presence of retinoic acid, the RAR ejects inhibitory components from its transcription complex and activates transcription of the target genes. Retinoic acid is very important in the rostro-caudal patterning of the hindbrain region and for development of many individual organs.

Body Plan Formation

General Body Plan

At the time of implantation, the epiblast of the mouse consists of a clump of cells, still expressing *Oct4* and other pluripotency genes. It soon becomes a deep cup-like structure, surrounded by visceral endoderm, extending into the blastocoelic cavity (Figure 5.9). The events of primary body plan formation take place within this cup and by means of interactions between the epiblast itself and the overlying visceral endoderm. In the human embryo, as in most mammals other than rodents, the epiblast remains flat and does not form a cup shape, but it is believed that the processes of regional specification are similar.

Initially the egg cylinder is exposed to a gradient of BMP from the extraembryonic ectoderm at the upper (proximal) end (Figure 7.5). This induces expression of Wnt3 in the proximal epiblast, so there is a proximal to distal gradient of both BMP and Wnt. These factors maintain a region of *Brachyury/T* expression at the proximal end of the epiblast and a region of *Hex* (encoding a non-Hox homeodomain protein) expression at the distal end of the visceral endoderm. At this stage, *Nodal* is expressed throughout the epiblast. Because it is still radially symmetrical this simple proximodistal pattern along the axis of the egg cylinder is not complex enough to generate a body plan. The key to how this occurs lies in cell movements which break the radial symmetry and lead to a new configuration with two signaling centers which can generate a two dimensional pattern in the egg cylinder epiblast (Figure 7.5). The cause of the cell movements is not entirely clear but probably owes something both to the pressure of cell division and to the active movements of cells via their filopodia. These forces cause the distal region to move proximally on one side and the proximal region to move distally on the other side. The orientation of this movement may depend on slight asymmetries derived from the early cleavage stages of the conceptus, but this is not certain and it may simply be random and initiated by tiny chance fluctuations, as in a typical symmetry breaking process. Coincident with the cell movements, the *Hex* domain, plus some additional visceral endoderm cells, start to secrete three inducing factors Lefty-1, Cer-l (Cerberus-like) and Dkk (Dickkopf), forming a signaling center called the anterior visceral endoderm (AVE). The three inducing factors are all inhibitors. Lefty-1 inhibits Nodal signaling; Dkk is an inhibitor of Wnt signaling; and Cerberus-like is an inhibitor of Wnt, BMP and Nodal signaling. Their combined effect is to reduce Nodal, Wnt and BMP signaling on the side occupied by the AVE, which is now defined as the rostral (anterior in mouse nomenclature) side. In the remainder of the epiblast, where these factors are still active, the cells undergo an epithelial-mesenchymal transition. This depends on FGF signaling by FGF4 and 8, which upregulate the *Snail* gene, encoding a transcription factor which represses expression of the *E-cadherin* gene. E-cadherin is the principal cell adhesion molecule at this developmental stage and, in its absence, the cells become mobile and migrate toward the midline of what is now the caudal (in mouse terms, posterior) part of the epiblast to form a cell mass called the primitive streak (Figure 7.5a). Expression patterns of various inducing factors and transcription factors are shown in Figure 7.C.1.

Gastrulation

The next set of cell movements is called gastrulation. Gastrulation of some sort occurs in all animal embryos and is the process whereby the early embryo, consisting of a sheet or a clump of cells, becomes rearranged into three germ layers: the endoderm, mesoderm and ectoderm. In mammals and other higher vertebrates, gastrulation occurs through the primitive streak, which is a condensation of cells forming in the caudal part of the epiblast. Cells from either side then move through the streak and exit to form a

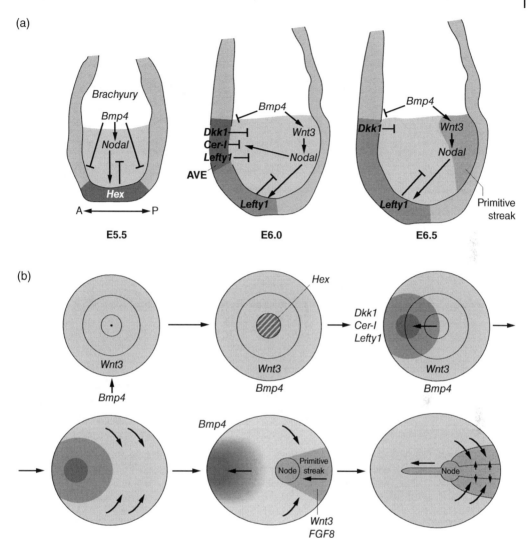

Figure 7.5 Generation of the AVE and primitive streak in the mouse embryo. (a) Shows events from the side. (Modified from Takaoka, K. and Hamada, H. (2012) Cell fate decisions and axis determination in the early mouse embryo. Development 139, 3–14. (b) Shows events from above, with the cup shaped egg cylinder presented as a flat disc.

lower layer, the definitive endoderm, and a middle layer, the mesoderm (Figure 7.6 a,b). Cells also move from the streak rostrally to form a prechordal plate. At the same time a condensation of cells called the node forms at the rostral end of the streak itself (Figure 7.6 c,d). In the later stages of gastrulation the node regresses, moving caudally along the streak, and structures of the body plan emerge as condensations of cells in its wake.

The production of Nodal is autocatalytic because one of the *Nodal* enhancers is activated by Smad 2/3, which are components of the Nodal signal transduction pathway. So as the streak forms, Nodal signaling progressively increases. This means that there is a temporal gradient of response by the cells that pass through the streak. The first cohort forms extraembryonic mesoderm (comprising the mesodermal part of

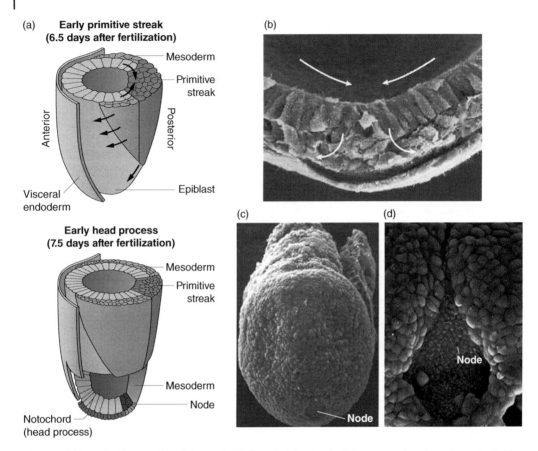

Figure 7.6 The primitive streak and the node. (a) The primitive streak of the mouse showing schematically the cell movements leading to the formation of the mesoderm, endoderm, node and head process. (b) Scanning electron micrographs of the mouse streak. (c, d) Scanning electron micrographs of the mouse node at low and high power. (Sources: (a) From http://www.mun.ca/biology/desmid/brian/BIOL3530/DB_03/fig3_24.jpg. (b) From Arnold, S.J. and Robertson, E.J. (2009) Making a commitment: cell lineage allocation and axis patterning in the early mouse embryo. Nature Reviews. Molecular Cell Biology 10, 91–103. (c, d) From Vogan, K.J. and Tabin, C.J. (1999) Developmental biology: A new spin on handed asymmetry. Nature 397, 295–298. Reproduced with the permission of Nature Publishing Group.)

the chorion, the ventral blood islands and the allantois), the next cohort forms the lateral plate mesoderm, the next the paraxial mesoderm (later forming the segmented structures called somites), and lastly the definitive endoderm, prechordal plate, cardiac mesoderm, and the node itself. If Smad signaling is weakened by deletion of individual *Smad2* or *3* genes, then these structures fail to be formed in a concentration dependent manner. In other words, with increasing severity of mutations, the node is lost first, then the paraxial mesoderm, then the lateral plate, and so on.

Cells passing through the streak to form the lower layer, the definitive endoderm, displace the visceral endoderm and become a strip of definitive endoderm on the ventral side of the embryo. The node itself generates the notochord along the whole length of the body. The region rostral to the node, consisting of cells forming the rostral notochord and prechordal plate, is called the head process. The cells that move to the midline but do not enter the streak end up overlying the midline mesodermal structures as definitive ectoderm. The node and other midline structures produce the BMP inhibitors chordin

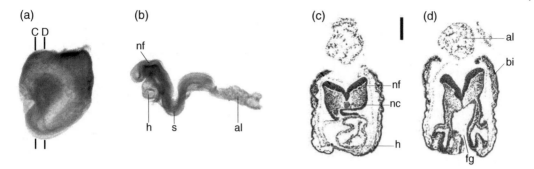

Figure 7.7 General body plan of the mouse embryo at 8.5 days. (a) 5 somite mouse embryo wrapped in its membranes (C, D show planes of section). (b) Shows a similar embryo stretched out rostrocaudally. nf: neural folds; h: heart; s: somites. (c,d) Sections as shown in (a). nf: neural folds; nc: notochord; h: heart; al allantois; bi: blood islands; fg: foregut portal. Scale bar 50 µm. (From Hart, A.H., Hartley, L., Sourris, K., Stadler, E.S., et al. (2002) Mixl1 is required for axial mesendoderm morphogenesis and patterning in the murine embryo. Development 129, 3597–3608. Reproduced with the permission of The Company of Biologists Ltd.)

and noggin, and these induce the adjacent ectoderm to become the neural plate, while the ectoderm above the peripheral parts of the streak becomes surface ectoderm, later to differentiate into the epidermis.

So far this process of body plan formation has generated a series of zones forming different cell types (Figure 7.7). In the ectoderm there is a neural and an epidermal territory. In the mesoderm there are notochord, prechordal plate, cardiac, paraxial and lateral plate zones. This pattern has emerged from response to a gradient of Nodal and of BMP activity with the central focus, the node, having high Nodal and low BMP. In essence it is a medio-lateral pattern, which in zebrafish or *Xenopus* would be called dorso-ventral. Onto this pattern now becomes superimposed the responses to gradients of Wnt and FGF which are high in the caudal region. These turn on the *Cdx* gene cluster, which in turn turns on the *Hox* genes, with more genes being active at higher signal concentrations. These factors are active in all three germ layers and impart a true rostro-caudal pattern to the embryo. The inductive processes generating the rostrocaudal pattern occur at the same time as the passage of cells through the streak and the regression of the node.

The end result is a three layered structure in which zones of cells committed to become

each of the main body regions are present in their definitive positions. The events described here are described for the mouse. But many of the genes and factors were discovered in other developmental model organisms including *Xenopus*, the chick and the zebrafish. The events are broadly similar for all vertebrates but the specific details, especially the geometry of cell movements, the exact role of each family of inducing factors, and the timing of events, do differ somewhat between groups. The overall similarity leads us to suppose that human body plan formation follows a course similar to that the mouse, except that the events take place in a flat epiblast rather than a cup-shaped one. Although experiments involving reimplantation of manipulated human embryos into the womb would be very unethical, it is becoming possible to investigate human development by other methods. For example RNAseq of individual cells can help to establish cell lineages and whole embryos can now be cultured in vitro to early post-implantation stages. Furthermore, investigation of the behavior of embryoid bodies or other embryo-like structures arising in vitro from pluripotent stem cells can also help to elucidate the steps of embryonic induction. So we can expect to see more detailed information about human early development in the near future, and this will establish just what is and what is not conserved from the mouse.

Embryo Folding

In the next stage of development the overall shape of the embryo changes considerably and it begins to look like an archetypal vertebrate animal. For these stages it is easier to consider the course of events in the human embryo from about 3–4 weeks of development (corresponding to about 7.5–9.5 days of mouse development). This is because in the egg cylinder configuration of the rodent embryo there occurs the complex process called turning (shown in Figure 5.10) which makes the description of events particularly complex, and the interpretation of histological sections rather difficult. Turning does not occur in humans or other types of embryo with a flat epiblast.

The first folding process is the closure of the neural plate to form a neural tube (Figure 7.8). This commences in the middle and proceeds both rostrally and caudally. Neural tube closure is rather susceptible to interference from a variety of genetic or environmental causes and failures of complete closure at this stage are responsible for the familiar neural tube defects found in a small percentage of human births. The end result is a tube with the wide rostral end destined to form the brain, and the narrower caudal end destined to form the spinal cord.

The next folding process, overlapping in time with neural tube closure, is a folding of the whole embryo which raises it from the surrounding extraembryonic tissue (Figure 7.9). At the rostral end, the prospective heart region becomes tucked under the head and the lateral regions move inward to undercut the head. The result is that the rostal end of the embryo now projects out of the overall cell sheet. These folds also enclose the rostral part of the endodermal layer to produce a blind-ended foregut tube. A similar process occurs somewhat later at the caudal end, raising the tail bud above the layer of the cell sheet and generating a hind-

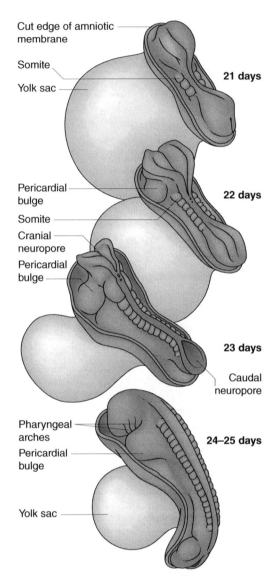

Figure 7.8 Neural tube closure. Here the human embryo is shown. The process begins in the middle and closure proceeds to rostral and caudal extremities. (From Larsen, W.J., Sherman, L.S., Potter, S.S. and Scott, W.J. (2001) Human Embryology, 3rd edition. Churchill Livingstone Inc., New York. Reproduced with the permission of Elsevier.)

gut tube. These movements continue, gradually constricting the attachment of the embryo to the surrounding cell sheet and incorporating the definitive endoderm into an actual gut lining. Finally the ventral attachment has constricted so much that it

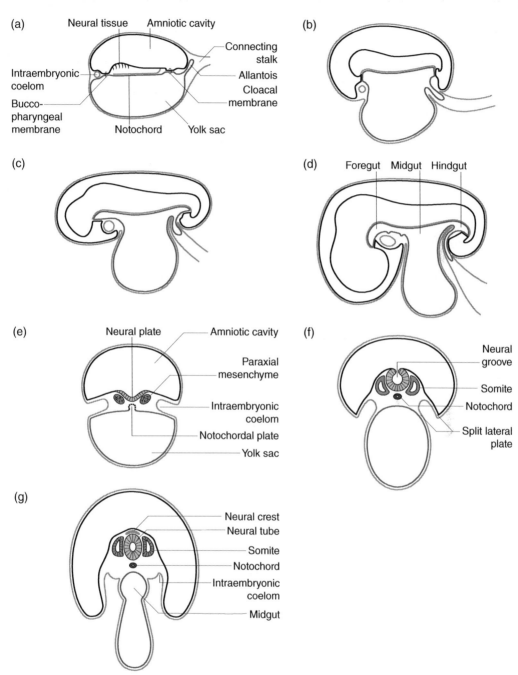

Figure 7.9 Embryonic folding. The human embryo is shown from about 18 to 28 days post-fertilization. (a)-(d) Sagittal sections. Rostral is to the left with the amnion above and the secondary yolk sac below the embryo. The head and tail ends become elevated from the cell sheet and the embryo closes around the ventral surface which becomes enclosed as a gut cavity. The residual opening of the midgut becomes the umbilical cord. (e)-(g) Transverse sections through the mid-body. (Modified from Grey's Anatomy 41st edn. 2015. Reproduced with the permission of Elsevier.)

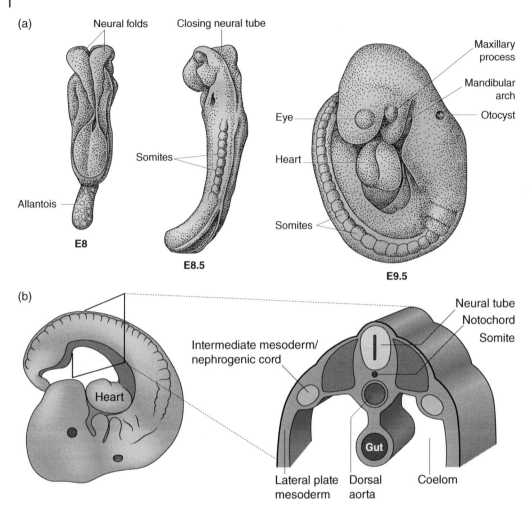

Figure 7.10 Whole mouse embryos at the general body plan stage. (a) External views. The E8 embryo is about 1 mm long and the E9.5 embryo about 2 mm crown-rump length. (From: Slack, J.M.W. (2013) *Essential Developmental Biology*, 3rd edn. Reproduced with the permission of John Wiley and Sons.) (b) Diagram to show structures in the trunk region. (Modified from http://www.stembook.org/sites/default/files/pubnode/45dfaa6a033d187df9d585bb17e6b50da0e0aaa2/Mouse_kidney_development/Davidson02.jpg.)

becomes a thin tube continuous with the mid- and hindgut, the umbilicus. This contains all three germ layers. It is covered with epidermis, the mesodermal part includes the principal blood vessels joining embryo to placenta, and the endodermal parts comprise the residual connection between gut and extraembryonic endoderm (vitellointestinal duct) and the allantois. The appearance of the mouse embryo at this "general body plan" stage is shown in Figure 7.10.

Further Reading

Akiyama, T. and Gibson, M.C. (2015) Morphogen transport: theoretical and experimental controversies. Wiley Interdisciplinary Reviews – Developmental Biology 4, 99–112.

Arnold, S.J. and Robertson, E.J. (2009) Making a commitment: cell lineage allocation and axis patterning in the early mouse embryo. Nature Reviews. Molecular Cell Biology 10, 91–103.

Bier, E. and De Robertis, E.M. (2015) BMP gradients: A paradigm for morphogen-mediated developmental patterning. Science 348, aaa5838.

Christian, J.L. (2012) Morphogen gradients in development: from form to function. Wiley Interdisciplinary Reviews – Developmental Biology 1, 3–15.

Deglincerti, A., Croft, G.F., Pietila, L.N., Zernicka-Goetz, M., Siggia, E.D. and Brivanlou, A.H. (2016) Self-organization of the in vitro attached human embryo. Nature 533, 251–254.

Dorey, K. and Amaya, E. (2010) FGF signaling: diverse roles during early vertebrate embryogenesis. Development 137, 3731–3742.

Ingham, P.W., Nakano, Y. and Seger, C. (2011) Mechanisms and functions of Hedgehog signaling across the metazoa. Nature Reviews. Genetics 12, 393–406.

Kimelman, D. and Martin, B.L. (2012) Anterior–posterior patterning in early development: three strategies. Wiley Interdisciplinary Reviews – Developmental Biology 1, 253–266.

Kopan, R. and Ilagan, M.X.G. (2009) The canonical notch signaling pathway: unfolding the activation mechanism. Cell 137, 216–233.

Ozair, M.Z., Kintner, C. and Brivanlou, A.H. (2013) Neural induction and early patterning in vertebrates. Wiley Interdisciplinary Reviews – Developmental Biology 2, 479–498.

Rhinn, M. and Dollé, P. (2012) Retinoic acid signaling during development. Development 139, 843–858.

Shen, M.M. (2007) Nodal signaling: developmental roles and regulation. Development 134, 1023–1034.

Slack, J.M.W. (2014) Establishment of spatial pattern. Wiley Interdisciplinary Reviews – Developmental Biology 3, 379–388.

Takaoka, K. and Hamada, H. (2012) Cell fate decisions and axis determination in the early mouse embryo. Development 139, 3–14.

van Amerongen, R. and Nusse, R. (2009) Towards an integrated view of Wnt signaling in development. Development 136, 3205–3214.

8

Organogenesis

This chapter deals with the development of some organ systems in the mammalian body which are particularly relevant to stem cell research. Their descriptive embryology has been known for both mouse and human for a long time. The molecular mechanisms have been established in recent decades using the mouse as the experimental model, with some assistance from work on chick, *Xenopus* and zebrafish embryos.

In particular, two groups of experimental method have been used to establish the molecular mechanisms of organogenesis. First, mouse knockouts can show the requirement for a particular gene in a process. If the process fails to occur in the absence of the gene it is presumed to be necessary. Often a gene is required several times in development and this may mean that a knockout embryo dies before the stage of interest. In such cases it is necessary to do organ-specific knock outs or overexpression using the techniques described in Chapter 3. Second, organ cultures can be set up in vitro for many of the systems considered here. The organ rudiment is dissected from a mouse embryo and grown in tissue culture medium, often with special substrates or extracellular matrix components (see Chapter 4). This enables inducing factors or specific inhibitors to be added to the culture. Furthermore it is often possible to separate epithelial and mesenchymal components of an organ rudiment by microdissection. This enables investigation of which component of the rudiment is responding to particular stimuli, and investigation of signals passing between epithelium and mesenchyme.

The descriptions given here are necessarily brief and it should be borne in mind that every system is actually much more complex than indicated here. It is usual for several inducing factors and signaling pathways to be involved in each developmental event, although one factor may predominate and be able to bring about appropriate effects on overexpression.

Nervous System

The central nervous system originates from the ectoderm of the embryo and comprises the brain and spinal cord. The region of the ectoderm which ends up overlying, or closely apposed to, the notochord and the prechordal plate becomes exposed to the BMP inhibitors: chordin and noggin. These suppress the formation of epidermis and induce the formation of the neural plate, characterized by expression of *Sox1* and *-2* and of other neural transcription factors. The neural plate is a keyhole shaped structure. The rostral end is wider and becomes the brain, the caudal end is narrower and becomes the spinal cord. As the node regresses, the neural plate folds up to form a tube, the folds starting in the center and proceeding both rostrally and caudally. As a result of neural tube closure the previous lateral edges become joined at the dorsal midline (see Figure 7.8).

The Science of Stem Cells, First Edition. Jonathan M. W. Slack.
© 2018 John Wiley & Sons, Inc. Published 2018 by John Wiley & Sons, Inc.
Companion website: www.wiley.com/go/slack/thescienceofstemcells

Apart from the morphogenetic movements involved in neural tube closure, the development of the neural tube involves two distinct types of process. There is regional specification, which creates a set of rostrocaudal and dorsoventral zones with different states of commitment. There is also differentiation of neurons and glia from the cells of the early neuroepithelium. The types of neurons and glia vary according to the region although most of the histological types are found throughout the CNS.

The Brain

The main parts of the developing human brain are shown in Figure 8.1a. The early brain has three swellings called the forebrain, midbrain and hindbrain, and these terms are also used to indicate the later anatomical regions arising from each swelling. Within the brain is a system of ventricles, derived from the original lumen of the neural tube, which are continuous with each other and with the spinal canal. The ventricles are lined with ependymal cells, which are cuboidal glial cells bearing cilia. Each ventricle contains a choroid plexus, consisting of capillaries surrounded by ependymal cells, which secrete the cerebrospinal fluid. Unless they arise in the dorsal or ventral midline, brain structures are paired, one on each side. Twelve pairs of cranial nerves arise from the brain. They supply various parts of the head and some parts of the upper thorax and may be motor or sensory or mixed in composition. In neuroanatomy, the term "nucleus" refers to a condensation of neurons, not to the nucleus of a single cell. Mouse and human brains contain generally similar parts but there are also large differences in proportion and scale (Figure 8.1b; Figure 8.C.1).

The forebrain comprises the telencephalon rostrally and the diencephalon caudally. The most rostral part of the telencephalon forms the olfactory bulb, the receptive area for the olfactory nerve (cranial nerve I). The olfactory neurons themselves arise from the olfactory placode, and the axons grow into the olfactory bulb. In mice the olfactory bulb is large, but in humans it is small and tucked below the brain. Rodents also have an additional vomeronasal organ in the nasal cavity connected to an accessory olfactory bulb, not found in humans.

The main structures formed by the telencephalon are the cerebral hemispheres, which arise as lateral evaginations separated by a medial longitudinal fissure, each containing a lateral ventricle. The dorsal part of each hemisphere becomes the cerebral cortex which is responsible for the higher brain functions including consciousness, perception, memory, and, in humans, thought and language. The main part of the cortex in mammals is called the neocortex and consists of six layers of neurons which arise from precursors adjacent to the ventricles. The cerebral cortex is, of course, relatively much larger in humans than in mice and during late fetal growth it increases greatly in surface area and becomes thrown into multiple folds (Figure 8.C.1). On the inner side of each hemisphere, within the medial fissure, forms a sausage-like body called the hippocampus (Figure 8.2a), just above the choroid plexus. This is later important in spatial navigation and in the consolidation of short to long term memories. It is notable in that its caudal portion, the dentate gyrus, is one of a very few areas of the brain containing neural stem cells which remain active through adult life. The ventral part of the cerebral hemispheres, called the striatal region, is much thicker than the dorsal part and contains folds called ganglionic eminences which are a source of migratory cells entering the neocortex (Figure 8.2b). Later, the striatal region forms the basal ganglia: the caudate nucleus, putamen and globus pallidus, which are involved in the selection of behaviors and movements. These have connections to the cortex, the thalamus, and the brainstem. In the adult brain the term striatum is used to refer to the caudate nucleus, putamen and some other structures.

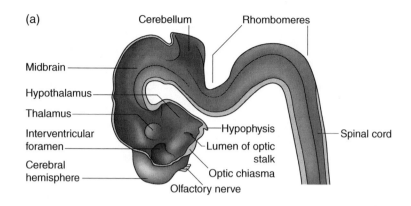

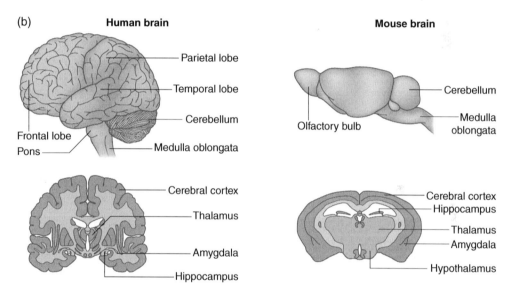

Figure 8.1 (a) Human brain at about 35 days from fertilization. (From: Hamilton Boyd and Mossman (1972) Human Embryology. Reproduced with permission.) (b) Comparison of structure of adult brains of human and mouse. (From: John F. Cryan, J.F. & Holmes, A. (2005) The ascent of mouse: advances in modelling human depression and anxiety. Nature Reviews. Drug Discovery 4, 775–790.)

The caudal part of the forebrain, the diencephalon, surrounds the third ventricle and produces several important structures. The thalamus, which later relays sensory information to the cerebral cortex, arises from the lateral walls. Later, as the cerebral cortex overgrows the diencephalon, the thalamus comes to lie adjacent to the basal ganglia (Figure 8.2b). The optic vesicles are lateral outgrowths of the forebrain that remain attached to the diencephalon. They become the neural parts of the eyes, comprising the sensory retina, pigmented retina and optic nerves (cranial nerves II). In the ventral part of the diencephalon arises the hypothalamus and the neural part of the hypophysis (= pituitary gland). The hypothalamus is later closely involved with control of the hypophysis. From the dorsal part of the diencephalon develops the epiphysis (pineal gland), and the choroid plexus of the third ventricle.

The midbrain cavity becomes narrowed to form the aqueduct, connecting the third and fourth ventricles. The dorsal part of the

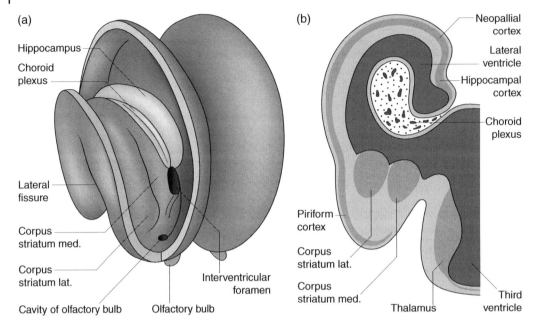

Figure 8.2 (a) Location of the developing hippocampus and corpus striatum within the cerebral hemisphere of a human embryo of about 10 weeks post-fertilization (embryo size 46 mm). (b) The developing basal ganglia viewed in transverse section. (From: Hamilton Boyd and Mossman, Human Embryology.)

midbrain forms the superior and inferior colliculi. The superior colliculus, known as the optic tectum in non-mammalian vertebrates, is the primary area for receipt of visual information from the optic nerves. The inferior colliculus is a center for receipt of auditory and balance information. The ventral part of the midbrain forms the nuclei of the third and fourth cranial nerves and the substantia nigra, an important center for motor control, reward seeking and learning. This is considered as one of the basal ganglia, along with those derived from the telencephalon, and contains GABAergic neurons connecting to the thalamus, and dopaminergic neurons connecting to the striatum.

The hindbrain comprises the cerebellum, later connecting to the thalamus and the brainstem and important in movement control; and the medulla oblongata which contains the remaining cranial nerve nuclei. The hindbrain shows a transient segmentation into seven rhomobomeres, of which the most rostral becomes the cerebellum.

Regional Specification of the CNS

Rostrocaudal

The rostrocaudal pattern of the neural tube and the later CNS depends initially on the gradients of FGF and Wnt which are high at the caudal end and low at the rostral end. These induce expression of *Cdx* genes which in turn induce expression of *Hox* genes. Because of the graded nature of the signals, the *Hox* genes are activated in a nested pattern such that all the genes are on in the tail bud and the rostral frontiers of expression vary such that the more 3′ the gene lies in the chromosome, the more rostral is its boundary (Figure 8.3). There are four Hox clusters in vertebrates which arose in evolution by a double duplication of the original cluster around the time of the origin of vertebrates. The equivalent genes in each cluster are known as paralogs and tend to have similar expression domains and functions.

The *Hox* gene expression domains extend into the future hindbrain. Rostral to this the initial subdivision of the neural plate depends

(a)

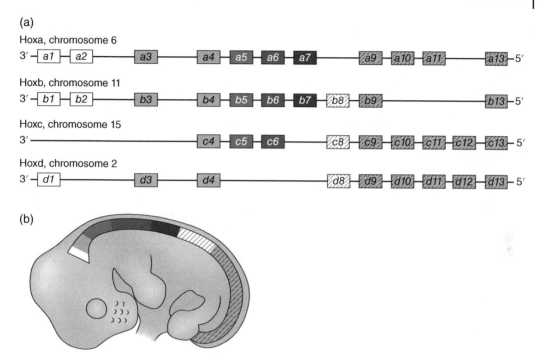

Hoxa, chromosome 6

3′ ⎡a1⎤—⎡a2⎤————⎡a3⎤———⎡a4⎤—⎡a5⎤—⎡a6⎤—⎡a7⎤————⎡a9⎤—⎡a10⎤—⎡a11⎤————⎡a13⎤—5′

Hoxb, chromosome 11

3′ ⎡b1⎤—⎡b2⎤————⎡b3⎤———⎡b4⎤—⎡b5⎤—⎡b6⎤—⎡b7⎤—⎡b8⎤—⎡b9⎤————⎡b13⎤—5′

Hoxc, chromosome 15

3′ ————————————⎡c4⎤—⎡c5⎤—⎡c6⎤————⎡c8⎤—⎡c9⎤—⎡c10⎤—⎡c11⎤—⎡c12⎤—⎡c13⎤—5′

Hoxd, chromosome 2

3′ ⎡d1⎤————⎡d3⎤———⎡d4⎤————————⎡d8⎤—⎡d9⎤—⎡d10⎤—⎡d11⎤—⎡d12⎤—⎡d13⎤—5′

(b)

Figure 8.3 Expression pattern of Hox genes in the mouse embryo neuraxis. (a) The 4 HOX gene clusters in the mouse. (b) Anterior expression limits of each paralog group of HOX genes. From: Box 5E p.215 in Wolpert, L. et al. (2015) Principles of Development. 5th edn. Oxford University Press, Oxford and New York. (Reproduced with the permission of Oxford University Press.)

on expression of two genes encoding non-*HOX* homeodomain transcription factors: *Otx2* and *Gbx2*. *Otx2* is expressed in the future fore- and midbrain while *Gbx2*, is expressed more caudally. There is a considerable subdivision of this simple rostrocaudal pattern which occurs during and shortly after neural tube closure. The forebrain is subdivided into telencephalon and diencephalon, in response to FGF from the anterior neural ridge. The rostral cerebral hemispheres are marked by expression of *Pax6* and the caudal part by *Emx2*. Regionalization of the caudal telencephalon also depends on Wnt and BMP from the cortical hem, which lies along the dorsal midline of the developing cortex. These signals are required for formation of the hippocampus.

The border between the midbrain and hindbrain, known as the isthmus, is marked by expression of FGF8 and WNT1 (Figure 8.4). On the rostral side these factors induce expression of the transcription factors Engrailed 1 and 2, defining the future midbrain. The tegmentum arises from the basal plate of the midbrain and requires both SHH from the ventral midline and FGF8 from the midbrain-hindbrain boundary for its formation. It later produces the dopaminergic neurons of the striatum. On the caudal side the FGF8 and WNT1 from the isthmus induce formation of the future cerebellum.

As mentioned above, the hindbrain becomes divided into seven segments called rhombomeres (Figure 8.5) of which the cerebellum forms from the most rostral, rhomobomere 1. The rhombomeres transiently form visible segments and the process of cell condensation depends on expression of Eph and ephrin adhesion molecules in alternate segments. Control of regional specification here is largely due to a local gradient of retinoic acid. This is produced from dietary vitamin A (retinol) by retinaldehyde dehydrogenase

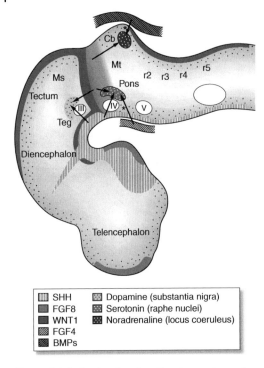

SHH

FGF8

WNT1

FGF4

BMPs

Dopamine (substantia nigra)

Serotonin (raphe nuclei)

Noradrenaline (locus coeruleus)

Figure 8.4 Inductive signals patterning regions of the CNS near the midbrain–hindbrain boundary. (Modified from Wurst, W. and Bally-Cuif, L. (2001) Neural plate patterning: Upstream and downstream of the isthmic organizer. Nature Reviews. Neuroscience 2, 99–108. Reproduced with the permission of Nature Publishing Group.) cb = cerebellum, Ms = mesencephalon, Mt = metencephalon, Teg = tegmentum, r2-r5 = rhombomeres, III, IV, V = cranial nerve roots. Signals marked * operate during gastrulation.

(RALDH2), expressed in the somites and the lateral plate of the trunk region. It is destroyed by CYP26, a cytochrome p450 enzyme, located in the fore- and midbrain. This establishes a caudal to rostral gradient of retinoic acid across the future hindbrain which induces expression of a unique combination of transcription factor genes in each rhombomere, including the *HOX* genes *b1-b4* and the gene encoding the zinc finger transcription factor KROX20. The knockout of *Cyp26A1* increases the level of retinoic acid and leads to caudalization of the hindbrain structures. Conversely the knockout of *Raldh2* decreases the level of retinoic acid and leads to a rostralization of the hindbrain.

The rostrocaudal organization of the spinal cord is less apparent in terms of gross morphology than that of the brain, but there are some differences, for example in the presence of certain types of motorneuron, and this is controlled by the nested expression of the *HOX* genes within the spinal cord.

Mediolateral

After about 10.5 days of mouse development, the spinal cord begins to differentiate into layers. The inner layer is the single layer of cells that becomes the ependyma lining the spinal canal. Outside this, the middle layer is called the mantle zone, composed of neuroepithelial

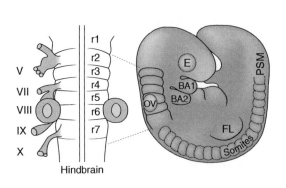

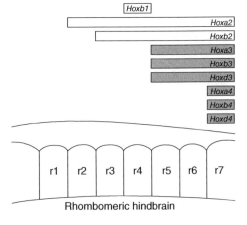

Figure 8.5 The rhombomeres of the hindbrain. Their individual character is determined by the nested expression of *HOX* genes, which is controlled by a local gradient of retinoic acid. (Modified from Alexander, T., Nolte, C. and Krumlauf, R. (2009) *HOX* genes and segmentation of the hindbrain and axial skeleton. Annual Review of Cell and Developmental Biology 25, 431–456.)

(a)

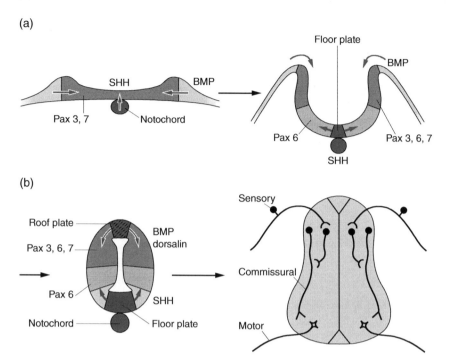

(b)

Figure 8.6 Dorsoventral patterning of the neural tube. (a) SHH from the notochord induces the floor plate in the overlying neural plate and this also secretes SHH. BMP from the epidermis induces the roof plate. (b) Inductive signals from the floor and roof plates induce zones of gene expression that later generate specific types of neuron. (Slack, J.M.W. (2013) Essential Developmental Biology, 3rd edn. Reproduced with the permission of John Wiley and Sons.)

cells and newly formed neurons. This will later become the gray matter. The outer layer is the marginal zone, consisting of the fiber tracts growing from the newly formed neurons, which will later be the white matter. In the brain the general arrangement is similar to the spinal cord, except that many cell populations migrate from the mantle zone through the marginal zone toward the surface, thus bringing layers of cells outside of the zone of fiber tracts.

Dorsoventral

Both the brain and spinal cord show considerable dorsoventral pattern. In the spinal cord the dorsal part is called the alar plate and the ventral part the basal plate. These are domains of different neuron types, with motorneurons ventrally and interneurons dorsally connecting to the sensory ganglia (Figure 8.6). This situation arises from a pair of morphogen gradients. Sonic hedgehog (SHH) comes from the notochord and the ventral midline structure called the floor plate. BMPs come from the epidermis flanking the lateral edges of the neural plate, which become dorsal on closure of the neural tube. The gradients initially activate transcription of *Pax6* in the ventral region and *Pax 3, -6* and *-7* in the dorsal region. These broad domains are further subdivided by expression of other transcription factors to produce a series of zones arranged dorsal to ventral. Each of these produces a particular repertoire of motorneurons or interneurons.

The Eye

The eyes are formed as outgrowths from the diencephalon called the optic cups (Figure 8.7). These grow out as vesicles which invaginate to form cups. The inner, invaginated, surface becomes the neural retina, composed of photoreceptors and ganglion cells.

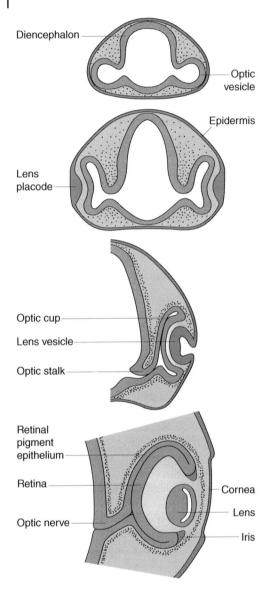

Diencephalon

Optic
vesicle

Epidermis

Lens
placode

Optic cup

Lens vesicle

Optic stalk

Retinal
pigment
epithelium

Retina

Cornea

Lens

Optic nerve

Iris

Figure 8.7 Development of the eye. The optic vesicle grows out of the diencephalon while the lens and cornea develop from the epidermis. The inner layer of the optic cup becomes the retina while the outer layer becomes the pigmented epithelium. (Hildebrand., 1995. Reproduced with the permission of John Wiley and Sons Inc.)

This sends axons back down the interior of the optic stalk into the brain making up the optic nerves, which project through the optic chiasma to the lateral geniculate nuclei and superior colliculi in the brain. Much of the optic projection crosses to the opposite side of the brain (contralateral) at the optic chiasma, although there is also some projection to the same side (ipsilateral).

The outer surface of the optic cup becomes the pigmented retina. The capsule of the eye is formed from mesenchymal cells of neural crest origin surrounding the optic cup. The lens and cornea arise from the ectoderm. They are formed from a thickened disc of cells, or placode, arising at the anterior margin of the neural plate. Like the neural crest, the epidermal placodes arise in areas of intermediate BMP signaling between the neural plate (low BMP) and the epidermis (high BMP). The position of specific placodes depends on a balance between FGF signaling from the anterior neural plate, BMP from the lateral region, WNT from the caudal region and SHH from the midline mesoderm. The whole eye territory, comprising optic vesicles and placodes, expresses the transcription factor PAX6, which is, remarkably, involved in eye development in all other types of animal, including lower invertebrates. When the optic cup contacts the optic placode, the lens invaginates forming a vesicle and the surface ectoderm which comes to overlay the lens becomes the corneal epithelium. The remainder of the cornea, comprising the stroma and the endothelium, arise from neural crest mesenchyme. Both the pigmented retina and the epithelium of the cornea are current targets for stem cell therapy.

The Neural Crest

Like the epidermal placodes, the neural crest is formed from the zone around the neural plate where the level of BMP inhibition is intermediate. Transcription factors active in the neural crest include FOXD3 (forkhead type), SNAIL and SLUG (Zn finger) and SOX9 and -10 (SRY related). Following neural tube closure, the crest comes to lie along the dorsal midline of the neural tube. The neural crest is characterized by an extensive migration of its cells, some of which can range all over the body, and by the wide range of differentiated cell types which they can become. The first step in the migration consists of the expression of SNAIL protein which represses expression of E-cadherin, hence reducing cell adhesion and promoting an epithelial-mesenchymal transition to a migratory behavior. In the course of their migration the crest cells show an affinity for the extracellular matrix components fibronectin and laminin, and they secrete proteases to assist in their progress. The pattern of migration is partly controlled by inhibition from the surroundings, for example the ephrins present in the posterior sclerotome of each segment, and also in rhombomere 3 and 5 of the hindbrain, inhibit crest migration. There are two main migration pathways in the trunk: a dorsolateral route under the epidermis, where the cells normally become pigment cells; and a ventrolateral route through the sclerotome to become the dorsal root ganglia and the other trunk derivatives.

The neural crest normally forms a remarkable range of cell types. These include:

- neurons and glia of the sensory and autonomic system;
- adrenal medulla and the calcitonin cells of the thyroid;
- pigment cells of the skin;
- skull bones and connective tissues of the head;
- part of the cardiac outflow tract.

The normal fate of different parts of the neural crest was originally established by orthotopic (same position in donor and host) grafts of tissue from quail to chick embryos (Figure 8.8a). The quail cells can be distinguished by the presence of a heterochromatin mass in the nuclei, or by immunostaining for species-specific cell surface proteins. More recently studies have used localized marks of the vital dyes DiI or DiO. The cranial neural crest, which later expresses SOX9 rather than SOX10, differs considerably from the trunk neural crest in that it generates a large amount of mesenchyme which becomes most of the skull and the soft tissues of the face. The trunk neural crest does not form skeletal tissues but both cranial and trunk crest form various types of neuron.

When, and how, do neural crest cells become committed to form their specific differentiated progeny? Labeling of single cells in the neural folds both before and during migration indicates that at least some of them generate clones containing several different cell types, in other words at this stage they are multipotent. When crest cells in culture are exposed to specific inducing factors, there is a clear ability to control the pathway of differentiation. BDNF (bone derived neurogenic factor) induces sensory neurons; BMPs induce autonomic neurons; endothelin 3 induces pigment cells and enteric neurons; neuregulin induces Schwann cells and TGFβ induces smooth muscle (Figure 8.8b). A role for these factors is also suggested by the phenotypes of mouse knockouts for the factors concerned, which tend to be defective in specific types of neural crest progeny. It is however also found that labeling of single crest cells at an early stage sometimes only gives rise to a single differentiated cell type, even though the progeny spread out and are exposed to more than one environment. So there is probably also an autonomous element to cell differentiation as well as that controlled by inducing factors.

There has been some debate about whether multipotent neural crest cells persist into postnatal life, in other words whether there is a neural crest stem cell population that survives long term. This is still unresolved, but there are some cells found in the skin (SKPs) which can under some circumstances show multipotency similar to the embryonic neural crest cells.

(a)

(b)

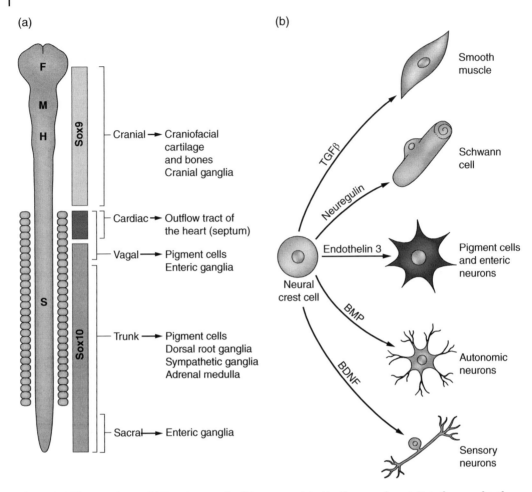

Figure 8.8 The neural crest. (a) Structures and cell types populated by the neural crest. Data from grafts of quail to chick embryos. F: forebrain; M: midbrain; H: hindbrain; S: spinal cord. (b) Differentiation behavior of neural crest cells exposed to different inducing factors. (Slack, J.M.W. (2013) *Essential Developmental Biology*, 3rd edn. Reproduced with the permission of John Wiley and Sons.)

Epidermis

After the closure of the neural tube the whole embryo is covered by surface ectoderm which will form the epidermal layer of the skin. The surface ectoderm is a simple one cell thick epithelium with a temporary periderm layer on the outside. In later gestation (mouse 14d; human from 11 weeks) this simple epithelium becomes multilayered, with a cuboidal basal layer and a series of progressively more flattened (squamous) upper layers (Figure 8.9). This transition depends on the activity of the *p63* gene,

encoding a transcription factor necessary for the formation and maintenance of all squamous epithelia in the body. During the transition the cell divisions change from symmetrical in the plane of the epithelium, to asymmetrical and vertical, i.e. one daughter remains in the basal layer while the other enters the next layer. As with other examples of asymmetric division, this process depends on localization of the PAR complex within each cell (see Chapter 9). As the epidermis matures, the cells become known as keratinocytes, expressing specific keratin genes in different layers. For example, K14 is

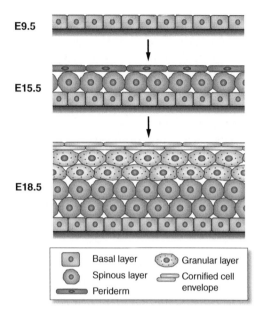

E9.5

E15.5

E18.5

	Basal layer		Granular layer
	Spinous layer		Cornified cell envelope
	Periderm		

Figure 8.9 Stratification of the mouse epidermis. The first step is the expression of *p63* in the surface ectoderm at about E8.5 and the upregulation by p63 protein of *K14*, encoding keratin 14 which remains expressed in the basal layer. Suprabasal cells initially arise by tangential divisions, express *K1,* cease dividing and mature into granular keratinocytes.

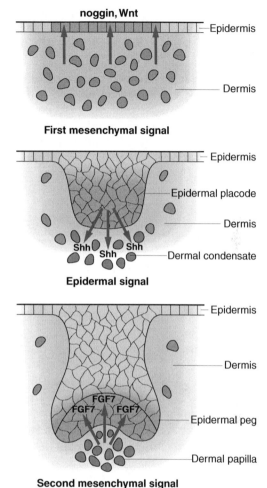

noggin, Wnt

Epidermis

Dermis

First mesenchymal signal

Epidermis

Epidermal placode

Dermis

Shh

Shh

Shh

Dermal condensate

Epidermal signal

Epidermis

Dermis

FGF7

FGF7

FGF7

Epidermal peg

Dermal papilla

Second mesenchymal signal

Figure 8.10 Embryonic development of hair follicles. (Slack, J.M.W. (2013) Essential Developmental Biology, 3rd edn. Reproduced with the permission of John Wiley and Sons.)

expressed in the basal layer and K10 in the upper layers. Keratins are one of the classes of intermediate filament proteins important in maintaining cell structure. In the postnatal organism the basal layer is the only layer containing dividing cells, and some of these are epidermal stem cells, which still depend on *p63* for their properties.

As well as the stratified epidermis, the surface ectoderm also generates a number of specialized structures: sweat glands, sebaceous glands and hair follicles. In recent years a lot has been learned about these, particularly about the hair follicles.

Hair Follicles

Hair follicles (see Figure 2.C.2) are formed through a set of reciprocal interactions between the surface ectoderm and the underlying mesenchyme, which becomes the dermal layer of the skin (Figure 8.10). The first signal comes from the mesenchyme and is a self-organizing pattern of Wnt activation and BMP inhibition set up by a lateral inhibition process (see Chapter 7 for symmetry breaking and Chapter 9 for lateral inhibition). This occurs about E14.5 of mouse development and establishes foci of cells from the surface epidermis committed to become epidermal placodes. In humans the first epidermal hair placodes appear at

7 weeks. The placodes emit sonic hedgehog (SHH) and provoke the formation of condensations of cells in the mesenchyme, which will become the dermal papillae of the later hair follicles. The papillae emit FGF7 and other factors which provoke the downgrowth of cells from the placode to surround the papilla and form the hair bulb. The evidence for this complex process comes from several sources. Mouse knockouts for key components show arrest of hair follicle development at the appropriate stage. Overexpression of the Wnt inhibitor Dickkopf can change the spacing of follicles, providing evidence for a lateral inhibition mechanism for placode formation. Grafting of the dermal papilla into the epidermis can, in favorable circumstances, provoke formation of new follicles.

Although the hair bulb contains the cells that support growth of the fiber, the true stem cells of the hair follicle are located higher up, in a region called the bulge (see Chapter 10). They arise in late gestation (mouse E17.5) as a population of SOX9-positive cells. The hair follicles are closely associated with sebaceous glands, which have their own stem cells and secrete lubricant onto the growing hairs.

Sweat glands arise around E17.5 and depend on BMP and FGF signaling from the mesenchyme and on a suppression of SHH activity.

Mammary Glands

Mammary glands develop from the ventral epidermis (Figure 8.11). The first indication is the formation of "milk lines", a slight thickening of the epidermis running from axilla to groin (mouse E10.5; human 4 weeks). This becomes a set of mammary placodes (5 pairs in mouse, one pair in humans). A mass of dense mammary mesenchyme appears underneath each placode. The epithelial cells grow inwards to form a bud and by E18.5 have formed a rudimentary duct system embedded in a dermal fat pad. The outer part of the placode becomes the nipple. The process, at least for placodes 2 and 3 in the mouse, is initiated by secretion of FGF10 from the somites. This induces Wnt expression in the milk lines and the later placodes, and expression of transcription factors including TBX3. Secretion of PTHrH (parathyroid hormone-related hormone) from the placode is responsible for induction of the mammary mesenchyme. This produces BMP which stimulates the ductal growth and the branching of the epithelial bud.

The mammary gland is unusual among organs in that most of its development occurs postnatally. The embryonic development is the same in males and females, but at commencement of puberty (from 4 weeks after birth in mouse, about 12 years in

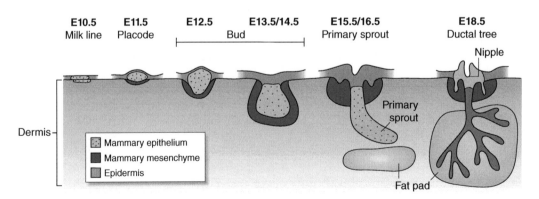

Figure 8.11 Development of the mammary glands in the mouse embryo. (From: Robinson, G.W. (2007) Cooperation of signalling pathways in embryonic mammary gland development. *Nature Reviews Genetics* 8, 963–972. Reproduced with the permission of Nature Publishing Group.)

human) there is a specific stimulation of further development in females. Increased gonadotrophin from the anterior pituitary stimulates release of estrogens from the ovaries. This causes a renewal of growth and branching activity in the mammary epithelium. The growth of the epithelium at this stage is estrogen-dependent, and in mice knockout of the estrogen receptor α (ERα) prevents further development. One effect is the upregulation of amphiregulin (an EGF-like factor) in the epithelium which stimulates growth in both epithelium and mesenchyme. There is also an effect of growth hormone (GH) via the mesenchyme. This causes release of IGF1 which stimulates growth of the epithelial cells. During this period the growth rate exceeds that of the rest of the body, so that the mammary glands become relatively larger. During pubertal growth, terminal end buds form which burrow into the fat pad, assisted by matrix metalloproteinases from the mesenchyme. These have an outer undifferentiated layer of cells and multiple inner layers. As they grow some of the inner cells apoptose to generate a duct lumen. Trailing cells of the bud become myoepithelial cells on the outer surface of the duct system.

The third phase of mammary development occurs during pregnancy and lactation and is described in Chapter 10.

Somitogenesis

The mesodermal germ layer initially forms the notochord, somites and lateral plate. These are arranged mediolaterally during gastrulation and as the embryo body folds develop they become rearranged to run from dorsal to ventral (Figure 8.12). In the embryo the notochord has an important signaling role as a source of SHH, and also has a morphogenetic function stiffening the embryo and helping to drive its elongation. The somites are segmented structures, discussed here. The lateral plate forms various organs including the limb buds, kidney and gonads,

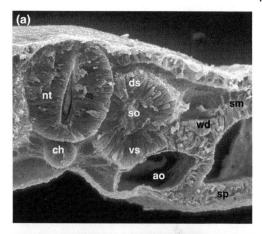

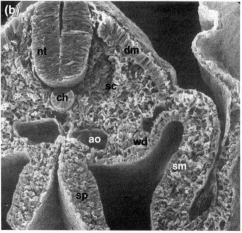

Figure 8.12 Scanning electron micrographs showing the disposition of the main axial structures in the chick embryo. (a) 2 days incubation. (b) 3 days incubation. ao aorta, ch notochord, dm dermomyotome, ds dorsal somite, nt neural tube, sc sclerotome, so somite cavity, sm somatopleure, sp splanchnopleure, vs ventral somite, wd Wolffian duct. (From: Christ, B., Huang, R. and Scaal, M. (2004) Formation and differentiation of the avian sclerotome. Anatomy and Embryology 208, 333–350. Reproduced with the permission of Springer.)

blood vessels and blood cells. Much of the lateral plate becomes divided by a cavity, called the coelom, into an inner splanchnic layer and an outer somatic layer. The somatic mesoderm with its overlying epidermis is often known as the somatopleure.

The presence of segmented somites is a prominent feature of all vertebrate embryos. They develop in the mesoderm in rostral to caudal sequence on either side of the noto-chord. Because they are clearly visible and can be counted, the somite number is often used as an indication of the developmental stage of the embryo. The somites are initially formed as cell condensations, they tran-siently become epithelial vesicles and later become divided into two cell masses. The central mass is the sclerotome and becomes the vertebrae, with each vertebra formed from the anterior part of one somite and the posterior part of the adjacent somite. The outer part is the dermomyotome, which forms the striated muscles of the trunk and the limbs, and the dermal layer of the skin in the dorsal part of the body. In lower vertebrates, such as *Xenopus* and zebrafish, the bulk of the somite becomes the myotome and the other parts are rela-tively very small.

The Somite Oscillator and Gradient

The development of somites consists of two clear phases: the formation of the segmented pattern, and the development of the scler-otome and dermomyotome within each somite. It was long suspected that the forma-tion of a perfectly regular series of structures might depend on some sort of clock or oscil-lator and indeed this is the case (Figure 8.13). The cells of the presomitic mesoderm express *Hes* genes, encoding basic helix-loop-helix (bHLH) type transcription factors, whose expression oscillates with a periodicity of a few hours (2 hours in mouse; 6–8 hours in

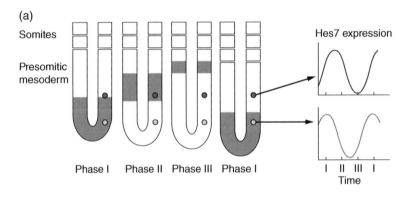

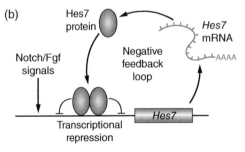

Figure 8.13 Mechanism of segment formation during somitogenesis. (a) The anterior end of the presomitic mesoderm becomes segmented every 2 hours in mouse embryos, forming a bilateral pair of somites. In each cycle, *Hes7* expression is initiated at the caudal end and appears to propagate rostrally because it is following the oscillatory time course illustrated on the right. A caudal to rostral gradient of FGF8 maintains the oscillations and as its overall level declines the oscillations stop in rostral to caudal sequence. When this occurs each group of cells with high *Hes7* increases its cell adhesivity and becomes the next somite. (b) Negative feedback mechanism for the oscillation of *Hes7* expression. (From: Kageyama, R., Niwa, Y., Isomura, A., Gonzalez, A. and Harima, Y. (2012) Oscillatory gene expression and somitogenesis. Wiley Interdisciplinary Reviews – Developmental Biology 1, 629–641. Reproduced with the permission of John Wiley & Sons.)

human). The basic oscillator mechanism depends on a negative feedback and delay. In the mouse, HES7 represses its own transcription, but the delay in transcription and translation, and the short lifetime of both the message and the protein, means that the repression is cyclical. HES7 also represses transcription of *Delta*, thus causing an oscillation of Notch activation in the surrounding cells. Synchronization of the oscillator between cells is achieved through *Lunatic fringe (Lfng)*, which is upregulated by HES7 and encodes a glycosyl transferase which modifies Notch and facilitates Notch signaling. Notch activation upregulates *Hes7* in neighboring cells thus synchronizing the *Hes7* oscillator between cells.

In addition there is a caudal to rostral gradient of FGF8 which is needed for continuation of the oscillations. The FGF gradient gradually declines so the oscillations stop in a smooth sequence from rostral to caudal as cells fall below the threshold level of FGF signaling. This effectively converts a temporal oscillation of *Lfng* mRNA into a spatial one, such that in each cycle Notch and FGF can drive expression of a transcription factor gene, *Mesp2*, in a one segment wide strip of presomite mesoderm. MESP2 represses *Snail* which represses genes for cell adhesion molecules such as integrins and cadherins. The result of this double inhibition is an increase in cell adhesion and the formation of an epithelial somite from the region of *Mesp2*

activity. It is now known that the FGF signaling is also oscillating, due to another negative feedback loop involving dephosphorylation of the ERK-phosphate. But it is still the gradual decline of the FGF8 gradient that confers the polarity and the sequence for the process of somite formation. The initially formed epithelial somite is characterized by expression of the paired-box transcription factor PAX3.

Subdivision of the Somites

The development of the sclerotome, myotome and dermatome from the epithelial somite is controlled by inducing factors from the surrounding structures (Figure 8.14). The sclerotome, which forms the cartilage, and later the bone of the vertebrae, is induced by SHH coming from the notochord and the floor plate of the neural tube. Any part of the epithelial somite can form sclerotome if combined with one of these tissues or exposed to SHH. Expression of *Pax1* and *Pax9* is induced and these upregulate *Nkx3.2* (*Bapx1*) which is a master gene for cartilage differentiation. At the same time, *Pax3* is repressed and a downregulation of N-cadherin causes the cells to undergo an epithelial to mesenchymal transition. The outer part of the sclerotome, formed furthest from the SHH signal, is known as the syndetome. It later generates tendons and is characterized by expression of the bHLH transcription factor Scleraxis. Each vertebra arises from the anterior half of one sclerotome

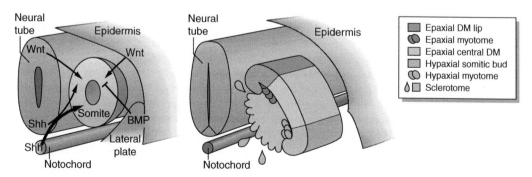

Figure 8.14 Subdivision of the somite by signals from the surrounding tissues. DM = dermomyotome. (Slack, J.M.W. (2013) Essential Developmental Biology, 3rd edn. Reproduced with the permission of John Wiley and Sons.)

and the posterior half of the next. The morphological character of the vertebrae (cervical, thoracic, lumbar etc.) is determined by the expression of *HOX* genes at that particular rostrocaudal level. Loss of function mutants of *HOX* genes tend to convert vertebrae into a more rostral type, and gain of function mutations to a more caudal type.

By the time that the sclerotome has appeared, the remainder of the somite is known as the dermomyotome. The myotome, which forms the skeletal muscle of trunk and limbs, is often depicted in the center of the mature somite, but it actually arises as two distinct primordia, the epaxial myotome on the dorsal side and the hypaxial myotome on the ventral side. Cells from these regions migrate to form layers of myoblasts under the central dermomyotome. The induction of the two myogenic regions arises from different signals. The epaxial myotome requires an early exposure to SHH, and a later exposure to WNT from the dorsal neural tube. The hypaxial myotome requires WNT7A from the dorsal epidermis. Myoblasts from both regions express *Pax7* in addition to *Pax3* and their division is maintained by FGF signaling.

Because of the process of myogenesis is ongoing, there is never really a distinct region of the somite that can be called the dermatome. The dorsal dermis does arise from cells of the dermomyotome, while the remainder of the dermis comes from mesoderm of the ventrolateral body wall.

Myogenesis

The myoblasts are elongated mononucleate dividing cells. Following a reduction of FGF signaling they become committed to form striated muscle by expression of the myogenic transcription factor MYF5. This is one of a group of bHLH type myogenic transcription factors that control expression of the characteristic proteins of striated muscle including actin, specific isoforms of myosin, and the sarcoplasmic reticulum ATPase. The other myogenic factors are MyoD, Myogenin

and MRF4 (= MYF6). These are expressed sequentially during myogenesis, but have overlapping biochemical properties. The combined knockout of *Myf5 + MyoD + Mrf4* lacks most embryonic muscle.

In the dermomyotome, the central myoblasts remain in situ to form the muscles of the dorsal body wall while the lateral ones undergo migration, under the influence of hepatocyte growth factor (HGF, =Scatter Factor) from the lateral plate, to form the muscles of the limbs and of the ventral body wall. Later differentiation of multinucleated striated muscle fibers involves cell fusion, described in Chapter 9.

The Kidney

The region of mesoderm ventral to the somites is called the intermediate mesoderm. From this develops both the gonads and the kidney. Because kidney transplantation is so important in human medicine, and there is a perpetual shortage of kidneys, the prospect of using stem cells to repair damaged kidneys, or to make completely new ones, has prompted much interest. Higher vertebrates (the amniotes) have three kidneys: a pronephros, which is vestigial, a mesonephros, which has a segmental structure but very limited function confined to embryonic life, and the metanephros, or definitive kidney. Lower vertebrates, such as *Xenopus* and the zebrafish, do not have a metanephros, and their definitive kidney is the mesonephros.

The metanephros arises from two distinct components of the intermediate mesoderm through a reciprocal inductive interaction. The nephric duct (= Wolffian duct) arises in the trunk region and grows caudally to the cloaca. Near the cloaca, a side bud is emitted, called the ureteric bud. This grows into the surrounding intermediate (nephrogenic) mesenchyme and undergoes repeated branching (Figure 8.15a). This branching behavior depends on an inducing factor called glial derived neurotrophic factor

(GDNF). This factor is not, in this case, derived from glia, but from the nephrogenic mesenchyme. Its receptor, RET, is present on the ureteric bud and stimulation activates both the ERK and the PI3K signaling pathways (see Figure 7.4). The branched ducts arising from the ureteric bud later become the collecting ducts of the kidney. They contain mostly cells called principal cells, responsible for water reabsorption, together with some intercalated cells, which regulate pH. Like many other examples of cell differentiation, formation of the intercalated cells depends on the Notch system (see Chapter 9).

The nephrons of the kidney arise from the nephrogenic mesenchyme, under the influence of the branching ureteric bud (Figure 8.15b). The main inducing factor from the bud is thought to be WNT9B, and the competence of the mesenchyme depends on expression of a Zn finger transcription factor Wilms' Tumor 1 (WT1, named after the tumor which arises from its loss of function mutation). Other important factors from the bud are BMP7 and FGF2. Following the initial steps of tubule induction, further development depends on WNT4 secreted by the nephrogenic mesenchyme itself.

During development of the nephron, a dense cap of cells forms around the bud tip which expresses the homeodomain transcription factor SIX2. The cap undergoes a mesenchymal to epithelial transition and grows to form the Bowman's capsule and the kidney tubule, fused to

(a)

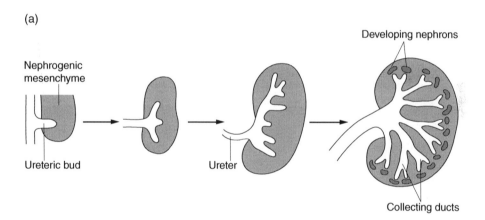

(b)

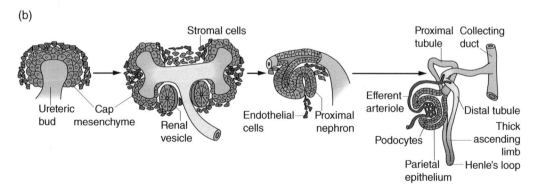

Figure 8.15 Development of the metanephric kidney. (a) Growth and branching of the ureteric bud into the surrounding nephrogenic mesenchyme. (b) Induction and development of an individual nephron. (Adapted from Dressler., 2009. Reproduced with the permission of Company of Biologists Ltd.)

and continuous with, the collecting system arising from the ureteric bud. The capsule contains specialized cells called podocytes that filter the blood. The tubule contains several other cell types with specific functions of reabsorption or secretion. The formation of new nephrons terminates just after birth in the mouse, coincident with the loss of cells expressing *Six2*. These cells are often called stem cells in the kidney literature, but as they only persist for a few cell cycles during embryonic development, they are better considered as progenitor cells.

Blood and Blood Vessels

Blood

Without blood and blood vessels no animal of any size could survive. Furthermore no grafted tissue can survive unless it rapidly acquires a vascular system. Effective tissue-engineered tissues and organs will need to be supplied with a vascular system so this is a topic of particular interest to stem cell researchers.

In the mouse the embryonic development of blood commences in the yolk sac at about 7.5 days gestation when a number of blood islands appear in the mesoderm. They are surrounded by endothelium but initially there is no circulatory system for them to join. These primitive blood islands contain a different repertoire of cells from the later definitive blood. Most of the primitive blood cells are erythrocytes which are larger than the definitive erythrocytes, and retain their nuclei (Figure 8.C.2). There are also some megakaryocytes (the cells that form blood platelets) and a population of primitive macrophages precursors that generate the microglia of the brain and other tissue resident macrophage types.

There has been much controversy about whether the cells of the primitive blood islands contribute progeny to the stem cells of the definitive blood. In the mouse it is dif-ficult to be certain, but in *Xenopus* it is possible to label the primitive and definitive blood by microinjection into different parts of the early embryo and this shows that the two populations are distinct. The definitive blood arises from the endothelium of the dorsal aorta and other central arteries (Figure 8.16). This has been shown by lineage labeling experiments and also directly observed by time lapse filming of the dorsal aorta in both mouse and zebrafish. In mammalian development the main site of hematopoiesis in the mid-gestation embryo is the liver, and in the late gestation and postnatal animal it is the bone marrow. The labeling experiments show that the hematopoietic cells of the fetal liver and the postnatal bone marrow are progeny of the cells originally arising from the central arteries. Definitive blood cells are often said to arise from the "AGM" region. This relates to early experiments which located their origin to a region of the lateral mesoderm including the aorta, gonads and mesonephros, but in fact the origin is just from the aorta and other central arteries.

The first time of appearance of real hematopoietic stem cells (HSCs) in the mouse embryo is known from experiments where cells are transplanted to lethally irradiated mice (see Chapter 10). HSCs are able to lodge in the bone marrow of the host and, in time, produce a new population of all the cell types of the blood and immune system which repopulate the host and prevent it from dying of the radiation. Such experiments indicate that before about E10.5 there are no HSC in the embryo. From E10.5 the first repopulating cells are detected in the region of the dorsal aorta and can be lineage labeled with endothelial cell markers. This shows that the primitive blood islands do not contain HSC, whereas the region of the dorsal aorta does contain them from the time that definitive hematopoiesis is first observed.

The three transcription factors SCL (bHLH type), LMO2 (LIM domain) and GATA1 (Zn finger) are all required both for primitive and definitive hematopoiesis, and form a complex

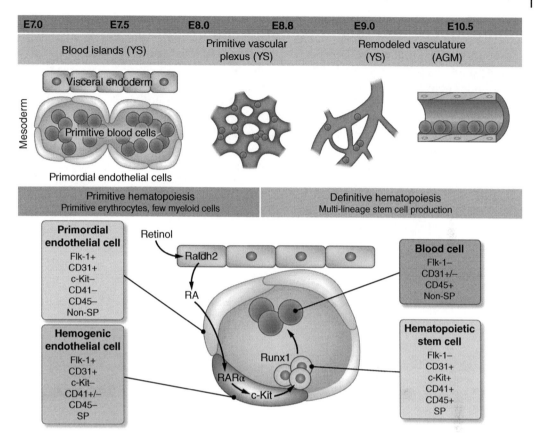

Figure 8.16 The switch from primitive blood island formation to definitive hematopoiesis from central artery endothelium, a process requiring retinoic acid. Surface antigens used to identify the various cell types are shown. (From: Marcelo, K.L., Goldie, L.C. and Hirschi, K.K. (2013) Regulation of endothelial cell differentiation and specification. Circulation Research 112, 1272–1287. Reproduced with the permission of Wolters Kluwer Health, Inc.) RA = retinoic acid, Raldh2 = retinaldehyde dehydrogenase, RARα = retinoic acid receptor, SP = side population.

for regulating gene activity. The mouse knockouts of all three factors die in mid-gestation with hematological defects. The transcription factor RUNX1 (runt domain) is expressed in both primitive and definitive hematopoietic cells, but mouse knockouts are defective only in definitive hematopoiesis, indicating that it is not required for primitive hematopoiesis.

Blood Vessels

Although the primitive blood islands of the yolk sac are surrounded by endothelial cells, the blood vessels of the embryo originate mainly from the splanchnic mesoderm of the lateral plate. Some vessels in the head arise from neural crest mesenchyme and in the heart from the endocardium. The primitive cells from which capillaries are assembled are called angioblasts. They arise from the splanchnic mesoderm in response to a signal from the endoderm, which can be replaced by treatment with FGF2. The angioblasts assemble themselves into a primitive vascular plexus under the influence of vascular endothelial growth factor (VEGF). VEGF-A which is the most active member of this family, binds to a receptor Fetal Liver Kinase-1 (FLK-1, = VEGFR2, KDR), which activates several intracellular transduction pathways. Mice lacking FLK-1 form angioblasts but not vessels, showing the importance of VEGF signaling for the assembly of vessels.

Soon after their formation it is apparent that the primitive vascular plexi have distinct arterial and venous regions. Arterial capillaries express the ephrin B2 adhesion molecule and venous ones express the complementary Eph B4. Possession of these complementary adhesion molecules enables the arterial and venous capillaries to meet and fuse with each other. High levels of VEGF induce production of the Notch ligand Dll4. The resulting stimulation of Notch induces ephrin B2 and represses EphB4, leading to artery formation. Lower level of VEGF lead to

expression of another transcription factor, COUPTFIII, which defines venous development (Figure 8.17). At least in the zebrafish a dorsal-ventral gradient of VEGF arises as a result of SHH signaling from the notochord and floor plate. This determines that the dorsal aorta arises in a dorsal position and has arterial character, whereas the cardinal veins arise more ventrally and have venous character.

Initially capillaries are solid rods of cells. They form a lumen through a symmetry breaking process whereby PAR3 (see Chapter 10) is attracted to the external (basal) side, repelling the protein podocalyxin

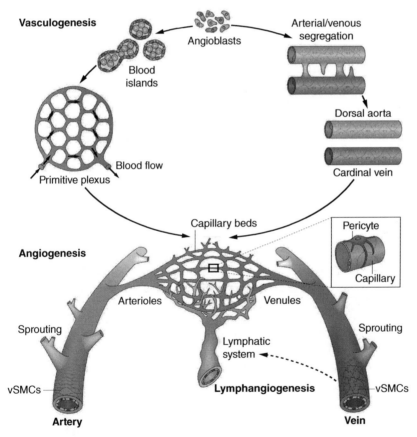

Figure 8.17 Blood vessel development. De novo formation is called vasculogenesis while sprouting from existing vessels is angiogenesis. Complementarity of adhesion molecules between arterial and venous capillaries enables their fusion into capillary beds. The lymphatics originate from the venous system. vSMCs = vascular smooth muscle cells. (From: Herbert, S.P. and Stainier, D.Y.R. (2011) Molecular control of endothelial cell behaviour during blood vessel morphogenesis. Nature Reviews. Molecular Cell Biology 12, 551–564. Reproduced with the permission of the Nature Publishing Group.)

to the inner, apical side. The junctional protein VE-cadherin, specific to endothelia, is shifted basally and the repulsion of podocalyxin molecules opens the lumen internally. Small capillaries remain as simple tubes of endothelium while the larger blood vessels develop outer layers. Pericytes are undifferentiated cells also formed from the splanchnic mesoderm, and recruited to the newly formed vessels, and smooth muscle cells are also recruited to invest the newly formed arteries.

The Heart

The heart arises from mesoderm at the rostral end of the body. Fate mapping experiments on chick embryos show that cells lateral to the node migrate through the primitive streak to form lateral mesoderm territories during gastrulation. These then migrate rostrally on both sides. When the embryo begins to fold, the prospective heart becomes visible as a cardiac crescent at the rostral end of the blastoderm (Figure 8.18). As the head fold forms, the crescent becomes tucked under the head. The actual specification of the heart rudiments occurs just before appearance of the cardiac crescent and depends on inductive signals from the rostral endoderm, especially BMP and FGF8. Evidence that specification occurs at this stage is that extirpation or transplantation of tissue within the crescent leads to subsequent heart defects, whereas this is not the case at earlier stages. The cardiac crescent itself is characterized by expression of several transcription factors including NKX2.5 (homeodomain), GATA4-6 (Zn finger), MEF2c (MADS box) and TBX5 (T box).

The early heart rudiments on each side of the body have the form of tubes. As these become tucked below the head, they fuse into a single midline tube. This early heart tube is the precursor of both atria and the left ventricle. Cells continue to be added at the rostral end to augment the tube and to add territories destined to become the right ventricle and the outflow tract. These cells come from the second heart field: a population lying medial to the cardiac crescent, which is characterized at early stages by expression of ISLET-1 instead of NKX2.5 (Figure 8.18). A third population of cells come from the neural crest. They express the transcription factor TBX1 and contribute to the outflow tract and the septum forming within it that later divides the pulmonary from the aortic circulation. The primitive heart tube has four layers: the endocardium inside, an extracellular layer called the cardiac jelly, the myocardium, which becomes the actual cardiac muscle, and the pericardium which becomes a thin connective tissue layer around the outside.

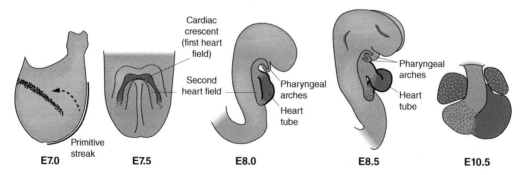

Figure 8.18 Heart development in the mouse embryo. The cardiac crescent, or first heart field, forms mainly the atria and the left ventricle, while the second heart field forms mainly the right ventricle and outflow tract. (Rosenthal and Harvey, 2010. Reproduced with permission of Academic Press.)

The heart commences its physiological functions at an early stage. The primitive heart tube begins autonomous pulsation from E8.5 in the mouse or 4 weeks in human. The heart is asymmetrical because of inherent difference between left and right side dependent ultimately on an asymmetrical expression of Nodal on the left side, which occurs somewhat after the formation of the node itself. Asymmetry initially becomes manifest as a looping of the heart tube to the right side. Different transcription factors are expressed in different territories: the prospective atria preferentially express COUP TFIII and TBX5, the prospective right ventricle dHAND and the prospective left ventricle eHAND. The later development of the heart involves a complex sequence of morphogenetic events that convert the simple tube into the four-chambered mammalian heart. These include the looping, which brings the atria to the rostral side, remodeling, the migration of blood vessel insertion sites on the surface of the heart, and the formation of septa to separate the chambers. Errors in these processes are not uncommon, resulting in a significant number of congenital heart defects in human infants (about 1% of live births). Because of the importance of these, a considerable amount of research has been conducted into the causes, which are, in many cases, mutations in the transcription factors controlling cardiac development.

The Gut

The endoderm becomes the epithelial lining of the gut tube and the respiratory system, plus the various organs that bud off the gut, including the liver and pancreas. The definitive endoderm arises as the bottom layer of the embryo during gastrulation (see Chapter 5), forming a strip of epithelium along the midline, flanked by the extraembryonic visceral endoderm. The definitive endoderm becomes transformed into a tube by the process of body folding (Figure 8.19). It commences as a lifting of the head end of the embryo above the surrounding blastoderm. Since all three germ layers participate, this means that at the rostral end a blind-ended cavity arises lined with definitive endoderm. This is the foregut. The folding continues progressively to more caudal levels bringing more and more of the definitive endoderm into the foregut tube, with less and less of it being open to the yolk sac. At the caudal end, a similar process starts later, resulting in the formation of a hindgut. The body folding continues until the whole embryo has everted from the original blastoderm and the open part of the endoderm is reduced to a small region forming a canal to the yolk sac lumen, called the vitellointestinal duct. This duct, along with the principal blood vessels to the placenta, and the allantois, becomes wrapped up into the umbilical tube.

The formation of the coelomic cavity within the lateral plate mesoderm separates the gut tube from the body wall. The gut tube now consists of a lining of endoderm surrounded by mesenchyme from the splanchnic mesoderm. It is suspended in the coelomic cavity by dorsal and ventral mesenteries, composed of splanchnic mesoderm, of which the dorsal mesentery remains in place permanently. The process of head folding brings the future heart into a midventral position. Between the heart and the vitellointestinal duct is a mass of mesoderm called the septum transversum. This divides the coelomic cavity into thoracic and abdominal regions and later contributes to the diaphragm.

The endoderm of chick and mouse embryos has been fate mapped by applying small marks of DiI to the early endoderm which is the lower layer of the blastoderm, and locating the labeled cells in the gut tube at a later stage. The results are complex but there is an approximate maintenance of rostrocaudal polarity, i.e. the rostral organs of the gut tube arise from the rostral part of the endoderm and vice versa. There is also a very pronounced mismatch between the fate map of the endoderm and its

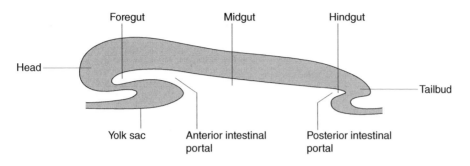

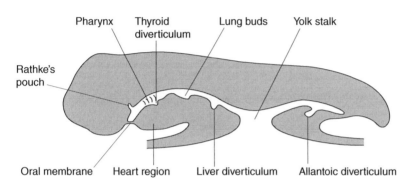

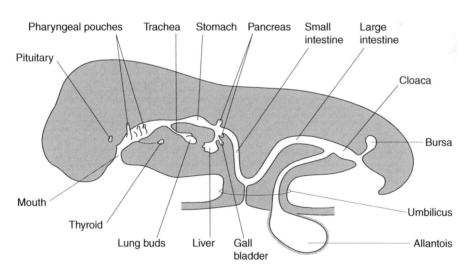

Figure 8.19 Formation of the regions of the gut in a higher vertebrate animal. The bursa is not found in mammals. (Hildebrand, 1995. Reproduced with the permission of John Wiley and Sons.)

associated splanchnic mesoderm. This arises from a progressive displacement between endoderm and mesodermal layers during the formation of the gut tube. This means, contrary to general belief, that the regional pattern of the endoderm cannot be simply "printed" by means of inductive signals from the mesoderm because the mesoderm that ends up as part of, say, the stomach, is initially present at a much more caudal level than the prospective stomach endoderm.

Regional Specification of the Endoderm

The initial formation of the endoderm depends on Nodal signaling, and mouse embryos lacking *Nodal* have no endoderm. Transcription factors important in the early endoderm include SOX17, FOXA1,2, and GATA 4,5,6. The first regional specification occurs during gastrulation and is a response to the caudal to rostral gradients of Wnt and FGF that also pattern the nervous system (Figure 8.20). These signals result in a nested expression of transcription factors, such as FOXA3 caudal to the liver, CDX2 in the prospective intestine, HNF4 in the prospective stomach and intestine. In addition many of the *HOX* genes are expressed, in both endo-

derm and mesenchyme layers, generally with expression in the caudal region and a boundary of expression at a specific body level. Retinoic acid, from the somitic mesoderm, is required in the foregut region, and the knockout of *Raldh1a* lacks stomach, lungs and dorsal pancreas.

It used to be thought that much of the pattern of the gut was derived from signals from the splanchnic mesenchyme. There are indeed such signals but there is also considerable signaling from the epithelium to the mesenchyme, so the situation tends to be quite complex for each gut region. The mesenchyme itself resolves into four layers, lamina propria, muscularis mucosa, submucosa and smooth muscle. This radial pattern is due to SHH secreted by the early gut tube. In addition to

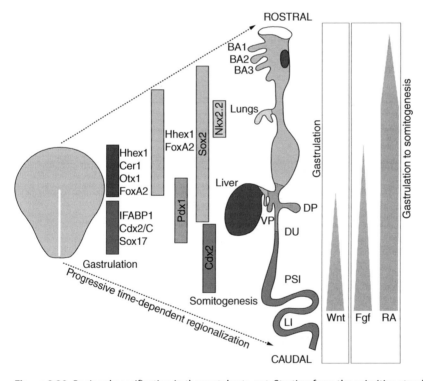

Figure 8.20 Regional specification in the vertebrate gut. Starting from the primitive streak stage the endodermal epithelium becomes subdivided in response to caudal-rostral gradients of Wnt, FGF and retinoic acid. The approximate domains of expression of various transcription factors are shown (IFABP is not a transcription factor but a fatty acid binding protein). BA: branchial arch; DP: dorsal pancreas; DU: duodenum; LI: large intestine; PSI: posterior small intestine; VP: ventral pancreas. (Modified from: Kraus, M.R.C. and Grapin-Botton, A. (2012) Patterning and shaping the endoderm in vivo and in culture. Current Opinion in Genetics and Development 22, 347–353. Reproduced with the permission of Elsevier.)

producing instructive signals for gut patterning, the mesenchyme is also the source of trophic signals such as FGF10, which are needed for growth and morphogenesis of the gut and especially for its outgrowths such as the lung buds, pancreas and cecum.

The Intestine

The mammalian small and large intestine are among the best studied and understood tissue-specific stem cell systems, so it is worth noting how the intestine develops. The mouse gut tube is fully formed by the processes described above by about 9 days of gestation. It is initially lined by a simple epithelium which, over the next 5 days, becomes pseudostratified, then stratified, and then columnar. This change in histology is accompanied by a pronounced elongation of the intestine which, like other cell intercalation processes, requires non-canonical Wnt signaling initiated by Wnt5A. Comparable events in human development occur from about 4 to 9 weeks post-fertilization.

After the epithelium has become columnar, mesenchyme invades to form nascent villi in a rostral to caudal sequence (Figure 8.21). Crypts initially form from cavities in the epithelium. Hedgehog signaling is important for the early establishment of crypts and villi. SHH and Indian Hedgehog (IHH) are both expressed in the early intestinal epithelium, but have opposite effects: loss of SHH causes villous overgrowth while loss of IHH causes formation of fewer, smaller, villi. BMPs are expressed in the mesenchyme invading the villi. Deletion of a BMP receptor, BMPR1A, from the epithelium, causes formation of ectopic crypts, suggesting that BMP signaling promotes villus rather than crypt formation. The transcription factor HNF4α, expressed in early stomach and intestinal epithelium, is needed for intestinal development; knockout in the embryo preventing the crypt-villus pattern from developing at all. The crypt-villus pattern is normally established in mice by E18.5, and in humans by 12–13 weeks post-fertilization. One issue that remains unclear is the precise role of canonical Wnt signaling in the formation of the crypts. This is of critical importance in maintaining the stem cells and proliferative behavior of postnatal crypts (see Chapter 10). However β-catenin activity only appears in the epithelium after villus

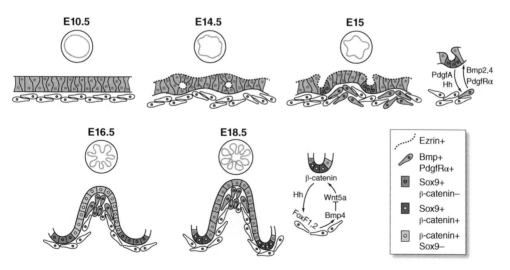

Figure 8.21 Development of crypts in the duodenum of the mouse embryo. Some of the cell interactions are shown. Ezrin (=villin 2) is required for the formation of microvilli on the absorptive cells. (From: Spence, J.R., Lauf, R. and Shroyer, N.F. (2011) Vertebrate intestinal endoderm development. *Developmental Dynamics* 240, 501–520. Reproduced with the permission of John Wiley and Sons.)

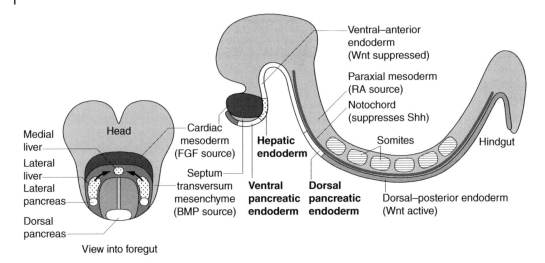

Ventral–anterior endoderm (Wnt suppressed)

Paraxial mesoderm (RA source)

Notochord (suppresses Shh)

Medial liver

Head

Cardiac mesoderm (FGF source)

Hepatic endoderm

Somites

Hindgut

Lateral liver

Lateral pancreas

Septum-transversum mesenchyme (BMP source)

Ventral pancreatic endoderm

Dorsal pancreatic endoderm

Dorsal–posterior endoderm (Wnt active)

Dorsal pancreas

View into foregut

Figure 8.22 Specification of mouse embryo liver and pancreas. On the left is shown a view into the foregut of a mouse embryo at E8.25. The territories indicated are not yet specified. Arrows indicate movement of lateral regions toward the ventral-medial region. On the right is a sagittal view of a later embryo showing the positions of the newly specified liver and pancreas tissue domains. Signals and cell sources that pattern the endoderm are shown.

emergence. By birth, Wnt signaling, and cell proliferation, is confined to the intervillus regions.

The Pancreas

The pancreas derives from two buds which arise in different ways (Figure 8.22). The dorsal pancreatic bud comes from the dorsal midline of the epithelium. It arises at about E9.5 in the region where the notochord contacts the epithelium, and suppresses SHH production locally. This effect can be mimicked by treatment with activin or FGF. The ventral pancreatic bud arises about 2 days later. Its formation also involves suppression of SHH, although here this is not due to the notochord but to something else. Retinoic acid is needed for formation of both the pancreatic buds. Once formed, the buds grow rapidly, stimulated by FGF10 from the mesenchyme, and generate branched structures. The two buds then fuse together to form a single organ, and the duct of the ventral bud becomes the main pancreatic duct.

The best known pancreatic transcription factor is the LIM-homeodomain factor PDX1. This is expressed in a ring around the duodenum encompassing the territories of the two buds. Its knockout does form rudimentary buds but they grow slowly and do not mature. The transcription factor HB9 is expressed a little earlier and is dependent on retinoic acid. The *Hb9* knockout lacks the dorsal bud but the ventral bud is still present. The factor PTF1 has three subunits, of which p48 is needed for formation of the ventral bud and the differentiation of exocrine tissue in the dorsal bud. The terminal differentiation of the exocrine and endocrine cells in the pancreas will be described in Chapter 9.

The Liver

The liver arises as a ventral diverticulum of the foregut endoderm (Figure 8.19). The cells separate from the epithelium and grow as individual cells into the region of mesenchyme called the septum transversum. This is a source of BMP and the adjacent cardiac mesoderm is a source of FGF. Both these factors are needed for the formation of the liver from the endoderm and for the suppression of formation of the ventral pancreas. The early cells of the liver are hepatoblasts: bipotent cells which can form either hepatocytes or biliary epithelial cells. Important transcription factors active in the early liver are C/EBPα, HNF1α and β, HNF4α, HNF6, and PXR.

Just between the liver and the ventral pancreas lies another bud which forms the extrahepatic biliary system: comprising the gall bladder, cystic duct, and extrahepatic bile ducts. This bud requires persistent expression of SOX17, which is mutually antagonistic to PDX1, ensuring that the ventral pancreatic and biliary buds remain distinct from one another. Expression of SOX17 is maintained by the bHLH factor HES1. Knockout of *Hes1* causes loss of SOX17 and persistence of PDX1, leading to the formation of ectopic pancreas in the biliary system.

Further Reading

Neural

Baggiolini, A., Varum, S., Mateos, José, M., et al. (2015) Premigratory and migratory neural crest cells are multipotent in vivo. Cell Stem Cell 16, 314–322.

Blaess, S. and Ang, S.-L. (2015) Genetic control of midbrain dopaminergic neuron development. Wiley Interdisciplinary Reviews-Developmental Biology 4, 113–134.

Copp, A.J., Greene, N.D.E. and Murdoch, J.N. (2003) The genetic basis of mammalian neurulation. Nature Reviews. Genetics 4, 784–793.

Dessaud, E., McMahon, A.P. and Briscoe, J. (2008) Pattern formation in the vertebrate neural tube: a sonic hedgehog morphogen-regulated transcriptional network. Development 135, 2489–2503.

Gammill, L.S. and Roffers-Agarwal, J. (2010) Division of labor during trunk neural crest development. Developmental Biology 344, 555–565.

Guillemot, F. (2007) Spatial and temporal specification of neural fates by transcription factor codes. Development 134, 3771–3780.

Kulesa, P.M., Bailey, C.M., Kasemeier-Kulesa, J.C. and McLennan, R. (2010) Cranial neural crest migration: New rules for an old road. Developmental Biology 344, 543–554.

Milet, C. and Monsoro-Burq, A.H. (2012) Neural crest induction at the neural plate border in vertebrates. Developmental Biology 366, 22–33.

O'Leary, D.D.M., Chou, S.-J. and Sahara, S. (2007) Area patterning of the mammalian cortex. Neuron 56, 252–269.

Ozair, M.Z., Kintner, C. and Brivanlou, A.H. (2013) Neural induction and early patterning in vertebrates. Wiley Interdisciplinary Reviews – Developmental Biology 2, 479–498.

Wurst, W. and Bally-Cuif, L. (2001) Neural plate patterning: Upstream and downstream of the isthmic organizer. Nature Reviews. Neuroscience 2, 99–108.

Xu, Q. and Wilkinson, D.G. (2013) Boundary formation in the development of the vertebrate hindbrain. Wiley Interdisciplinary Reviews – Developmental Biology 2, 735–745.

Epidermis

Fuchs, E. (2007) Scratching the surface of skin development. Nature 445, 834–842.

Howlin, J., McBryan, J. and Martin, F. (2006) Pubertal mammary gland development: Insights from mouse models. Journal of Mammary Gland Biology and Neoplasia 11, 283–297.

Koster, M.I. and Roop, D.R. (2007) Mechanisms regulating epithelial stratification. Annual Review of Cell and Developmental Biology 23, 93–113.

Lu, C.P., Polak, L., Keyes, B.E., Fuchs, E., 2016. Spatiotemporal antagonism in mesenchymal-epithelial signaling in sweat versus hair fate decision. Science 354.

Robinson, G.W. (2007) Cooperation of signaling pathways in embryonic mammary gland development. Nature Reviews Genetics 8, 963–972.

Sennett, R. and Rendl, M. (2012) Mesenchymal–epithelial interactions during hair follicle morphogenesis and cycling. Seminars in Cell and Developmental Biology 23, 917–927.

Somitogenesis and Myogenesis

Buckingham, M. and Vincent, S.D. (2009) Distinct and dynamic myogenic populations in the vertebrate embryo. Current Opinion in Genetics and Development 19, 444–453.

Christ, B., Huang, R. and Scaal, M. (2007) Amniote somite derivatives. Developmental Dynamics 236, 2382–2396.

Kageyama, R., Niwa, Y., Isomura, A., Gonzalez, A. and Harima, Y. (2012) Oscillatory gene expression and somitogenesis. Wiley Interdisciplinary Reviews-Developmental Biology 1, 629–641.

Pourquie, O. (2011) Vertebrate segmentation: from cyclic gene networks to scoliosis. Cell 145, 650–663.

Kidney

Costantini, F. and Kopan, R. (2010) Patterning a complex organ: branching morphogenesis and nephron segmentation in kidney development. Developmental Cell 18, 698–712.

Dressler, G.R. (2006) The cellular basis of kidney development. Annual Review of Cell and Developmental Biology 22, 509–529.

Michos, O. (2009) Kidney development: from ureteric bud formation to branching morphogenesis. Current Opinion in Genetics and Development 19, 484–490.

Blood and Blood Vessels

Adamo, L. and García-Cardeña, G. (2012) The vascular origin of hematopoietic cells. Developmental Biology 362, 1–10.

Herbert, S.P. and Stainier, D.Y.R. (2011) Molecular control of endothelial cell behaviour during blood vessel morphogenesis. Nature Reviews. Molecular Cell Biology 12, 551–564.

Marcelo, K.L., Goldie, L.C. and Hirschi, K.K. (2013) Regulation of endothelial cell differentiation and specification. Circulation Research 112, 1272–1287.

Medvinsky, A., Rybtsov, S. and Taoudi, S. (2011) Embryonic origin of the adult hematopoietic system: advances and questions. Development 138, 1017–1031.

Poole, T.J., Finkelstein, E.B. and Cox, C.M. (2001) The role of FGF and VEGF in angioblast induction and migration during vascular development. Developmental Dynamics 220, 1–17.

Heart

Abu-Issa, R. and Kirby, M.L. (2007) Heart field: from mesoderm to heart tube. Annual Review of Cell and Developmental Biology 23, 45–68.

Buckingham, M., Meilhac, S., and Zaffran, S. (2005) Building the mammalian heart from two sources of myocardial cells. Nature Reviews Genetics 6, 826–837.

Chien, K.R., Domian, I.J. and Parker, K.K. (2008) Cardiogenesis and the complex biology of regenerative cardiovascular medicine. Science 322, 1494–1497.

Olson, E.N. (2006) Gene regulatory networks in the evolution and development of the heart. Science 313, 1922–1927.

Rosenthal, N. and Harvey, R.P. (2010) Heart Development and Regeneration. Academic Press, London.

Endoderm

Gittes, G.K. (2009) Developmental biology of the pancreas: A comprehensive review. Developmental Biology 326, 4–35.

McCracken, K.W. and Wells, J.M. (2012) Molecular pathways controlling pancreas induction. Seminars in Cell and Developmental Biology 23, 656–662.

Si-Tayeb, K., Lemaigre, F.P. and Duncan, S.A. (2010) Organogenesis and development of the liver. Developmental Cell 18, 175–189.

Spence, J.R., Lauf, R. and Shroyer, N.F. (2011) Vertebrate intestinal endoderm development. Developmental Dynamics 240, 501–520.

Zaret, K.S. and Grompe, M. (2008) Generation and regeneration of cells of the liver and pancreas. Science 322, 1490–1494.

Zong, Y. and Stanger, B.Z. (2012) Molecular mechanisms of liver and bile duct development. Wiley Interdisciplinary Reviews – Developmental Biology 1, 643–655.

Zorn, A.M. and Wells, J.M. (2009) Vertebrate endoderm development and organ formation. Annual Review of Cell and Developmental Biology 25, 221–251.

9

Cell Differentiation and Growth

The previous chapters have covered the account of mammalian development from the generation of gametes to the formation of organ rudiments. Developmental biology textbooks often stop at this point, but here we shall continue and look at some selected examples of how organ rudiments become the functional tissues and organs of the body. All tissue types grow and expand during development, but most of those dealt with here do not have real stem cells which persist through adult life. Several bona fide stem cell systems are dealt with in Chapter 10, but in the present chapter the only tissue types which have stem cells are skeletal muscle and the central nervous system. Those of skeletal muscle are a population of muscle satellite cells that can become new myofibers. In the postnatal mammalian brain most of the neurons cannot be renewed, but there are some small stem cell populations in specific areas. Otherwise there are no true stem cells although some tissue types maintain slow cell turnover by division of their functional differentiated cells.

Organs, Tissues and Cell Types

The three concepts of "organ", "tissue" and "cell type" are often confused, as when speaking of "muscle" without specifying whether what is meant is the whole anatomical muscle or just the multinucleated myofibers within a muscle. An actual muscle contains many other tissues and cell types, including connective tissue, blood vessels, nerves and macrophages. Gene expression studies are often conducted on pieces of whole organ despite the fact that these contain multiple tissues and cell types each with vastly different gene expression repertoires. This means that the gene expression patterns of organs are dominated by the most abundant mRNAs from the most abundant cell types in the sample. Worse, these data are then used for various purposes by theoreticians who may not appreciate the limitations of the information arising from the complexity of the initial samples. So it is really worthwhile to be clear about what is meant by organ, tissue or cell type in specific situations.

An organ is a named part of the body familiar from gross anatomy. The stomach, the kidney, or the lungs are all organs. The skin is also an organ although it has a less discrete character. Organs have an identifiable physiological function and always consist of several tissue types which in turn usually contain multiple cell types.

There is no clear definition of a "tissue" in the histological literature, but a tissue may usefully be regarded as the set of cell types originating from a single type of stem cell (or embryonic progenitor cell if the tissue in question does not have stem cells). Under this definition the small intestinal epithelium is a tissue. It is composed of multiple cell types, but they all come from one stem cell

The Science of Stem Cells, First Edition. Jonathan M. W. Slack.
© 2018 John Wiley & Sons, Inc. Published 2018 by John Wiley & Sons, Inc.
Companion website: www.wiley.com/go/slack/thescienceofstemcells

population. The small intestine as an organ comprises also the connective tissue layers, the blood vessels, the lymphatics, the nerve supply, and some patches of lymphoid tissue. In the liver the hepatocytes and the biliary system comprise one tissue, as they arise in late embryonic development from one type of progenitor: the hepatoblast. The liver as an organ also contains an abundant vascular system and numerous cells of different lineages: the Kupffer cells and hepatic stellate cells. Histology textbooks often classify tissues into five general types: epithelia, connective tissues, nervous tissues, muscle and blood. The last three are dealt with later, in this chapter and in Chapter 10, but some preliminary remarks on the first two are appropriate here.

Epithelia

Epithelia (singular: epithelium) are sheets of cells which may consist of one or many layers (simple or stratified). The cells may be flat (squamous), cuboidal or columnar (Figure 9.1a).

The cells of a simple epithelium have an apical-basal polarity. The apical surface abuts the lumen of the structure formed by the epithelium and often bears specializations such as cilia or microvilli. The basal side abuts a basement membrane (Figure 9.1b). Part of this, the basal lamina, is secreted by the epithelium itself and consists of the extracellular proteins laminin, type IV collagen, entactin and heparan sulfate proteoglycan. This is usually underlain by collagen fibers secreted by the adjacent connective tissue, the whole making up the basement membrane. The lateral surfaces of epithelial cells are attached to each other by means of junctional complexes. The tight junctions form an impermeable barrier, isolating the luminal from the basal side of the tissue. The adherens junctions and the desmosomes bind the cells together via calcium-dependent adhesion molecules called cadherins. The gap junctions, formed from proteins called connexins which contain small pores, enable transfer of small molecules between adjacent cells.

Many organs in the body have an epithelium as their principal component. This is obviously the case for those organs forming the gut (esophagus, stomach, small and large intestine) where the respective epithelia line the luminal surfaces, but is also true of many solid organs including the liver, the kidneys and the salivary glands. Many organs are glands with a secretory function. Glands often originate from a region of an epithelial sheet which invaginates into the surroundings. If it persists, the duct of the later gland shows the original position of this invagination. In the case of endocrine glands, which secrete their products into the bloodstream, the duct disappears in the course of development.

The term "epithelium" is a descriptive one and does not imply anything about embryonic origin. Epithelia may originate from any of the three embryonic germ layers, and from numerous different positions within them. Epithelia derived from the mesoderm are sometimes referred to as "endothelia" or "mesothelia".

Connective Tissues

The term "connective tissue" is used in two very different senses. In its wider usage it refers to all skeletal tissues: bone, cartilage, tendons, ligaments, and also adipose tissue. In its narrower usage it refers just to the fibrous tissue that fills the spaces between other structures. In the latter sense, connective tissue consists of individual cells called fibroblasts, embedded in a loose extracellular matrix. This consists of proteoglycans, hyaluronan, fibronectin, type I collagen, type III collagen (= reticulin) and elastin, which are secreted by the fibroblasts themselves (Figure 9.1c). Connective tissues are often considered to derive from the mesoderm of the embryo, but in fact most of the skeleton and loose connective tissue of the head is derived from the neural crest. In the embryo the tissue filling up the gaps between other structures is called "mesenchyme" (Figure 9.1d). Again, this is a descriptive term

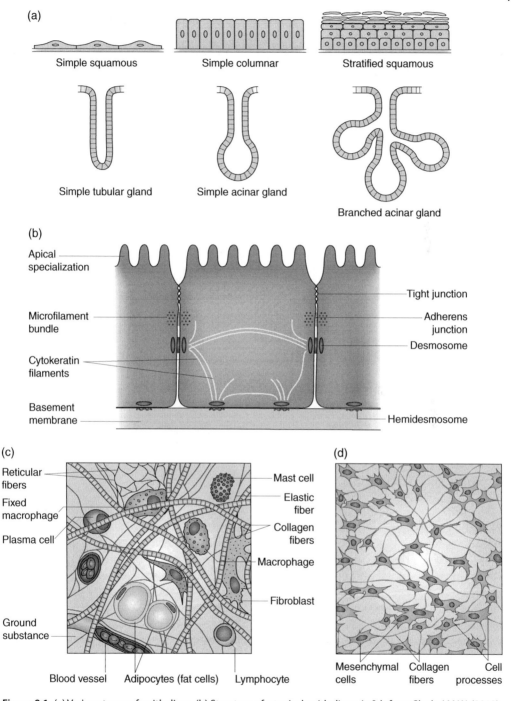

Figure 9.1 (a) Various types of epithelium. (b) Structure of a typical epithelium. (a & b from Slack, J.M.W. (2013) *Essential Developmental Biology*, 3rd edn. Reproduced with the permission of John Wiley and Sons.) (c) Typical mature connective tissue. (Modified from http://www.mhhe.com/biosci/ap/histology_mh/loosctfs.html.) (d) Loose mesenchyme as found in embryos. (Modified from http://www.mhhe.com/biosci/ap/histology_mh/loosctfs.html.)

and does not designate a specific embryonic origin. Mesenchymal cells have an irregular stellate appearance and their extracellular matrix consists largely of hyaluronic acid and glycosaminoglycans. Although of similar appearance, different regions of the mesenchyme have different developmental commitments, forming the various skeletal structures and the masses of adipose tissue as well as the loose connective tissue of the postnatal organism. Another related term is "stroma". This refers to the non-epithelial part of an organ or a tumor, much of which consists of connective tissue. It is used for adult rather than embryonic tissues.

Cell Differentiation

Regulation of Gene Activity

Different cell types express different repertoires of genes so regulation of gene expression is obviously central to their nature. Regulation of gene expression is mostly exerted at the stage of transcription of DNA to messenger RNA, but there are also some translational controls, exerted by regulatory proteins that bind to mRNAs, and by micro RNAs that impede translation or destabilize mRNA.

Key to the control of transcription are the transcription factors, which are proteins that control the activity of other specific genes. They usually have two important parts: an effector region and a DNA-binding region (Figure 9.2a). The effector region is often rich in acidic amino acids and activates transcription by interacting with the general transcription complexes present in all cells. The DNA-binding region determines the specificity of the transcription factor by binding to specific sequences in the regulatory regions of the target genes. For example, T-box transcription factors like BRACHYURY bind to the sequence TCACACCT in the DNA. The regulatory sequences may be next to the binding site for the general transcription complex, in the region known as the promoter,

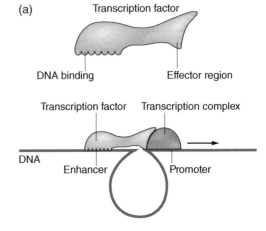

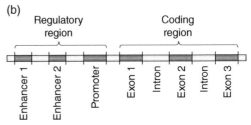

Figure 9.2 (a) Schematic view of a transcription factor binding to an enhancer sequence. (b) Typical gene structure showing multiple regulatory regions. (Slack, J.M.W. (2013) *Essential Developmental Biology*, 3rd edn. Reproduced with the permission of John Wiley and Sons.)

or they may be distant, then being called enhancers. Genes involved in development tend to have rather complex regulatory regions with a number of binding sites for transcription factors, and specific combinations of factor may be necessary to activate the transcription complex (Figure 9.2b). Because the presence of a specific repertoire of transcription factors in a cell determines the pattern of gene expression, the genes whose activity defines states of developmental commitment, discussed in Chapter 7, mostly encode transcription factors.

In addition there is an important level of gene regulation exerted by the chromatin (Figure 9.3a). Chromatin comprises the genetic DNA plus the many proteins that control its structure and accessibility to regulatory factors. The most important

(a)

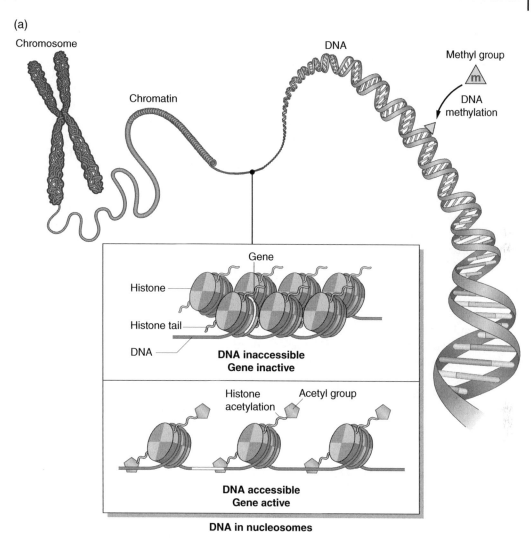

DNA in nucleosomes

(b)

Inheritance of DNA methylation

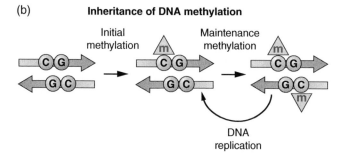

Figure 9.3 (a) Chromatin, showing opening of the structure by histone acetylation. (New drawing based on those by National Institutes of Health.) (b) How DNA methylation patterns are inherited through DNA replication and cell division. (Slack, J.M.W. (2013) Essential Developmental Biology, 3rd edn. Reproduced with the permission of John Wiley and Sons.)

chromosomal proteins are the histones, which are basic proteins highly conserved across all eukaryotic organisms, but there are also many other non-histone chromosomal proteins with critical roles. Most of the DNA is incorporated into structures called nucleosomes, each containing eight histone molecules. The accessibility of the DNA to transcription factors is obviously important for their action and to facilitate it the nucleosomes are in a state of dynamic equilibrium, being continually assembled and disassembled. Opening of the chromatin structure to make the DNA accessible is promoted by the acetylation of lysine groups in the histones, which reduces the number of positive charges displayed and hence weakens the association of the histones with the negatively charged phosphate groups of the DNA. Processes of gene upregulation usually involve the action of histone acetylases to open up the chromatin region concerned. Conversely, the methylation of DNA recruits histone deacetylases which reduce histone acetylation and hence accessibility of the gene. This is why DNA methylation is often associated with repression of gene activity.

DNA methylation is found at the 5-position of cytosine residues in GC sequences. The CG dimer is generally underrepresented in the mammalian genome but higher levels are found in about 40,000 "CG islands" which are generally undermethylated and are often found in the vicinity of gene promoters. Methylation of CG cytosines is achieved by DNA methyl transferase enzymes of which there are several types. Some of them are de novo methylases that can insert methyl groups on a previously unmodified CG. Others are maintenance methylases, that methylate the CG which is paired in the double-stranded DNA molecule with an already methylated CG. DNA methylation offers a very simple mechanism for the inheritance of states of cellular commitment through DNA replication and cell division. This is because when a methylated site is replicated, it becomes a hemimethylated site and is thereby a substrate for the maintenance

methylase (Figure 9.3b). So long as maintenance methylase is present, the hemimethylated site becomes fully methylated, with a methyl group on the C of the CGs on both DNA strands. Through this mechanism a methylated CG will remain methylated however many times the cell divides. In the absence of maintenance methylase, passive DNA demethylation will occur to the extent of 50% in each replication cycle. In addition to the process of methylation maintenance, some specific de novo methylation and demethylation events also occur, as with the setting of the sex-specific imprints in developing germ cells (see Chapter 5).

Another class of chromatin modification important in the control of gene expression is the methylation of histone molecules. This occurs on the N-terminal "tails" of the histones exposed on the surface of nucleosomes. Lysines or arginines can be methylated and the modifications are associated either with boosting gene activity (e.g. histone 3 lysine 4 trimethylation), or inhibiting it (e.g. histone 3 lysine 9 dimethylation). The effects of histone methylation are exerted by the recruitment of other proteins that stimulate or inhibit transcription.

There is evidence that the states of histone acetylation and methylation can also be propagated through cell division, although the mechanism is not so simple as that of DNA methylation. Histones remain associated with the daughter DNA strands after replication and it may be that pre-existing modifications can be repeated on the newly inserted histones in the presence of the relevant enzymes.

Because the state of the chromatin can be maintained through DNA replication and cell division, both in terms of DNA methylation and probably also in terms of histone acetylation and methylation patterns, we can regard the developmental history of cells in the adult as being recorded in their chromatin. Erasure of such chromatin states may be achieved by the methods for inducing pluripotency, namely somatic cell nuclear transfer into an oocyte, or overexpression of specific transcription factors to generate iPS cells. The low efficiency of these procedures show that complete reprogramming is difficult, and, in fact, despite their pluripotent behavior, many iPS cell lines still preserve partial chromatin signatures of their original parent cells and show a preferential tendency to re-differentiate in this direction.

At a macroscopic level, active regions of chromatin are called euchromatin and inactive regions heterochromatin. Heterochromatin is tightly condensed, making access to DNA difficult for transcription factors. The regions around the centromeres and telomeres of chromosomes are heterochromatic, as is the entire X-chromosome, which becomes inactivated in female cells as discussed in Chapter 5.

Lateral Inhibition

One of the most important generic processes of developmental biology is lateral inhibition (Figure 9.4a). This is important in all cases where one type of cell, A, differentiates from a background of another type, B, to generate a scattered population of A embedded in a sheet of B. Lateral inhibition occurs during

neuronal development, the formation of secretory cells in the intestine, the formation of endocrine cells in the pancreas, and many other examples. The starting situation is a sheet of cells that are all the same. Because of the small number of molecules of certain types present in a cell, not least the genes themselves, there are always small random fluctuations between apparently identical cells in terms of the levels of gene activity or the amounts of particular gene products present. Imagine that one substance, conventionally called the "activator", stimulates its own production, and also stimulates the production of another substance called the "inhibitor". The inhibitor in turn inhibits activator production. Imagine further that the inhibitor is freely diffusible between cells while the activator is confined to the cell of production. A sight random increase in activator level in one cell will lead to an excess production of both activator and inhibitor in that cell. But the inhibitor level soon falls by diffusion, leading to a situation where the activator exceeds the inhibitor in that cell while, because of its diffusion, the inhibitor exceeds the activator in the surrounding cells. This means that the production of

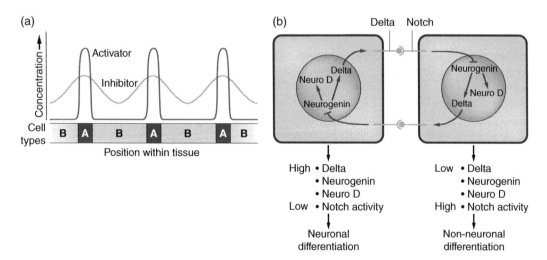

Figure 9.4 (a) The principle of lateral inhibition. The activator promotes synthesis of itself and of the inhibitor. The inhibitor inhibits production of the activator, and is diffusible. (b) Lateral inhibition via the Notch system. (Slack, J.M.W. (2013) *Essential Developmental Biology*, 3rd edn. Reproduced with the permission of John Wiley and Sons.)

activator and inhibitor will further increase in the original cell, while the production of activator (and consequently inhibitor) is suppressed in the surrounding cells. After a while the situation will reach a steady state where the original cell has high activator and inhibitor and is exporting inhibitor to the surroundings. The surrounding cells have low activator and although they produce little inhibitor themselves, the inhibitor continues to be supplied by diffusion. It only remains to suppose that the activator can regulate specific gene activity to explain the formation of one specific cell type, type A, embedded in a background of another, type B.

In the 1990s it was found that this model did operate to control the differentiation of neurons from the embryonic neuroepithelium. The molecular basis for lateral inhibition is formed by the Notch signaling system. As described in Chapter 7, Notch is a cell surface receptor. Its ligands are also cell surface molecules, called Delta and Jagged. When Notch binds to its ligand, it becomes cleaved by an intramembranous protease to generate a free intracellular domain (NICD). This then combines with transcription factors of the CSL (=Su(H)) class, enters the nucleus, and regulates specific genes. The Notch system resembles the theoretical model except that the components are not freely diffusible but are tethered to the cell surface, so the range of the system is limited by the range of cell contacts. The activator is a transcription factor, in the case of neurogenesis it is NeuroD, a bHLH factor promoting neuronal differentiation. The formation of NeuroD (the activator), and Delta or Jagged (the inhibitor), is repressed by Notch signaling. So the way the system works is that a slight random increase in NeuroD leads to more Delta, this stimulates Notch on adjacent cells and represses formation of NeuroD and Delta in those cells. Eventually scattered cells will express high levels of NeuroD and Delta and the cells in between will express little or none but experience high levels of intracellular Notch signaling (Figure 9.4b). The high NeuroD cells will then become neurons while the others remain as undifferentiated neuroepithelium.

Evidence for this system was first found in *Drosophila* where the loss of function mutation of *Notch* gave 100% neuron formation from the ventral neurogenic region of the embryo. Experiments on *Xenopus* later showed that primary neurogenesis from the neural plate was stimulated by inhibition of Notch and reduced by overexpression of Notch ligands. It has also been found in mice that neurogenesis from the telencephalon is reduced by stimulation of Notch signaling, and the proportion of undifferentiated subventricular cells is increased.

Subsequently it has been found that the formation of secretory cells in the intestine and of endocrine cells in the pancreas occurs by a very similar mechanism. The key transcription factors are ATOH1 (= MATH1) in the intestine and Neurogenin 3 in the pancreas. Some aspects of the Notch system and its role in lateral inhibition remain uncertain. One issue is whether the initial perturbations are really random, or depend on the intrinsic oscillatory character of the system as displayed during somitogenesis (see Chapter 8). Another issue is the range of the Notch ligands, which may be increased through the ramification of fine cellular processes to reach non-neighboring cells.

Asymmetrical Cell Division

Stem cells do not have to divide in an asymmetrical manner but they often do so. The best understood example of asymmetric cell division is not of a stem cell but of the early stages of the nematode worm *Caenorhabditis elegans*. This organism is one of a small number of "model organisms" commonly used in developmental biology. It is characterized by an invariant sequence of unequal cell divisions in the early stages of development. The mechanism underlying the process was discovered by isolating maternal effect mutations which abolished the asymmetry. The mutations were known as "partitioning defective", and the genes as *par* genes.

The gene products act as two complexes. They are initially uniformly distributed but following fertilization they sort themselves to the anterior and posterior parts of the egg respectively (Figure 9.C.1). The anterior complex is formed from PAR-3 and -6, which are PDZ type proteins, and PKC-3 (= aPKC), an atypical protein kinase C. The posterior complex consists of PAR-1, a kinase, and PAR-2, a RING finger type protein. The complexes diffuse freely in the cell cortex, the region immediately underlying the plasma membrane. However they are antagonistic to each other. PKC-3 phosphorylates PAR-1 and -2, causing them to dissociate from the cortex and thereby reduces their presence in the anterior. Likewise, PAR-1 phosphorylates PAR-3, causing it to dissociate from the cortex and reducing its presence in the posterior. Evidence for the mutual antagonism is found in the behavior of mutants: when either complex is reduced in activity the other expands its domain.

The establishment of polarity in the zygote depends on the position of sperm entry. Initially the PAR-3/6/PKC-3 complex is present all over the cortex while the PAR-1/2 complex is present in the cytoplasm. The sperm introduces a centrosome which establishes a radiating set of microtubules and also initiates a flow of actomyosin away from the point of sperm entry. Both these events help establish the polarization. PAR-2 binds to the microtubules which inhibits its phosphorylation by PKC-3 and enables it to reach the nearby cortex, recruit PAR-1, and eject PAR-3 by phosphorylation. Thus the point of sperm entry becomes the posterior and the opposite side becomes the anterior of the fertilized egg.

How does this cell polarity affect subsequent differentiation? It is because the PAR complexes are associated with determinants and control their localization with respect to the cell division planes (Figure 9.C.1). A determinant is any intracellular substance that can affect the subsequent pathway of differentiation. In early embryos determinants are often maternal mRNAs encoding key transcription factors, but they may also be proteins or miRNAs that can affect gene expression in some less direct way. There are many determinants in *C. elegans* embryos, and as an example consider MEX-5. This is an RNA binding protein which regulates translation of specific mRNAs. Its distribution depends on phosphorylation by PAR-1, and on dephosphorylation by a uniformly distributed phosphatase, PP2A. MEX-5 mobility is increased by phosphorylation and since it is phosphorylated by PAR-1 in the posterior it accumulates in the anterior where it is less mobile. The action of MEX-5 is to repel various components which then end up in the posterior. These include PIE-1, which represses transcription in the early germline; and P granules, which are droplets containing many proteins and mRNAs involved with germ cell formation. The end result is a zygote which contains determinants for somatic development in the anterior and for germline development in the posterior, and when cleavage occurs into an anterior AB cell and a posterior P_1 cell, the two different lineages are established (Figure 9.C.1f).

Does the lesson of *C. elegans* tell us anything about mammalian stem cells? In principle it shows that localization can arise from a symmetry breaking process following a minor asymmetric perturbation, in this case sperm entry. In terms of actual molecular components there is some evidence for the involvement of PAR homologs in asymmetric cell divisions of vertebrates. For example PAR-3 is expressed in the radial glia (neural stem cells) of the zebrafish embryo forebrain. It is localized in an apical (ventricular) position and sequesters an E3 ubiquitin ligase called Mindbomb (Figure 9.5). Mindbomb increases the activity of Delta-like ligands stimulating the Notch pathway. On cell division, the apical daughter cell displays more Delta than the basal one, and so stimulates more Notch signaling in the basal daughter. This causes the basal daughter to persist as a self-renewing stem cell while the apical daughter becomes a neuron. The evidence of the mechanism is that clones of cells without PAR-3 mostly

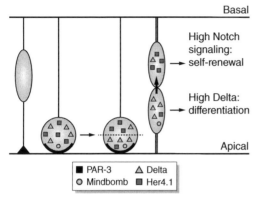

Basal

High Notch
signaling:
→ self-renewal

High Delta:
→ differentiation

Apical

■ PAR-3   △ Delta
◎ Mindbomb ■ Her4.1

Figure 9.5 Involvement of the PAR system in the unequal division of radial glia cell in the developing zebrafish brain. PAR-3 sequesters Mindbomb on the apical side and this generates an asymmetry of Notch signaling between apical and basal daughters. (Modified from Dong, Z., Yang, N., Yeo, S.-Y., Chitnis, A. and Guo, S. (2012) Intralineage directional notch signaling regulates self-renewal and differentiation of asymmetrically dividing radial glia. Neuron 74, 65–78. Reproduced with the permission of Elsevier.)

divide symmetrically, and that loss of function mutants of *mindbomb* have an early excess of neuron formation. This example nicely brings together the effects of the Notch system leading to lateral inhibition and of the PAR system leading to cellular polarization and unequal division.

Neurogenesis and Gliogenesis

Neurons and Glia

Neurons have a cell body, a dendritic tree and an axon (Figure 9.6). They are electrically excitable and typically connect with many other neurons via their axon, which makes synapses with the dendrites or cell bodies of other neurons. The cell body, or soma, is responsible for most of the protein synthesis, with the products being moved up and down the axon by anterograde and retrograde transport mechanisms. Cell bodies are rich in ribosomes and in rough endoplasmic reticulum (called Nissl bodies in neurons). Neurons

contain a special sort of intermediate filament called neurofilaments, composed of class IV intermediate filament proteins. The axons may be myelinated, which increases the conduction rate of action potentials, or unmyelinated. Myelin consists of multiple layers of plasma membrane wrapped around the axon by an oligodendrocyte (in the CNS) or a Schwann cells (in the PNS). The principal types of neuron are the motor neurons, which have a long unbranched axon, the sensory neurons, which have a branched axon, and the interneurons, which have a short axon. However each of these comprise many sub-types, both in terms of morphology, the neurotransmitter deployed, and their gene expression. Neurons typically maintain a membrane potential of about 70–80 mV (negative inside). This arises from the action of ion pumps moving Na, K and Cl ions, together with unequal back-diffusion through ion channels. Excitatory signals received from other neurons via synapses lead to a drop in the magnitude of the membrane potential and, through opening of voltage sensitive ion channels, this initiates a major depolarization or reverse polarization of the cell which propagates along the axon. This is called an action potential and may excite other neurons that are contacted by synapses. There are also inhibitory signals which lead to an increase in membrane potential and make the neuron more refractory to initiation of an action potential. Neurons normally release just one neurotransmitter at their synapses, and the following substances are commonly found as neurotransmitters in the CNS: glutamate (excitatory), γ-amino butyric acid (GABA, inhibitory), acetyl choline (ACh), dopamine and serotonin (5-hydroxytryptamine). The firing patterns of different neuron types are very diverse.

A few neuron types of particular interest in stem cell biology are as follows. Spinal motorneurons are found in the ventral horns of the spinal cord innervating muscles and glands. They secrete acetyl choline at their synapses. α-motorneurons connect to normal muscle fibers and β and

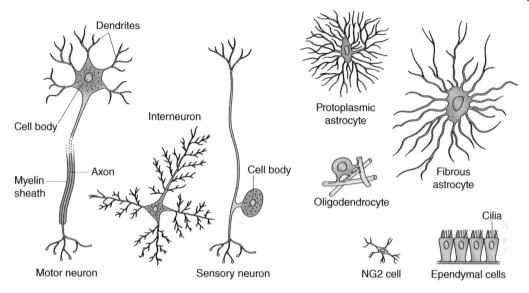

Figure 9.6 Cell types of the central nervous system, neurons are shown on the left and glia on the right. Microglia are not shown as they are not really glia but immune cells of hematopoietic origin. (Slack, J.M.W. (2013) Essential Developmental Biology, 3rd edn. Reproduced with the permission of John Wiley and Sons.)

γ-motorneurons to the intrafusal stretch sensitive fibers. Sometimes the cortical neurons connecting to the spinal motorneurons are also classed as motorneurons. These are pyramidal cells in the motor cortex. Both types of motorneuron, cortical and spinal, may be lost in amyotrophic lateral sclerosis which is a progressive and fatal disorder. Dopaminergic neurons make up 3–5% of the substantia nigra (Latin for "black substance") and project to the basal ganglia and cortex. These cells may have a dark color due to neuromelanin, a by-product of dopamine synthesis, which gives the substantia nigra its name. They express the transcription factors LMX1b, PITX3 and NURR1, which control genes required for dopamine synthesis. Dopaminergic neurons are lost in Parkinson's disease, and their replacement is an important goal of stem cell research. Medium spiny neurons are small cells making up a high proportion of neurons in the striatum. They have a large dendritic tree and contain receptors for glutamate and aspartate. They are GABAergic and innervate cells in the thalamus and elsewhere. These are the first cells to be lost in Huntingdon's disease.

About 50% of cells in the human brain are not neurons but glia (Figure 9.6). Astrocytes provide structural support, control water and ion balance, and can modulate signaling between neurons. There are two main types. The fibrous astrocytes are star shaped with long unbranched processes and mostly located in the white matter. They contain many glial filaments made up of glial fibrillary acidic protein (GFAP), and express aldehyde dehydrogenase 1L1 (ALDH1L1) among other markers. Fibrous astrocytes can proliferate on injury and form glial scars in the CNS. Protoplasmic astrocytes are of mossy appearance with many short and highly branched processes, and are mostly found in gray matter. They express the complex carbohydrate antigen A2B5. In rodents, the astrocytes are generated in late fetal development and grow their processes in postnatal weeks 1–4.

Oligodendrocytes are the cells which produce myelin and invest the neuronal axons in myelin sheaths. They express myelin basic protein (MBP), myelin gene regulatory factor (MRF) and proteolipid protein 1. There is some plasticity of myelination in adult life, requiring continuing production and renewal of oligodendrocytes from oligodendrocyte precursor cells.

A fourth type of glial cell are the ependymal cells which line the ventricles. They have cilia and microvilli on the luminal side and secrete cerebrospinal fluid. Certain other types of cell which are called glia are not really glia at all. The radial glia found in the embryo are actually precursor cells of neurons and mature glia. The microglia of the CNS are macrophage-like immune cells derived from the hematopoietic system in early development which are permanently resident in the CNS.

Neurogenesis

Because the generation of neurons from pluripotent stem cells is a very important practical goal, it is worth considering what happens normally during mammalian neurogenesis (Figure 9.7a). The early neuroepithelium is a simple columnar epithelium. In terms of ultrastructure the side which later abuts the brain ventricles (the ventricular surface) is apical while that which abuts the exterior (the pial surface) is basal. As it is conventional to show the ventricular surface at the bottom of a figure, this means that neuroepithelia are normally depicted upside down relative to other epithelia where the apical side is shown at the top.

In lower vertebrates, there is substantial neurogenesis from the open neural plate, but this is minimal in mammals. In the mouse the main periods of neurogenesis are E9–12 in the spinal cord, E11–16 in fore- and midbrain, and E12-postnatal in the hindbrain and retina. Similar to lower vertebrates, experimental procedures that activate the Notch signaling pathway tend to suppress neurogenesis and prolong the period of multiplication of undifferentiated cells. After neural tube closure the cells undergo a process of "interkinetic migration", which involves movement of the nucleus from the basal (pial) side where they reside during G1 and S phases of the cell cycle, to the apical (ventricular) side for the G2 and M phases (Figure 9.7b). This gives the appearance of stratification to the epithelium although for a while the cells remain attached at both ends. Shortly after this the so-called radial glial cells arise, which continue to span the distance from ventricular to pial surface (Figure 9.7c). Although these cells express some glial markers, such as GLAST (Glu-Asp transporter) and GFAP (glial fibrillar acidic protein) they are not really glia, but are progenitor cells for neurons, astrocytes, oligodendrocytes and ependymal cells. They are polarized, with a concentration of prominin (= CD133), which is a cell surface glycoprotein; nestin, which is an intermediate filament protein; and brain lipid-binding protein (BLBP), at the ventricular end. The proteins Numb and Numblike are also present in an apical position and are retained when the radial glial cell buds off new neurons. The fate of radial glia in parts of the CNS can be observed by Cre labeling of reporter mice using adenoviral delivery of Cre to the pial surface. Because only the radial glia contact the pial surface they can be infected and transport the virus to the nucleus where recombination and labeling occurs. Such experiments show that the radial glia are precursors for neurons, intermediate progenitors, adult neural stem cells, astrocytes, oligodendrocytes and ependymal cells.

The neurons that bud off the radial glial cells arise near the ventricular surface and migrate up the radial glia until they reach their definitive position. This means that the formerly single layered neuroepithelium now acquires three layers. In the ventricular zone is the region of neuronal generation. Near the pial surface is a marginal zone devoid of cell nuclei, and in the center is a mantle zone composed of neurons migrating from the ventricular zone. The spinal cord and the brainstem maintain this three-layered appearance, but in some parts of the brain it becomes more complex. In the telencephalon a population of intermediate or "basal" progenitors appears. These are not really basal (i.e. pial) but they form a subventricular zone separated from the ventricular surface.

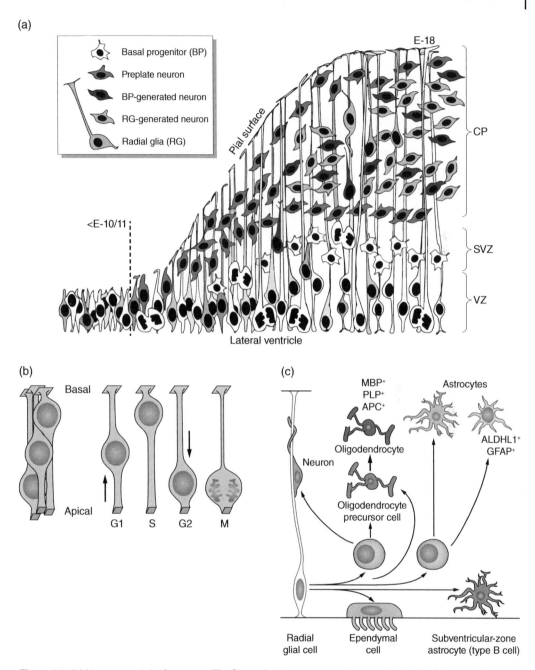

Figure 9.7 (a) Neurogenesis in the mouse. The figure depicts neurogenesis occurring in the future cerebral cortex of a mouse embryo between E10, when the neuroepithelium is just one cell thick, to E18 when the cortical plate is formed. BP = basal progenitor; RG = radial glia; CP = cortical plate; SVG = subventricular zone; VZ = ventricular zone. (b) Interkinetic migration, showing movements of nuclei and change of cell shape through the cell cycle. (c) Products of division of radial glia. Type B cells are considered to be adult neural stem cells. ALDH1L1: aldehyde dehydrogenase 1 family member L1; APC, adenomatous polyposis coli; GFAP, glial fibrillary acidic protein; MBP, myelin basic protein; PLP, proteolipid protein 1. (*Sources:* (a) Malatesta et al., 2007. Reproduced with permission from Springer. (b) From Miyata, T., Okamoto, M., Shinoda, T. and Kawaguchi, A. (2015) Interkinetic nuclear migration generates and opposes ventricular-zone crowding: insight into tissue mechanics. Frontiers in Cellular Neuroscience. 8, 1–11. (c) Modified from Rowitch, D.H. and Kriegstein, A.R. (2010) Developmental genetics of vertebrate glial-cell specification. Nature 468, 214–222. Reproduced with the permission of Nature Publishing Group.)

They express the transcription factor TBR2 (= Eomesodermin) and also Deltalike1 which maintains Notch signaling in the parent radial glia cells, helping to maintain their undifferentiated state. In the cerebral cortex the intermediate progenitors generate neurons which migrate through the mantle zone to form further layers on its pial side. This builds up the cortical plate which eventually consists of six layers of neurons. Additional cells (GABAergic interneurons) enter the cortical plate by lateral migration from the ganglionic eminences at the base of the cerebral hemispheres. In the cerebellum a different process occurs whereby cells migrating from the ventricular surface set up a new generative zone near the pial surface.

The developing human CNS also contains radial glia. They differ slightly from those of the mouse, for example, GFAP is expressed from the time of their first formation at 5–6 weeks gestation. Neurogenesis in the cerebral cortex occurs from then until about 20 weeks gestation, after which glia are formed. Evidence that human radial glia can form neurons has been obtained by sorting them from telencephalon of human fetuses, then culturing them in vitro, and showing that they can produce neuronal, glial or mixed clones. Although in vitro behavior may not be the same as in vivo, this is at least consistent with a similar multipotency to that shown by the mouse radial glia.

Gliogenesis

Radial glia not only generate neurons during development but also glial cells. In each part of the CNS there is a process which controls the switch from neurogenesis to gliogenesis. This is complex, but one important element is the presence of cardiotrophin, secreted by newly formed cortical neurons. This is a LIF-like factor (see Chapter 6 for LIF). Experimental application of cardiotrophin increases the production of glia, and knockout of the receptors gives rise to a glial deficiency. In the neurogenic phase the transcription factors NGN1 and -2 and MASH1

suppress glial genes, while in the gliogenic phase the transcription factor NF1 (nuclear factor 1) suppresses neurogenic genes. A few days after birth in mice the radial glia disappear, mostly differentiating to ependymal cells that line the ventricles, and also to some astrocytes and oligodendrocyte precursors. Cell labeling studies show that glial clones tend to be either astrocytes or oligodendrocytes, but not both.

It should be borne in mind that the control of the cell types formed in different parts of the CNS depends in large measure on the regional specification of those parts, as briefly described in Chapter 8. The intimate association between regional specification and cell differentiation is very clearly shown in the ventral part of the spinal cord where different zones are established in response to the gradient of SHH originating from the floor plate and notochord (Figure 9.8). The p3 domain is defined by expression of NKX2.2 and produces V3 interneurons and oligodendrocytes. Dorsal to this the pMN

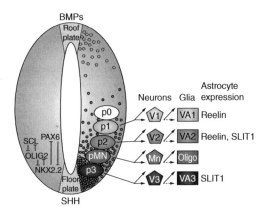

Figure 9.8 Effect of regional specification in the ventral spinal cord on the types of cell differentiation. On the left are shown transcription factors whose expression is induced by the gradient of SHH, together with their mutually inhibitory interactions. On the right are shown five resulting zones of cell differentiation. Mn = motorneurons; Oligo = oligodendrocytes. (From: Rowitch, D.H. and Kriegstein, A.R. (2010) Developmental genetics of vertebrate glial-cell specification. *Nature* 468, 214–222. Reproduced with the permission of Nature Publishing Group.)

domain is defined by expression of OLIG2 and produces motorneurons and oligodendrocytes. There is a mutual inhibition of expression between NKX2.2 and OLIG2 that sharpens the boundary between these domains. Loss of either factor causes the domain boundary to shift and the cell types produced to change accordingly.

Postnatal Cell Division

Once neurons have differentiated then they remain permanently post-mitotic. If a pulse of BrdU is given to an embryo then those neurons formed shortly after the label has been incorporated into DNA will retain it for life because it is not diluted by further replication and division. By contrast, those that have already differentiated are not labeled because they do not make DNA, and those that continue to divide will dilute out the label so it becomes no longer visible after a few cell cycles. The time of the last S-phase prior to neuronal differentiation is known as the "birthday" of the neuron and much work has been done cataloging the birthdays of different neuronal types in different parts of the CNS. Both neurons and glia may migrate considerable distances after their initial formation and the building of the final neuroanatomy depends on the assembly of

various cell layers and the formation of axonal connections between different regions.

Mature oligodendrocytes are also post-mitotic, but there is a population of glial precursor cells, sometimes misleadingly called oligodendrocyte precursor cells (OPCs), which continue to proliferate throughout life. They initially derive from the ventral part of the CNS but later from the dorsal part. They express the cell surface chondroitin sulfate proteoglycan NG2, the receptor PDGFRα, and the transcription factor SOX10. Unlike mature oligodendrocytes, they have some neuron-like properties, being able to respond to glutamate by depolarization and generation of an action potential. Mature astrocytes in the CNS are normally quiescent but can be stimulated to divide by injury.

Adult Neurogenesis

Although most neurogenesis is completed before birth, there is a small amount of new neuron formation throughout life, located in specific parts of the brain. This has been detected by labeling with DNA precursors to detect persistent cell division and also by cell lineage labeling. In the mouse, persistent neurogenesis occurs in the subventricular zone of the lateral ventricles and in the dentate gyrus of the hippocampus (Figure 9.9). The subventricular cells become olfactory

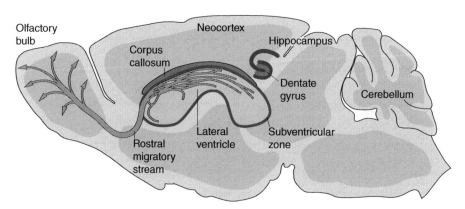

Figure 9.9 Location of adult neurogenesis in the mouse brain. Stem cells are found in the subventricular zone of the lateral ventricles and in the dentate gyrus of the hippocampus. (Zhao et al., 2008. Reproduced with the permission of Elsevier.)

neurons and migrate rostrally into the olfactory bulb. The dentate gyrus cells become glutaminergic granular neurons. Cell lineage labeling experiments show that the adult neural stem cells arise from the radial glia of the embryo. In mice, the experimental inhibition of adult neurogenesis causes a reduction of odor discrimination and of pattern separation ability. These are functions of the brain regions showing adult neurogenesis and indicates that it is of functional importance.

In the subventricular zone the neural stem cells are elongated, extending one process to a blood vessel and another, bearing a cilium, to the ventricular surface (Figure 9.10a). This potentially enables the cells to respond to conditions both in the blood and the cerebrospinal fluid. At least in the mouse, they generate transit amplifying cells, which in

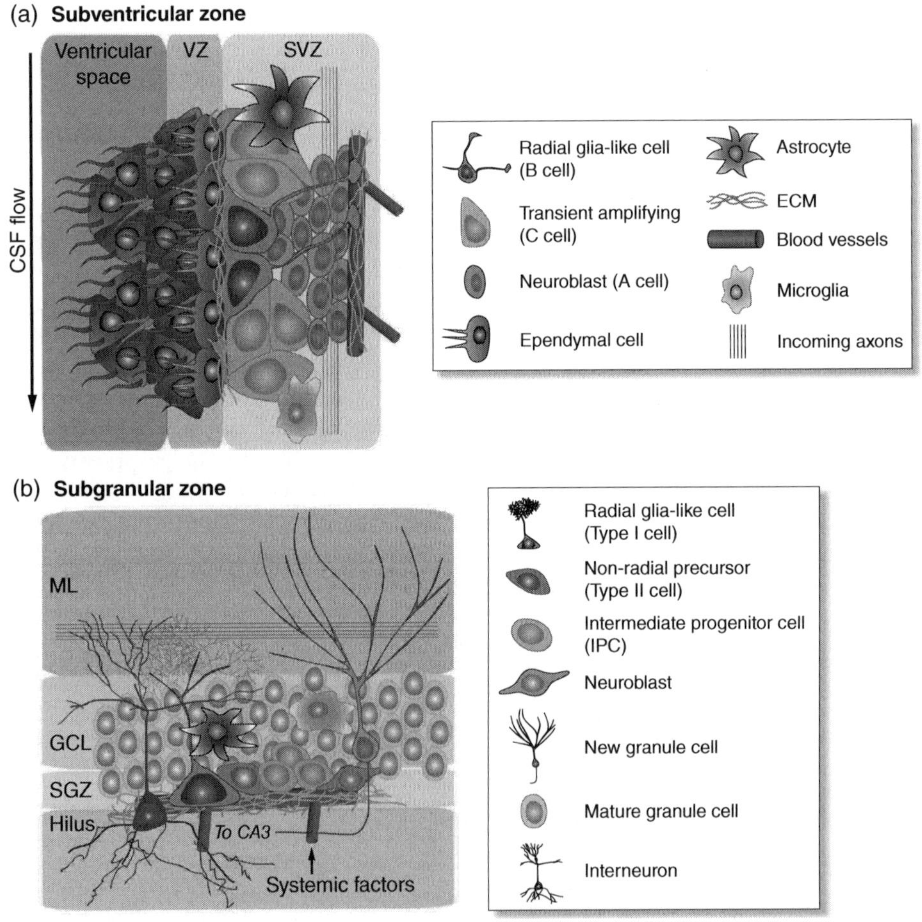

(a) Subventricular zone

(b) Subgranular zone

Figure 9.10 Stem cell niches in the adult mouse brain. (a) Subventricular zone. The B cells are considered to be stem cells, with processes contacting both the cerebrospinal fluid and the blood vessels. CSF = cerebrospinal fluid; VZ = ventricular zone; SVZ = subventricular zone. (b) Dentate gyrus of the hippocampus. The Type 1 cells are considered to be stem cells. ML = molecular layer; CGL = granule cell layer; SGZ = subgranular zone; CA3 = cornu ammonis 3 (another region of the hippocampus). (From: Bond, Allison, M., Ming, G.-l. and Song, H. (2015) Adult mammalian neural stem cells and neurogenesis: five decades later. Cell Stem Cell 17, 385–395. Reproduced with the permission of Elsevier.)

turn generate neuroblasts. These migrate to the olfactory bulbs where they migrate radially and eventually differentiate into olfactory interneurons.

The arrangement of the stem cell niche in the hippocampus is different. Here the stem cells are radial glia-like in shape and lie between the inner granule layer and the hilus (Figure 9.10b). They give rise to intermediate progenitors which become neuroblasts. These migrate tangentially along the subgranular zone, then become immature neurons and move radially into the granule layer, differentiating as granule neurons.

In humans there is also ongoing adult neurogenesis in the same regions of the brain, but the situation does differ from that in the mouse. In humans few if any new olfactory neurons are produced. Instead most of the new neurons produced in the subventricular zone end up in the striatum. There may possibly also be some endogenous neurogenesis in the striatum. Estimation of neuronal turnover using the ^{14}C dilution method (see Chapter 2) indicates that in a population of striatal interneurons there is about 2.7% cell replacement per year. In the dentate gyrus of the hippocampus a higher proportion of cells is subject to turnover than is found in mice, the rate according to ^{14}C dilution being about 1.75% per year.

Neurospheres

There is an interesting in vitro model for neuronal stem cells in the form of the neurosphere culture system. This is established from fetal or postnatal explants of CNS which are dispersed and cultured in a medium containing EGF and FGF. Neurospheres are cell clumps of about 0.3 mm diameter which grow and divide and can be cultured indefinitely in suspension. If dissociated into single cells, a few percent of these will reform neurospheres, thus indicating stem cell behavior. When neurospheres are plated onto a laminin surface in the presence of serum they differentiate to form both neurons and glia. Neurospheres may be cultivated from regions of the CNS that are known not to contain stem cells functioning in the intact organism. This again illustrates the fact that stem cell behavior depends on the cell environment and that specific cells in vitro may behave differently from the same cells in vivo.

Skeletal and Cardiac Muscle

Skeletal Muscle

An anatomical muscle has a hierarchical structure (Figure 9.11). The basic cellular unit is the myofiber, which is a highly elongated multinucleated cell surrounded by a

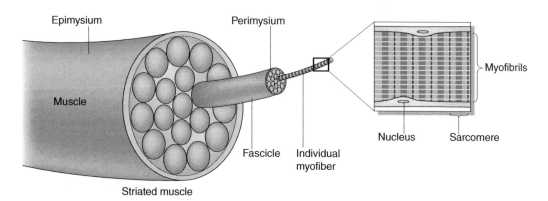

Figure 9.11 Skeletal muscle. The basic units are the myofibers. These are bundled in fascicles which are grouped into a whole muscle. (Slack, J.M.W. (2013) Essential Developmental Biology, 3rd edn. Reproduced with the permission of John Wiley and Sons.)

basal lamina. Myofibers are surrounded by delicate connective tissue called the endomysium. Several myofibers are packed into a bundle (fascicle) surrounded by perimysium, and many fascicles make up one anatomical muscle surrounded by a connective tissue sheath, the epimysium. Muscles are supplied with a vascular system by capillaries investing individual myofibers, and with a nervous system comprising motor axons innervating myofibers and sensory axons innervating the stretch-sensitive muscle spindles, made up of modified myofibers called intrafusal fibers.

The contractile structures found within each myofiber are called myofibrils and each myofiber contains many myofibrils arranged in register to give the characteristic striated appearance. Myofibrils are about 1 μm diameter and consist of repeating units of 2.2 μm called sarcomeres. The ends of the sarcomeres are called Z-discs, and to these are attached thin filaments composed of actin. Thick filaments composed of myosin interdigitate with the thin filaments. Each myosin molecule consists of a dimer of two heavy chains and two pairs of dissimilar light chains. There is a globular head region projecting from the thick filament that makes contact with the adjacent thin filament. Myosin is a motor protein and when supplied with ATP will migrate along the actin filament, driven by cyclic movements of the head region. The myofibrils are invested by a modified endoplasmic reticulum called the sarcoplasmic reticulum and also by a network of T-tubules formed by invagination of the plasma membrane, or sarcolemma. They are aligned with pairs of terminal cisternae of the sarcoplasmic reticulum to form triads. Flanking each complex of myofibril, sarcoplasmic reticulum and T-tubules are mitochondria. Their numbers vary depending on the myofiber type. Slow, type I fibers have an abundance of mitochondria and the oxygen storage protein myoglobin, and function mostly via oxidative metabolism. They tend to be found where recurrent or continuous movement is required. The predominant myosin heavy chain type is MHCIβ. Fast fibers have fewer mitochondria and little myoglobin but abundant glycogen. They function mostly using glycolytic metabolism to generate ATP and are found where short bursts of rapid contraction are required. In rodents fast fibers are classified as IIA, IID and IIB, with predominant myosin heavy chains MHCIIa, d and b respectively. In humans there is no MHCIIb.

Skeletal muscle contraction is initiated by receipt of action potentials from motor axons at specialized synapses between axon and myofiber called neuromuscular junctions. These are rich in acetyl choline receptors. Release of acetyl choline stimulates the receptors and causes depolarization of the myofiber, initiating a muscle action potential. Like the nerve action potential this is short in duration, about 2–3 msec. Because of the structure of T tubules the action potential propagates immediately throughout the whole myofiber. This causes opening of L type Ca channels in the T tubules. In the triad regions these are linked to Ca channels in the sarcoplasmic reticulum (ryanodine receptors), and when these open large amounts of stored Ca^{++} enters the cytoplasm. At rest, the myosin motor is inhibited because of obstruction of interaction with actin by the protein tropomyosin. The Ca^{++} binds to troponin, which is linked to tropomyosin and detaches it from the myosin. This permits the myosin to contact the actin and, so long as enough ATP is present, to drive the contraction.

Development of Skeletal Muscle

Most skeletal, or striated, muscle develops from the dermomyotome region of the somites. As described in Chapter 8, this buds off myoblasts which express myogenic transcription factors such as MYOD and MYF5, and are committed to become skeletal muscle. These continue to proliferate for a while and may also migrate, for example the myoblasts that form the abdominal muscles migrate to the ventral side of the body, while those that form the limb muscles migrate into the limb buds to form dorsal and ventral muscle masses. The myoblasts are attracted into the limb buds by hepatocyte growth factor (HGF = Scatter factor) secreted by the limb mesenchyme.

Much work on mammalian myogenesis has utilized the cell line C2C12. At least in these cells continued proliferation requires FGF, and FGF signaling maintains expression of the transcription factor MSX1, which inhibits differentiation. On withdrawal of FGF, cell division ceases and the cells become competent to fuse. Fusion requires various components including the integrin types β1, α3 and α9, the cell adhesion molecule M-cadherin, and the protease ADAM12 (Figure 9.12). The actual fusion process involves the actin cytoskeleton and essential components include various GTP exchange proteins and small GTPases.

In the development of the mouse embryo there are two phases of myoblast fusion. Primary myoblast fusion occurs around E11 generating mostly slow fibers, while secondary fusion occurs around E14.5–17.5 generating mostly fast fibers. In embryonic life the myosin heavy chains MHC-emb and MHC-neo are present, becoming replaced by the mature forms after birth. Initially cells of the epaxial and hypaxial myotome express PAX3, although lineage labeling shows that PAX3 positive cells can become capillary endothelium as well as muscle. If these cells are ablated then primary myogenesis fails. PAX7 is necessary for the formation of muscle satellite cells (see below) and thus for postnatal myogenesis, but it is dispensable for secondary myogenesis, presumably because of the overlapping role of PAX3. In the postnatal growth period of the mouse there is continued myoblast fusion to preexisting fibers, increasing their length and girth. In mouse the myofiber number does not increase after P7. But the volume of myofibers does continues to increase to adult size (Figure 9.C.2), and this size may be further augmented by exercise.

In humans there are three phases of myoblast fusion, which occur between about the 7th and 23rd week of gestation. New myofiber formation and myoblast fusion with existing fibers is finished by about 5 months gestation.

Muscle Satellite Cells

Once nuclei have entered a multinucleate fiber, they become post-mitotic. So the cells of striated muscle are post-mitotic, but striated muscle considered as a tissue is still capable of regeneration from muscle satellite cells. These are small cells found within the basal lamina of myofibers (Figure 9.13). They are characterized by expression of the transcription factor PAX7 and are responsible for postnatal muscle regeneration. Satellite cells become visible from about E16.5, as basal laminas form around the fibers. In mouse, satellite cells are dividing and their progeny fusing with myofibers until about P21, after which the remaining satellite cells (about 5 per fiber) become quiescent. Slow fibers have more satellite cells than fast fibers. In adult life satellite cells are normally quiescent and express PAX7, but not the myogenic transcription factors. Following toxic or traumatic damage to their fiber, the satellite cells become activated. They divide, upregulate expression of myogenic transcription factors, and start to migrate. They can fuse with each

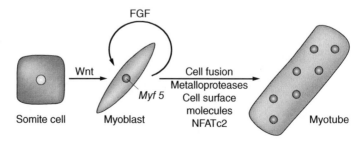

Figure 9.12 Myoblast fusion in the embryo. (Slack, J.M.W. (2013) *Essential Developmental Biology*, 3rd edn. Reproduced with the permission of John Wiley and Sons.)

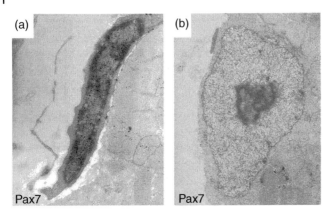

(a) Pax7 (b) Pax7

Figure 9.13 Nucleus of muscle satellite cell (a) and myofiber nucleus (b) viewed by transmission electron microscopy. PAX7 is immunostained with gold beads and is only present in the satellite cell nucleus. The specimen is *Xenopus laevis* tadpole tail muscle. (Author's photos.)

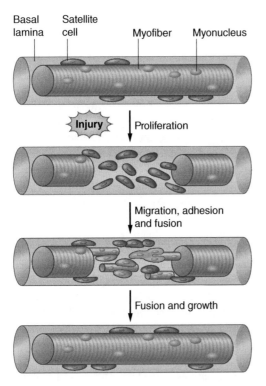

Basal lamina / Satellite cell / Myofiber / Myonucleus

Injury ↓ Proliferation

↓ Migration, adhesion and fusion

↓ Fusion and growth

Figure 9.14 Regeneration of adult mouse muscle over about 2 weeks from injury.

other to form small myotubes, which can fuse with each other, and with the ends of the original myofiber, to form a new fiber (Figure 9.14). Regenerated fibers can be identified because they have centrally positioned nuclei when viewed in section, rather than the peripherally positioned nuclei of mature fibers.

Continued Notch activity is required for the maintenance of the quiescent state of muscle satellite cells. It is possible to transplant satellite cells between animals, and if the host is defective in muscle regeneration, for example the *mdx* mouse, which lacks dystrophin and is a partial model for human muscular dystrophy, then the host muscle can be repopulated from the graft.

In mice, the transcription factor PAX7 is not needed for muscle development so long as PAX3 is present, but it is needed for satellite cell function in the young mouse. This is demonstrated by the fact that the *Pax7* knockout mouse is seriously compromised in postnatal muscle growth and regeneration. However, PAX7 is no longer needed after about P21 when the cell recruitment to myofibers has finished. This is shown by ablation of *Pax7* at different times using an inducible knockout system. At the same time, PAX7 does remain expressed permanently in satellite cells. This makes it possible to ablate the cells using controlled expression of diphtheria toxin. The mouse used is *Pax7-CreER/R26R-DTA*. This means that Cre-ER is driven by the *Pax7* promoter. When activated by tamoxifen the CreER will activate expression of *DTA*, encoding diphtheria toxin A, and this will kill the cells in which it is produced. If the cells are destroyed at any stage of life, then muscle regeneration does not occur. So the satellite cells are needed even if the *Pax7* gene itself is not.

Muscle satellite cells are often described as "facultative stem cells" because they persist for the lifetime of the animal and they can generate both myoblasts and further satellite cells on division, but do not feed continuous ongoing renewal of skeletal muscle. However, the satellite cell numbers do decline with age to about 25% of the starting numbers (mouse), so the regenerative ability in this case is less than perfect.

Cardiac Muscle

The functional cells of the heart are cardiac muscle cells, or cardiomyocytes (Figure 9.C.3). They are responsible for the lifelong rhythmic contraction without which death would be rapid. The heart as an organ contains many cells other than cardiomyocytes, being about 50% connective tissue as well as containing an essential vascular and nervous supply. The cardiomyocytes themselves are surrounded by a delicate connective tissue and have a very abundant blood supply.

The contractile apparatus of cardiomyocytes is similar to that of skeletal myofibers. The sarcomeres form a branching three dimensional network through the cytoplasm and there are numerous mitochondria. Cells have just one or two nuclei. They are joined to each other by intercalated discs which are rich in gap junctions and transmit action potentials between cells, enabling them to behave as a functional syncytium (Figure 9.15). Cardiomyocyte action potentials are much longer in duration than those of skeletal muscle, about 200 msec compared to 2 msec.

Although the process of excitation coupling via T-tubules and sarcoplasmic reticulum is similar to skeletal muscle, a number of the key proteins are derived from different genes. MHCα (human gene *MYH6*) is found in all parts of the heart during development and mostly in the atria in the mature heart. MHCβ (human gene *MYH7*), which is also found in skeletal muscle, predominates in the ventricles of the mature heart. There is a specific cardiac subunit of troponin, cTnT, and a specific myosin light chain MLC2v (human gene *MYL2*) is expressed in the cardiac crescent in early development and then restricted to the ventricles. ALC2 (human gene *MYL7*) is found in the mature atria. The actual channel

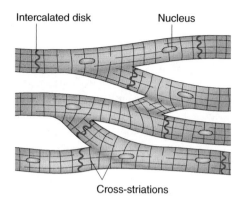

Figure 9.15 Organization of cardiac muscle. Cells may be mono or binucleate. They are joined by intercalated discs to form branched networks. (Slack, J.M.W. (2013) *Essential Developmental Biology*, 3rd edn. Reproduced with the permission of John Wiley and Sons.)

subunit of the L type Ca channel in the heart is CaV1.2 (human gene *CACNAIC*), which is not found in skeletal muscle.

Cell division among cardiomyocytes does extend after birth, for about one week in mice and 2 months in humans. At least in the mouse there is also a short burst of cell division due to thyroid hormone at about P14. But after this further growth is mostly by cell enlargement which is substantial, comprising about a 30–40 fold increase in volume. Such a large increase in size for a cell normally requires more than a diploid complement of DNA to fuel the necessary transcription and protein production. In mice this is achieved by most cardiomyocytes becoming binucleate during postnatal day 5–10. In humans there are some binucleate cells but most cells achieve the increase of DNA content by the single nucleus becoming polyploid, i.e. replicating the chromosomes without nuclear division, to 4n, 8n or even higher degrees of ploidy. The issue of whether there is any cell division after the postnatal period has been very controversial. The extensive polyploidy in humans, and the existence of DNA repair synthesis following damage, both mean that it is possible to observe DNA synthesis without actual cell multiplication. However the balance of evidence from cell labeling and ^{14}C dilution studies is in favor of a very small degree of cell multiplication throughout life. In the human heart perhaps half of the cardiomyocytes are replaced by cell division during the course of a whole lifetime. The persistence of cardiac progenitor cells after the embryonic period has also been very controversial. Some examples, such as cells expressing the *c-kit* gene, are probably derived from the blood and are not cardiac cells at all. But calculations from cell numbers in different sized hearts in humans, and from cell lineage labeling following induced infarction in mice, suggest that there may be a limited contribution to heart muscle from non-cardiomyocytes.

In summary, the postnatal growth behavior of skeletal and cardiac muscle is very different. Skeletal muscle fibers are completely post-mitotic but can regenerate from satellite cells. Cardiomyocytes are almost entirely post-mitotic, showing a very low level of proliferation and little or no regeneration from committed progenitors.

Endodermal Tissues

Cell Differentiation in the Pancreas

The pancreas (Figure 9.16) consists mostly of exocrine cells which secrete digestive enzymes into the intestine. The secretions are collected by a system of ducts, also considered to be exocrine cells. In addition, 1–2% of the organ consists of endocrine cells, mostly grouped into the Islets of Langerhans. These comprise α-cells (secreting glucagon), β-cells (secreting insulin), δ-cells (secreting somatostatin), PP cells (secreting pancreatic polypeptide) and ε-cells (secreting ghrelin). Like the exocrine cells they arise from the endoderm of the pancreatic bud. This may be shown by Cre labeling experiments in which the Cre is driven by a promoter active in the endodermal component of the bud, such as that for the *Pdx1* gene, and both exocrine and endocrine cells become labeled.

Exocrine differentiation depends on the transcription factor PTF1, which has three subunits of which one, P48, is pancreas-specific. The *p48* gene is expressed in the whole pancreatic bud in early development but is downregulated in developing endocrine cells. Commitment to ductal differentiation seems to occur just in the narrow time window E8.5–10.5, based on lineage labeling with *Pdx1–CreER*, in which recombination is initiated at different stages by doses of tamoxifen. The transcription factor SOX9 is important for the postnatal ductal phenotype but in early pancreatic bud development SOX9 is more widely expressed and it maintains expression of FGF receptors needed for response to the growth promoting factor FGF10 secreted by the pancreatic bud mesenchyme.

An early population of endocrine cells appears in the mouse pancreas from about E10.5 which produce more than one hormone and do not respond to glucose. Cell lineage labeling indicates that these early-appearing endocrine cells are not the progenitors of the definitive endocrine population. Definitive endocrine cell commitment depends on the transcription factor Neurogenin3 (NGN3). This plays a similar role in a lateral inhibition system to that played by NeuroD in neurogenesis (Figure 9.17a). Notch signaling suppresses formation of NGN3, while NGN3 promotes synthesis of Delta. Starting with tiny random fluctuations of NGN3 levels, the operation of the lateral inhibition system brings about a population of isolated cells expressing NGN3 and Delta, surrounded by other cells in which Notch signaling is active and synthesis of NGN3 and Delta is low. The NGN3-positive cells become endocrine cells while the NGN3-negative cells become exocrine (Figure 9.17b). Overexpression of *Ngn3*, or knockout of components of the Notch pathway, can drive a high proportion of pancreatic cells to become endocrine, while knockout of *Ngn3* suppresses endocrine cell formation. In vivo the maximal expression of *Ngn3* is during the period E13.5–15.5 and the main wave of endocrine cell differentiation occurs a little later.

It is still not entirely clear what causes the diversification of the endocrine precursors into the five types of mature endocrine cell.

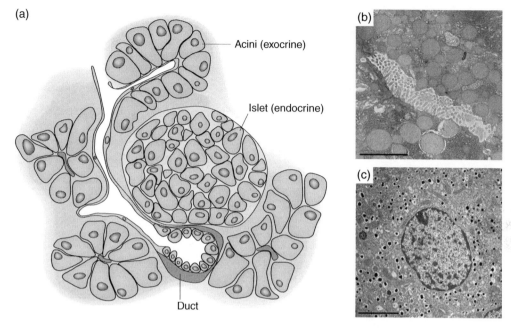

Figure 9.16 The mature pancreas. (a) Diagram of structure, showing exocrine acini and ducts and an endocrine islet. (b) Transmission EM picture of mouse exocrine cells, showing large secretory granules, abundant endoplasmic reticulum, and a luminal space into which project many microvilli. Scale bar 2 μm. (c) Transmission EM picture of a mouse beta cell, showing dense core granules containing insulin. Scale bar 4 μm. (*Sources:* (a): Slack, J.M.W. (2013). Essential Developmental Biology, 3rd edn. Reproduced with the permission of John Wiley and Sons. (b): Behrendorff, N., Shukla, A., Schewiening, E. and Thorn, P. (2009) Local dynamic changes in confined extracellular environments within organs. Proceedings of the Australian Physiological Society 40, 55–61. Reproduced with the permission of John Wiley and Sons. (c): Author's photo.)

However various transcription factors have been identified as essential. Formation of β-cells require the genes *Nkx6.1, Nkx2.2, Pax4, Pax6* and *MafB*. In addition proper maturation requires *MafA* and the re-expression of *Pdx1*. Formation of α-cells requires *Arx* and *Nkx2.2*. δ-cells require *Pax4* but the proportion of δ-cells also increases in the absence of *Arx*. Interestingly, the loss of *Nkx2.2, Pax4* or *Pax6* causes an increase in the proportion of ε-cells. Whether the choice between these various factors arises spontaneously or as the result of some local signaling process is not yet known.

In the normal pancreas the endocrine cells, of which the majority are β-cells, coalesce into islets of Langerhans. During growth of the young animal both exocrine and endocrine cells increase in number as the organ grows, but lineage labeling experiments indicate that this occurs by division of differentiated cells rather than from pancreatic stem cells. In adult life, normal cell turnover is very slow, although the exocrine tissue can regenerate rapidly following toxic damage, and the islets do increase in size during pregnancy. Labeling of the pancreas with DNA precursors indicates that cell turnover is by scattered divisions of differentiated cells, not fed from a stem cell population.

Production of functional β-cells is a major goal of stem cell biology. Although immunostaining for insulin is often used to identify β-cells, the presence of insulin not a sufficient criterion. A mature β-cell has a very high content of insulin, in the range 1–5% of the total cell protein. The insulin is found in secretory granules with a characteristic "dense core" appearance in the transmission electron microscope (Figure 9.16c). The secretion of insulin occurs in response to external glucose concentration. This

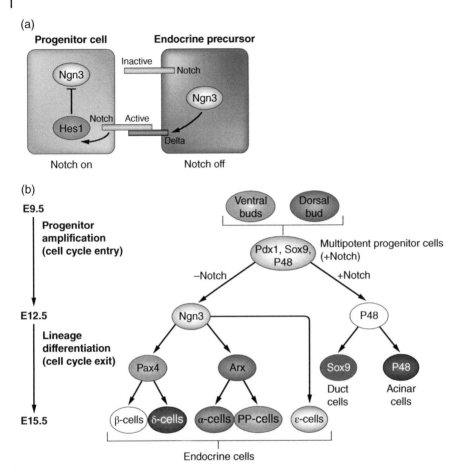

Figure 9.17 Lateral inhibition in the developing pancreas generates the endocrine precursors. (a) Cells with slightly more NGN3 cause more Notch signaling in adjacent cells, with repression of *Ngn3* transcription in these cells. High NGN3 cells become endocrine cells. (b) Putative cell lineage of pancreatic progenitors, showing some of the key transcription factors active at each stage. (From: Li, X.Y., Zhai, W.J. and Teng, C.B. (2015) Notch signaling in pancreatic development. International Journal of Molecular Science, E48. doi: 10.3390/ijms17010048.)

enters the β-cell with the help of the transporter GLUT2, it is then phosphorylated by glucokinase and enters the glycolytic pathway, raising the intracellular level of ATP. The ATP blocks potassium channels, retaining K^+ ions in the cell, and thus decreasing the magnitude of the (negative inside) membrane potential. This opens voltage-sensitive Ca^{++} channels, and the rise of intracellular Ca^{++} causes fusion of the secretory granules with the plasma membrane and the release of insulin. All of this system must be present and working correctly for a cell to be regarded as a β-cell. Immature β-cells, which are the type usually generated following directed

differentiation of pluripotent stem cells, have a low insulin content and do not show full glucose sensitivity. Such cells are therefore not suitable for clinical transplantation.

Cell Differentiation in the Intestine

The small intestinal epithelium is lined with columnar absorptive cells. In addition there are several cell types described as "secretory", comprising goblet, Paneth, enteroendocrine and tuft cells. All the cell types are formed from the intestinal stem cells near the base of the crypts, which will be considered in more detail in the next chapter. The formation of

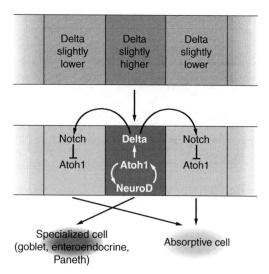

Figure 9.18 Differentiation of the secretory cell types in the intestinal epithelium controlled by lateral inhibition. (Modified from Slack, J.M.W. (2013) Essential Developmental Biology, 3rd edn. Reproduced with the permission of John Wiley and Sons.)

the secretory population is once again controlled by a lateral inhibition system based on Notch signaling. This time the key transcription factor is ATOH1 (= MATH1), a bHLH type factor, whose expression commits the cells to secretory differentiation. ATOH1 promotes the synthesis of Delta, which stimulates Notch in the surrounding cells. Notch-ICD de-represses the transcription of *Hes1*, and HES1 inhibits the transcription of *Atoh1* and thereby of *Delta*. So a small random increase in ATOH1 in one cell will stimulate Notch signaling in adjacent cells, driving down their production of Delta. The original cell will have less Notch signaling and therefore the level of ATOH1 will increase. Eventually this results in a scattered population of high ATOH1 cells (the future secretory cells) surrounded by a larger population of low ATOH1 cells (the future absorptive cells) (Figure 9.18).

Cell Differentiation in the Liver

The liver is a large organ with many vital functions. It is a major center of intermediary metabolism, including the synthesis of cholesterol, urea and glycogen. It is a source of serum proteins, especially of albumin; of hormones including insulin-like growth factor 1 (IGF1), angiotensin and thrombopoietin; and it secretes bile into the intestine. It also detoxifies drugs and toxins using the cytochrome P450 system. All these functions are exerted by hepatocytes, which are the principal cell type. These are epithelial cells whose basolateral surfaces contact the blood and whose apical surfaces contact the bile canaliculi. A histological section of liver shows a rather uniform appearance, but the organ does have a precise substructure composed of plates of hepatocytes arranged in lobules (Figure 9.19). Lobules are roughly hexagonal in shape with a central vein in the middle. At each apex is a portal triad, composed of branches of the hepatic artery, bringing oxygenated blood from the lungs; the hepatic portal vein, bringing nutrient-loaded blood from the intestine; and a bile duct. In between the portal triads and the central vein run blood sinusoids, lined by specialized sinusoidal endothelium. The hepatocytes are arranged along these sinusoids, and because the endothelia have gaps (fenestrations), the hepatocytes are in direct contact with the blood. Other cell types found in the sinusoids are the Kupffer cells, which are a type of macrophage, and hepatic stellate cells, which store vitamin A and are involved in fibrosis. The small bile canaliculi abut the apical surfaces of the hepatocytes but near the portal triad they become lined with biliary epithelial cells, also called cholangiocytes. The biliary system comprises a series of ducts, starting with the small ducts of individual portal triads, which fuse to form larger and larger ducts until they coalesce into the hepatic duct. This leads to the intestine. On the way there is a side branch, the cystic duct, leading to the gall bladder, which is a bile storage organ, and, more distally, fusion with the pancreatic duct to form the common bile duct entering the duodenum. The hepatic duct, cystic duct, gall bladder and common duct are known as the extrahepatic biliary system and originate from a distinct endodermal bud adjacent to the hepatic and ventral pancreatic buds (see Chapter 8).

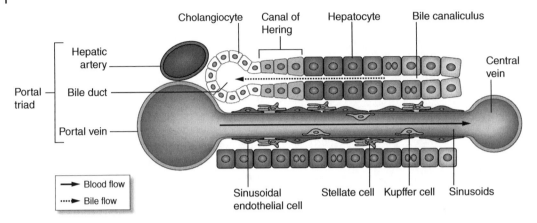

Figure 9.19 Structure of the liver lobule. Hepatocytes are arranged along blood sinusoids running between the portal vein and the hepatic vein. The basal sides of hepatocytes face the sinusoid and the bile canaliculi are formed from apposed apical surfaces. (From: Miyajima, A., Tanaka, M. and Itoh, T. (2014) Stem/progenitor cells in liver development, homeostasis, regeneration, and reprogramming. Cell Stem Cell 14, 561–574. Reproduced with the permission of Elsevier.)

Hepatocytes and Cholangiocytes

The liver bud in the embryo is composed of hepatoblasts, bipotential cells capable of forming both hepatocytes and biliary epithelial cells. Around E15.5 in the mouse some of these congregate around the branches of the future portal vein to form ductal plates, which express the transcription factor SOX9 (Figure 9.20). Clonal labeling using *Sox9-Cre* mice show that the ductal plate cells can form both bile duct cells and hepatocytes. In vitro experiments indicate that TGFβ signaling is needed for the formation of the ductal plates, and knockout studies show that Notch signaling is needed for the formation of cholangiocytes, the signal in this case being Jagged2 from periportal mesenchyme. The transcription factors HNF6(= OC1) and OC2 are needed for biliary and hepatocyte development. In the absence of both, the heptoblasts differentiate as mixed phenotype cells. The function of HNF6 and OC2 is to cause production of TGFβ inhibitors that enable the formation of a gradient of TGFβ activity across the ductal plates such that the high level induces differentiation of cholangiocytes on the portal vein side while the low level allows differentiation of hepatocytes on the parenchymal side.

The Kupffer cells are derived from the hematopoietic system, while the stellate cells come from the mesothelium, which is the mesoderm-derived layer lining body cavities. This is shown by labeling with *Mesp1-Cre*, expressed in the mesothelium. The liver is the main center for hematopoiesis at fetal stages. The hematopoietic cells secrete the cytokine oncostatin M, which may help to promote hepatocyte differentiation. Full differentiation of the hepatocytes is not achieved until after birth, especially in terms of the cytochrome P450 detoxification systems. By this stage the hepatocyte differentiated state depends on a small group of transcription factors which activate each other's expression in a cross-catalytic web: C/EBPα, HNF1β, FOXA2, HNF4α1, HNF6 and LRH1 (Nr5a2).

The hepatocytes of the liver are not all identical but become organized in a pattern of metabolic zonation along the portal triad-central vein axis of each lobule. The cells near the portal triad are specialized for gluconeogenesis and for urea synthesis, thereby exhibiting high levels of glucose-6-phosphatase and carbamoyl phosphatase. The cells near the central vein are specialized for glycolysis and for glutamine

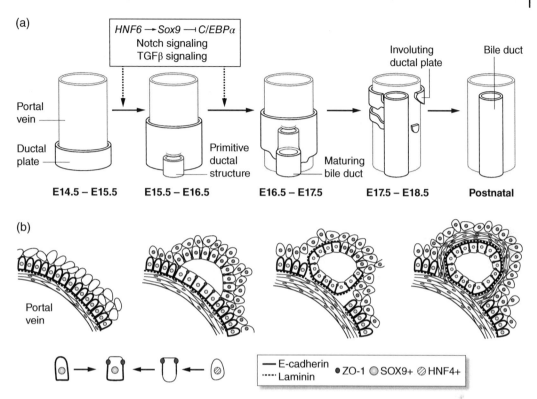

Figure 9.20 Origin of bile ducts during liver development in the mouse. (a) The ductal plates arise as a single layer of hepatoblasts around the portal vein and then bud off tubes that become bile ducts. SOX9 promotes ductal development while C/EBPα and β promotes hepatocyte development. (b) Schematic sections through the ductal plate. E-cadherin is a cell adhesion molecule; laminin is found in basement membranes; ZO-1 is found in tight junctions; HNF4 is one of the hepatocyte transcription factors. (Modified from Antoniou, A., Raynaud, P., Cordi, S., Zong, Y., et al. (2009) Intrahepatic bile ducts develop according to a new mode of tubulogenesis regulated by the transcription factor SOX9. Gastroenterology 136, 2325–2333. Reproduced with the permission of Elsevier.)

synthesis, showing high levels of glucokinase and of glutamine synthase (Figure 9.C.4). This pattern becomes established in late embryogenesis and depends on the Wnt system. There is a gradient of Wnt signaling across the lobule with the high level at the perivenous end and the low level at the periportal end. Upregulation of Wnt signaling by liver specific ablation of the inhibitor APC, using *TTR-CreER × floxed APC*, causes perivenous markers to be expressed all over the lobule. Conversely, liver specific inhibition of the Wnt system, by adenoviral expression of the Wnt inhibitor DKK1, causes periportal markers to be expressed across the lobule.

Liver Growth and Regeneration

The liver continues to grow after birth by division of all of its cell types. Most attention has not surprisingly been devoted to the hepatocytes. Saturation labeling with [3]HTdR following by long term observation of the subsequent label dilution indicates that in the mouse hepatocytes divide on average about every 100 days, and cholangiocytes about every 50 days. Probably because of the need to maintain a high protein synthesis capability, hepatocytes are, like cardiomyocytes, prone to become binucleate, and also to become polyploid. In the mouse about 60% of cells are polyploid by 3 months and 80% in the adult. About 50–60% cells are binucleate, some having two diploid nuclei and others having two tetraploid nuclei. In humans the level of polyploidy is 20–45%, and of binuclearity about 30%. Polyploid cells can divide but the daughters are usually aneuploid and incapable of further division, so long term cell renewal inevitably depends on the minority of diploid cells. Proposals for the existence of a hepatic stem cell population living in the portal triad and feeding hepatocytes across the lobule towards the central vein have not been supported by subsequent investigations. There is no concentration of cell division at any position across the lobule, and the shapes of clones visualized by retroviral labeling indicate a random rather than an oriented pattern of clonal expansion.

Although the liver shows very slow normal cell turnover and does not have stem cells, it is capable of vigorous regeneration following damage. Surgical removal of 1–2 liver lobes is followed by a rapid growth of the remaining lobes to restore the missing volume. Classical parabiosis experiments, in which the circulatory systems of two animals were experimentally joined together, showed that the signal for regeneration was present in the blood. In other words an animal with an intact liver starts to regenerate if it shares the circulation and blood supply of another animal with a partially resected liver. The nature of this signal is still unclear but probably includes hepatocyte growth factor (HGF), bile acids, tumor necrosis factor (TNF) and interleukin 6 (IL6) from the damaged liver. Hepatocyte division is elevated within 24 hours, along a portal to venous gradient, and is followed by division of biliary and endothelial cells. HGF is essential for regeneration and it is derived from stellate and endothelial cells within the regenerating liver. The original volume is restored in about 1 week in the rat (2 weeks in human). Initially the size of each liver lobule is increased, but this adjusts back to normal over several weeks, presumably by cell migration and some respecification of the metabolic zonation pattern. The reason for the cessation of regeneration when organ volume is restored is not entirely understood but one component is probably TGFβ from stellate cells, which has an inhibitory effect on hepatocyte proliferation.

Many studies in which hepatocytes are labeled before regeneration show that the new hepatocytes are derived from pre-existing hepatocytes and not from stem cells. In fact the overall growth capability of hepatocytes is very high. As many as five successive partial hepatectomies carried out on a rat are followed by complete restoration of volume. This involves an expansion of the residual liver tissue by a factor of 18. Even more impressive is the behavior of hepatocytes transplanted as cells. There is a very useful mouse transplantation model afforded by mice deficient in the enzyme fumaryl acetoacetate hydrolase (FAH). This enzyme is part of the pathway for tyrosine catabolism. Mice lacking the *FAH* gene die because of liver failure caused by the accumulation of the toxic fumaryl acetoacetate. They can be saved by treatment with a drug called NTBC (or nitisinone) which inhibits a previous step in the pathway leading to accumulation of a less toxic intermediate. Transplantation of normal hepatocytes into an immunodeficient *FAH⁻* mouse, followed by withdrawal of NTBC, leads to a rapid colonization of the liver by the graft (Figure 9.C.5). Up to six serial transplants of this type have been carried out with an estimated 86 doublings of the donor hepatocytes.

For many years it has been known that certain toxic and carcinogenic stimuli to the rodent liver can elicit formation of a novel cell population from the portal triad regions called "oval cells". These have some properties of embryonic hepatoblasts such as production of α-fetoprotein. They have been considered as "facultative stem cells" which can regenerate the liver when the division of hepatocytes is blocked, for example after treatment of rats with the toxin 2-acetylaminofluorene. Recent cell lineage labeling with *Sox9-CreER* mice have shown that the oval cells originate from the SOX9-positive small bile ducts. However such studies have also shown that they do not contribute to regenerated hepatocytes, which arise, as in normal regeneration, from pre-existing hepatocytes. Since the cell damage models available in mice and rats are different, the nature of the "oval cells" may also be different, and so this conclusion might not hold for rats or indeed humans. However in mice it seems clear that in all the regeneration models hepatocytes come from pre-existing hepatocytes.

Transdifferentiation and Direct Reprogramming of Cell Type

In the late 1990s there was a spate of reports to the effect that grafts of hematopoietic stem cells from one animal to another could repopulate a variety of tissue types in addition to the blood. These included the brain, muscles, liver, and many epithelial structures. The results gave rise to some curious theories of continuous cell renewal of all parts of the adult body fed by stem cells in the bone marrow. They also probably helped to establish the myth of the miracle stem cell therapy: the cells that could be grafted into a patient which would seek out any area of decay and rebuild it to restore full function. However further investigation showed that the results were incorrect, either being due simply to lodging of grafted cells in various organs, or to fusion of grafted cells with cells of the host. There are a few genuine examples of transdifferentiation in nature but, within the mammalian body, tissue-specific stem cells generally only produce the cell types of their own tissue. Moreover, as is well-known, grafts of pluripotent stem cells into animals do not repopulate damaged tissues, instead they form the tumors called teratomas which will eventually kill the host (see Chapter 6).

Although the transdifferentiation of normal mammalian cells is generally not possible just by altering their environment, it is possible to force cells to change type by introduction of selected transcription factors. These are usually small subsets of the transcription factors known to be important in normal development. For example, neuron-like cells have been generated by introducing the transcription factors BRN2 (= POU3f2), ASCL1 and MYT1L, into fibroblasts. These have a typical neuronal morphology and gene expression pattern. They can generate action potentials and form synapses in vitro. A more sophisticated cocktail, FOXG1, SOX2, ASCL1, DLX5, and LHX6, can convert fibroblasts specifically into GABAergic neurons resembling telencephalic interneurons. These have firing patterns comparable to cortical interneurons, form functional synapses, and release GABA. Cardiomyocyte-like cells have been generated by introducing GATA4, MEF2c, and TBX5, and undergo spontaneous contraction like bona fide cardiomyocytes. Hepatocyte-like cells have been generated by introducing FOXA3, HNF1A, and HNF4A. These cells display functions characteristic of mature hepatocytes, including cytochrome P450 enzyme activity and biliary drug clearance. Upon transplantation into mice with concanavalin-A-induced acute liver failure and fatal metabolic liver disease due to fumarylacetoacetate dehydrolase (FAH) deficiency, they restore the liver function.

Generally it is harder to do direct reprogramming with human than with mouse cells, and it is harder the further apart the starting cells are from the desired cell type in terms of developmental history. For example, cardiomyocyte-like cells are easier to make from cardiac fibroblasts which are derived from the original heart-forming territory of the mesoderm, than from dermal fibroblasts. However, there are also some examples of cross germ layer transformations such as fibroblast to neuron or hepatocyte. In the cocktails of transcription factor used, at least one needs to be a "pioneer factor" that can locate its target sequences in chromosomal DNA even when these are embedded in closed chromatin. Some studies have incorporated Cre-based markers of known progenitor cell types to find whether the process of reprogramming involves going "backwards", and thus becoming a normal progenitor, or "jumping sideways", and missing out any normal progenitors. The results indicate that progenitors are not formed so the process really is a "jump" from one cell type to another.

The essence of direct reprogramming is that the process operates a bistable switch. As a result of the transient action of the introduced transcription factors, a whole ensemble of genes is turned on or off and the new state is stable and self-maintaining in the absence of the original stimulus. So studies of this type need to show that the factors causing the transdifferentiation do indeed act in a transient manner and that the new cell phenotype is stable in their absence. This is usually done by using a doxycycline-inducible vector to express the gene products (see Chapter 3), and to show that the new cell types persist after removal of the doxycycline. In cases where the new cell type is dividing such evidence is convincing, but it is less certain where the new cell type is postmitotic, such as neurons, cardiomyocytes or beta cells. The reason is that proteins can persist for a long time in non-dividing cells so the introduced factors may not be properly removed, and also the differentiation markers used to characterize the new cell type may persist long after their production has stopped.

Detailed studies of gene regulatory networks in transdifferentiated cells have shown that they are never exact copies of the normal cell type. However they may nonetheless be stable, and they may be close enough to the normal phenotype for the purpose intended, for example in vitro studies of drug metabolism, or cell transplantation therapies. One persistent problem, also found with the directed differentiation of cells from pluripotent stem cells discussed below, is the tendency to create immature rather than mature forms of the differentiated cells: for example β-cells that are non-glucose responsive or hepatocytes without full detoxification

functions. Overcoming this problem will require further work.

Differentiation Protocols for Pluripotent Stem Cells

What are the lessons of developmental biology for the design of an effective way to make a specific cell type from human pluripotent stem cells? In principle the answer is simple. The cells should be exposed to a sequence of inducing factors at appropriate concentrations and durations of treatment to mimic the series of signals encountered during normal development from the epiblast to the final differentiated cell type required. The reality is more difficult. Pluripotent stem cells grow as aggregates, so even supplying a pure factor to a dish of cells results in different cells receiving different stimuli, depending on the size of the aggregate, and their position within it. This causes a heterogeneity of response to the applied factors, and means that the next cycle of treatment will be of a diverse cell population. Furthermore, once development is commenced by removing pluripotent cells from their normal growth medium, they display a lot of self-organizing potential. They may readily form signaling and responding regions within the aggregate, and this generates further new cellular territories as a result of local interactions. Because of the inevitable presence of endogenous inducing factors generated by such processes, inhibiting these may be just as important as adding new ones. Small molecule agonists and antagonists may be preferred to the protein inducing factors themselves because of better standardization and better penetration into cell aggregates. The monitoring of progress through a protocol is normally carried out by looking for the expression of key transcription factors, or other markers indicative of specific intermediate states of developmental commitment. Some pluripotent cell lines have been made carrying fluorescent protein reporters of useful markers to assist with protocol development. This makes monitoring very easy because the cells change color with each of the steps for which a reporter is present.

Over the last 10–15 years many new protocols for directed differentiation have been devised and it has become clear that it can be relatively easy to coax the cells into becoming immature or juvenile versions of the cell types required, but much harder to get fully mature cells. This is very important for such targets as cardiomyocytes, or hepatocytes or β-cells, where the immature cells do not display the essential physiological properties that are needed. There has also been considerable difficulty in devising protocols that are robust enough to be repeated successfully in other labs. The existence of many irreproducible protocols has led workers to be very careful to try to ensure robustness when they publish new ones.

Workers in this area have tended not to try to drive differentiation by overexpression of specific transcription factors, and to prefer the use of soluble factors. The reason is a desire not to have to deal with the complexities of introducing new genetic material.

Since transient overexpression can now be achieved using the introduction of RNA or protein, it need not involve any heritable change to the cells, and it is likely that such methods will be used more in future.

If the differentiated cells are intended for clinical transplantation the whole process must be carried out under conditions of "good manufacturing practice" (GMP). This means that high standards of sterility and quality control must be adhered to. No animal products may be used as they could be the source of infectious agents. Very high standards of recording and documentation must be met for all materials and procedures used. Moreover the pluripotent cell line itself should have been prepared under GMP conditions. Compared with the normal mode of activity in an academic lab, GMP is very cumbersome and expensive. However it may be worth thinking about GMP well in advance in order to avoid having to go back several years and repeat a lot of preliminary work under GMP conditions. It is also worth noting that the regulations are somewhat different in the USA and Europe.

Further Reading

Organs, Tissues and Cell Types

Chitnis, A., Henrique, D., Lewis, J., Ishhorowicz, D. and Kintner, C. (1995) Primary neurogenesis in Xenopus embryos regulated by a homolog of the Drosophila neurogenic gene – Delta. Nature 375, 761–766.

Griffin, E.E. (2015) Cytoplasmic localization and asymmetric division in the early embryo of *Caenorhabditis elegans*. Wiley Interdisciplinary Reviews: Developmental Biology 4, 267–282.

Latchman, D.S. (2008) Gene Regulation: A Eukaryotic Perspective, 2nd edn. Springer, New York.

Meinhardt, H. and Gierer, A. (2000) Pattern formation by local self-activation and lateral inhibition. Bioessays 22, 753–760.

Ross, M.H. and Pawlina, W. (2016) Histology: A Text and Atlas: With Correlated Cell and Molecular Biology, 7th edn. Wolters Kluwer Health, Philadelphia.

Young, B., O'Dowd, G. and Woodford, P. (2014) Wheater's Functional Histology: A Text and Colour Atlas, 6th edn. Elsevier, Churchill Livingstone, Philadelphia.

Neurogenesis

Anthony, T.E., Klein, C., Fishell, G. and Heintz, N. (2004) Radial glia serve as neuronal progenitors in all regions of the central nervous system. Neuron 41, 881–890.

Bond, A.M., Ming, G.-l. and Song, H. (2015) Adult mammalian neural stem cells and neurogenesis: five decades later. Cell Stem Cell 17, 385–395.

Goldman, S.A. (2016) Stem and progenitor cell-based therapy of the central nervous system: hopes, hype, and wishful thinking. Cell Stem Cell 18, 174–188.

Götz, M. and Huttner, W.B. (2005) The cell biology of neurogenesis. Nature Reviews. Molecular Cell Biology 6, 777–788.

Hobert, O. (2011) Regulation of terminal differentiation programs in the nervous system. Annual Review of Cell and Developmental Biology 27, 681–696.

Jessberger, S. and Gage, F.H. (2014) Adult neurogenesis: bridging the gap between mice and humans. Trends in Cell Biology 24, 558–563.

Kageyama, R., Ohtsuka, T., Shimojo, H. and Imayoshi, I. (2008) Dynamic Notch signaling in neural progenitor cells and a revised view of lateral inhibition. Nature Neuroscience 11, 1247–1251.

Malatesta, P., Appolloni, I. and Calzolari, F. (2007) Radial glia and neural stem cells. Cell and Tissue Research 331, 165–178.

Rowitch, D.H. and Kriegstein, A.R. (2010) Developmental genetics of vertebrate glial-cell specification. Nature 468, 214–222.

Skeletal and Cardiac Muscle

Abmayr, S.M. and Pavlath, G.K. (2012) Myoblast fusion: lessons from flies and mice. Development 139, 641–656.

Bergmann, O., Bhardwaj, R.D., Bernard, S., Zdunek, S., et al. (2009) Evidence for cardiomyocyte renewal in humans. Science 324, 98–102.

Brack, A.S. and Rando, T.A. (2012) Tissue-specific stem cells: lessons from the skeletal muscle satellite cell. Cell Stem Cell 10, 504–514.

Fan, C.-M., Li, L., Rozo, M.E. and Lepper, C. (2012) Making skeletal muscle from progenitor and stem cells: development versus regeneration. Wiley Interdisciplinary Reviews – Developmental Biology 1, 315–327.

Laflamme, M.A. and Murry, C.E. (2011) Heart regeneration. Nature 473, 326–335.

Musunuru, K., Domian, I.J. and Chien, K.R. (2010) Stem cell models of cardiac development and disease. Annual Review of Cell and Developmental Biology 26, 667–687.

Relaix, F. and Zammit, P.S. (2012) Satellite cells are essential for skeletal muscle regeneration: the cell on the edge returns centre stage. Development 139, 2845–2856.

Rosenthal, N. and Harvey, R.P. (2010) Heart Development and Regeneration. Academic Press, London.

White, R., Bierinx, A.-S., Gnocchi, V. and Zammit, P. (2010) Dynamics of muscle fibre growth during postnatal mouse development. BMC Developmental Biology 10, 21.

Endodermal Organs

Afelik, S., Jensen, J., (2013) Notch signaling in the pancreas: patterning and cell fate specification. Wiley Interdisciplinary Reviews – Developmental Biology. 2, 531–544.

Carpentier, R., Suñer, R.E., van Hul, N., Kopp, J.L., et al. (2011) Embryonic ductal plate cells give rise to cholangiocytes, periportal hepatocytes, and adult liver progenitor cells. Gastroenterology. 141, 1432–1438.

Crosnier, C., Stamataki, D. and Lewis, J. (2006) Organizing cell renewal in the intestine: stem cells, signals and combinatorial control. Nature Reviews. Genetics 7, 349–359.

Mastracci, T.L. and Sussel, L. (2012) The endocrine pancreas: insights into development, differentiation, and diabetes. Wiley Interdisciplinary Reviews – Developmental Biology 1, 609–628.

Miyajima, A., Tanaka, M. and Itoh, T. (2014) Stem/progenitor cells in liver development, homeostasis, regeneration, and reprogramming. Cell Stem Cell 14, 561–574.

Rieck, S., Bankaitis, E.D. and Wright, C.V.E. (2012) Lineage determinants in early endocrine development. Seminars in Cell and Developmental Biology 23, 673–684.

Saisho, Y., Manesso, E., Butler, A.E., Galasso, R., et al. (2011) Ongoing β-cell turnover in adult nonhuman primates is not adaptively increased in streptozotocin-induced diabetes. Diabetes. 60, 848–856.

Si-Tayeb, K., Lemaigre, F.P. and Duncan, S.A. (2010) Organogenesis and development of the liver. Developmental Cell 18, 175–189.

Spence, J.R., Lauf, R. and Shroyer, N.F. (2011) Vertebrate intestinal endoderm development. Developmental Dynamics 240, 501–520.

van der Flier, L.G. and Clevers, H. (2009) Stem cells, self-renewal, and differentiation in the intestinal epithelium. Annual Review of Physiology 71, 241–260.

Cell Type Switching

Huang, P., Zhang, L., Gao, Y., He, Z., et al. (2014) Direct reprogramming of human fibroblasts to functional and expandable hepatocytes. Cell Stem Cell 14, 370–384.

Pang, Z.P., Yang, N., Vierbuchen, T., Ostermeier, A., et al. (2011) Induction of human neuronal cells by defined transcription factors. Nature. 476, 220–223.

Qian, L., Huang, Y., Spencer, C. I., Foley, A., et al. (2012) In vivo reprogramming of murine cardiac fibroblasts into induced cardiomyocytes. Nature 485, 593–598.

Slack, J.M.W. (2007) Metaplasia and transdifferentiation: from pure biology to the clinic. Nature Reviews. Molecular Cell Biology 8, 369–378.

Xu, J., Du, Y. and Deng, H. (2015) direct lineage reprogramming: strategies, mechanisms, and applications. Cell Stem Cell 16, 119–134.

Zhou, Q. and Melton, D.A. (2008) Extreme makeover: converting one cell into another. Cell Stem Cell 3, 382–388.

Directed Differentiation

Bruin, J.E., Rezania, A. and Kieffer, T.J. (2015) Replacing and safeguarding pancreatic β cells for diabetes. Science Translational Medicine 7, 316ps23.

Burridge, P.W., Keller, G., Gold, J.D. and Wu, J.C. (2012) Production of de novo cardiomyocytes: human pluripotent stem cell differentiation and direct reprogramming. Cell Stem Cell 10, 16–28.

Cahan, P., Li, H., Morris, S.A., Lummertz da Rocha, E., Daley, G.Q. and Collins, J.J. (2014) CellNet: network biology applied to stem cell engineering. Cell 158, 903–915.

Goldman, O. and Gouon-Evans, V. (2016) Human pluripotent stem cells: myths and future realities for liver cell therapy. Cell Stem Cell 18, 703–706.

Mummery, C.L., Zhang, J., Ng, E.S., Elliott, D.A., Elefanty, A.G. and Kamp, T.J. (2012) Differentiation of human embryonic stem cells and induced pluripotent stem cells to cardiomyocytes. Circulation Research 111, 344–358.

Murry, C. E., Keller, G. (2008) Differentiation of embryonic stem cells to clinically relevant populations: Lessons from embryonic development. Cell 132, 661–680.

Pagliuca, F.W., Millman, Jeffrey, R., Gürtler, M., et al. (2014) Generation of functional human pancreatic β cells in vitro. Cell 159, 428–439.

10

Stem Cells in the Body

It is now time to consider the nature of the tissue-specific stem cells: namely those stem cells that exist and function throughout life in the animal or human body. Recall the key features of the definition of stem cell behavior propounded in Chapter 1: the stem cell divides to renew itself, it produces differentiated cells, and it persists long term, usually for the lifetime of the organism. This definition means that only tissues undergoing continuous renewal really have stem cells. Those that do not self-renew or renew very slowly, like cardiac muscle or cartilage, do not have stem cells. One could enlarge the definition of a stem cell to include, for example, hepatocytes, which divide to generate further hepatocytes following toxic or surgical damage to the liver; or pancreatic exocrine cells which can rapidly replenish the exocrine population following toxic ablation. However such examples correspond to different types of biological behavior which may reasonably be given different names. They are of course very important and interesting but they are somewhat different from stem cell behavior. The neural stem cells found in the subventricular zone and in the hippocampus are real stem cells, but to avoid disruption to the flow of the text these have already been described in Chapter 9.

The Intestinal Epithelium

One of the best understood stem cell systems is that maintaining the intestinal epithelium. This lines the luminal surface of the gut all the way from the pyloric sphincter of the stomach to the anus. There is a significant morphological difference between the small intestine, responsible for most digestion and absorption of the food, and the large intestine, responsible for absorbing water and accommodating a large population of gut bacteria. The small intestine (Figure 10.1a) is lined with villi which massively expand the surface area for the purpose of absorption. In between the villi are crypts of Lieberkühn in which active cell division takes place (Figure 10.2). Cells differentiate as they leave the crypts, and migrate up the villi, eventually to die by apoptosis and to drop off into the lumen. The flux of cells is considerable and it is estimated that the epithelium, apart from the long lived Paneth cells, is renewed every 4–7 days in the mouse. There are several cell types making up the epithelium. The most numerous are the columnar absorptive cells, or enterocytes. Next most numerous are the goblet cells, containing large vesicles filled with mucins. In the crypt bases lie Paneth cells which contain granules and secrete lysozyme and cryptdins as a defense against infection. In addition there are several types of enteroendocrine cells, secreting hormones such as cholecystokinin, GLP1, PYY and secretin. Although most textbooks only mention these four principal cell types there are also three others: the cup cells, which are a type of absorptive cell; the M cells, which overlie patches of lymphoid tissue; and the tuft cells which bear a tuft of microvilli. Beneath the epithelium lies the

The Science of Stem Cells, First Edition. Jonathan M. W. Slack.
© 2018 John Wiley & Sons, Inc. Published 2018 by John Wiley & Sons, Inc.
Companion website: www.wiley.com/go/slack/thescienceofstemcells

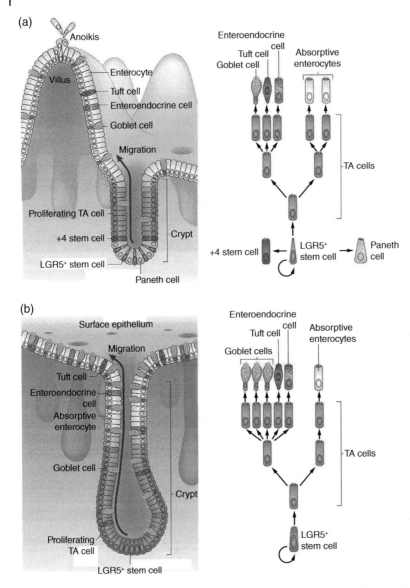

Figure 10.1 Organization of the small and large intestine with putative cell lineage diagrams. (a) Small intestine. (b) Large intestine. TA cells = transit amplifying cells. (From: Barker, N. (2013) Adult intestinal stem cells: critical drivers of epithelial homeostasis and regeneration. Nature Reviews Molecular Cell Biology 15, 19–33. Reproduced with the permission of Nature Publishing Group.)

lamina propria, a stroma containing blood vessels, lymphatics, loose connective tissue and some lymphoid tissue. The core of each villus also contains similar stromal tissues. Outside this lies a thin layer of smooth muscle called the muscularis mucosa. Often all of the tissues lying within the muscularis

mucosa are simply described as "the mucosa". Outside the muscularis mucosa lies more connective tissue, substantial layers of smooth muscle and a peritoneal lining on the surface. The small intestine is divided into the duodenum, jejunum and ileum. They are histologically fairly similar although the

(a)
1h BrdU

(b)
24h BrdU

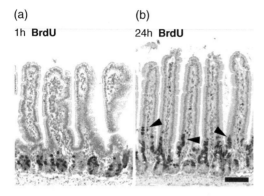

Figure 10.2 Cell division in the intestinal epithelium of a mouse. The mouse has been labeled with an injection of bromodeoxyuridine (BrdU), visualized by immunostaining (dark nuclei). Only epithelial cells in the crypts have incorporated BrdU, plus a few non-epithelial cells in the interior of the villi. (a) 1 hour after BrdU injection. (b) 24 hours after BrdU injection showing migration of labeled cells onto the villi (arrowheads). Scale bar 100 μm. (From: Algül, H., Schneider, M.R., Dahlhoff, M., Horst, D., et al. (2010) A key role for E-cadherin in intestinal homeostasis and paneth cell maturation. Plos One 5, e14325.)

duodenum has Brunner's glands in the submucosa which secrete mucin and bicarbonate, and the ileum contains lymphoid masses called Peyer's patches, associated with M cells in the epithelium. There are also differences in the types of enteroendocrine cells in the three regions.

The large intestine, or colon, is larger in circumference and histologically it differs from the small intestine in that there are no villi and no Paneth cells, the goblet cells are more abundant and the repertoire of enteroendocrine cells is different (Figure 10.1b). As in the small intestine, cell division occurs in the crypts and differentiated cells are exported to the luminal surface. The overall cell renewal rate in the large intestine is somewhat slower than the small intestine.

Intestinal Stem Cells

In many ways the intestinal epithelium currently represents the best understood tissue-specific stem cell system. The stem cells can be identified in vivo, their cell lineage can be visualized, their cell kinetics has been well

studied, they can be cultivated in vitro, and their mechanism of differentiation is fairly well understood.

Proliferation in the intestine is dependent on Wnt signaling. Conditional knockouts of Wnt pathway components will bring cell renewal to a halt, and overexpression of Wnt pathway components is associated with intestinal tumors. A key breakthrough came with the discovery of leucine-rich G protein-coupled receptor 5 (LGR5) as a stem cell marker (Figure 10.C.1a,b). This is a co-receptor for Wnt factors, responding to a Wnt agonist called R-spondin, and is expressed on about 12–14 cells at base of each crypt in mouse, interspersed with the Paneth cells. The *Lgr5* knockout is a neonatal lethal, not because of effects on the intestine but because effects on the tongue prevent suckling. The key evidence that the LGR5-positive cells are stem cells comes from CreER labeling experiments. In these, a transgenic mouse is made in which an *Lgr5* promoter drives a *CreER* gene. It also contains a reporter, such as *lacZ* or *GFP*, whose expression is activated by excision of a transcription stop sequence flanked by *loxP* sites. The cells expressing *Lgr5* continuously produce CreER. If tamoxifen is given to the mice, the CreER becomes active and excises the transcription stop sequence in front of the reporter. Those cells will then express the reporter thereafter, regardless of their state of differentiation (see Chapter 3 for a fuller description of this important method). The results in the intestine are as follows (see Figure 1.5 in Chapter 1; also Figure 10.C.1c). Shortly after the tamoxifen dose, some LGR5-positive cells themselves are labeled. A few days later, from each labeled cell at the crypt base a file of labeled cells may be seen extending up the crypt and up the neighboring villus. These files include cells of all the intestinal types found on the villi: absorptive, goblet and enteroendocrine. After a few days, the files extend right to the villus tips where they terminate, and also some Paneth cells at the crypt base become labeled. If the tamoxifen dose is a low single dose, the recombination is inefficient and only a few well separated stem cells become labeled. If tamoxifen doses are large and repeated then a high proportion of stem cells can be labeled leading to labeling of most of the epithelium. These results reveal several things. First, the stem cells are permanent. Once a clone has become labeled it persists long term. Second, the stem cells lie in the crypt base because this is where labeled cells appear first following the tamoxifen dose. Third, the stem cells can produce all the cell types of the epithelium. Even well-spaced clones contain all the cell types showing that they must arise from single stem cells. It also confirms what was previously deduced from [3]HTdR and BrdU labeling studies, that the cells originate in the crypt base, migrate up to the villus tips and then die. For most of the cell types the turnover time is about 5 days, but, unlike the rest of the epithelium, the Paneth cells have a longer lifetime of 6–8 weeks. In the large intestine the situation appears to be broadly similar. Although there are no actual Paneth cells, there are some CD24-positive cells which may serve the same function of providing a niche for the stem cells. CD24, which is also present on Paneth cells, is a glycosyl phosphatidyl inositol (GPI)-anchored sialoglycoprotein with important functions in the immune system.

The transcription factor that controls stem cell character is ASCL1, a bHLH type factor normally expressed in the stem cells, whose targets include the *Lgr5* gene. Its expression is induced by Wnt signaling and its gene is also auto-activating, so that once it has been turned on, it remains on (in other words it is a bistable switch, see Chapter 7). Conditional knockout of *Ascl1* will destroy the stem cells, although in vivo the rate of cell turnover is so high that the intestine rapidly becomes repopulated from the few surviving wild type crypts. Overexpression of *Ascl1* will increase the number of stem cells and give rise to hyperplasia and to crypt fission.

The situation has been somewhat complicated by the discovery of other stem cell markers which behave in a similar way to *Lgr5* in CreER experiments, but may be present on different cell populations. There has been a persistent suggestion of the existence of some radiation-resistant stem cells around the +4 position in the crypt (i.e. about four cells up from the crypt base) but this remains uncertain. These cells have been described as expressing BMI1, a component of the PRC1 polycomb complex regulating chromatin state; also LRIG1, a membrane receptor, telomerase, which rebuilds telomeres after cell division, and HOPX, an atypical homeoprotein. However it is now thought that all these markers are also expressed in the LGR5-positive crypt base cells, so the clonal labeling experiments not surprisingly give similar results. In addition there is CD133 (= Prominin), a transmembrane glycoprotein beloved of those who work with cancer stem cells, but this is actually expressed in the whole dividing population of the crypt.

Analysis of the effects of radiation damage indicate that a larger population of cells per crypt are capable of regenerating the crypt than normally act as stem cells in the steady state. There are about 30–40 regeneration-competent cells as opposed to 6–8 steady state stem cells estimated from cell kinetic measurements or 12–14 cells positive for LGR5. Indeed it is possible to treat transit amplifying cells with WNT3A to mimic normal niche conditions and thereby cause them to produce organoids in vitro, which is a behavior characteristic of stem cells (see below). Once again this fact emphasizes the importance of the local cell environment in determining whether or not specific cell populations do or do not behave as stem cells.

In the human intestine, the current lack of good antibodies for LGR5 makes it impossible to establish that it also serves as a stem cell marker in humans. However, cells from the crypt base isolated by sorting for the Wnt target Ephrin 2 will generate organoids in Matrigel expressing high levels of LGR5, so the situation is probably similar. A clonal labeling study using the loss of cytochrome c oxidase from mitochondria by somatic mutation (see Chapter 3) shows stem cell derived clones similar in behavior to those shown by LGR5 positive cells in the mouse.

The smallest individual clones occupy about 15% of crypt circumference, indicating that there are about six stem cells/crypt, as in mice.

In Vitro Culture

Tissue-specific stem cells are notoriously difficult to culture in vitro, but this has been achieved for intestinal stem cells in the form of organ cultures. Single LGR5-positive cells cultured in laminin-rich Matrigel in the presence of various factors including the Wnt agonist R-spondin, noggin, EGF and Notch ligand, will grow to become organoids resembling a mini-intestine with villi extending inward and crypt-like regions on the outside (Figure 10.3). In vivo it appears that the Paneth cells provide a niche for the stem cells by providing them with WNT3 and other Wnt factors. Removal of Paneth cells reduces the number of functioning LGR5-positive cells, and inclusion of Paneth cells in the organoid cultures enables them to grow in the absence of exogenous Wnt, although R-spondin is still required in the medium. LGR5-positive cells from the mouse colon will also produce organoids in Matrigel. The normal source of R-spondin in vivo is probably from the underlying stromal tissue and not from the epithelium itself.

Clonality of Intestinal Crypts

It first became possible to examine the clonal structure of animal tissues when mouse aggregation chimeras and X-inactivation mosaics were studied. Aggregation chimeras are embryos formed by aggregating together two morulae of different genetic strains to form a single embryo. The resulting mice consist of a mixture of cells from the two strains. X-inactivation mosaics are females heterozygous for a marker on the X-chromosome (described in Chapter 5). In both cases the whole body consists of two cell populations which can be distinguished using suitable antibodies or

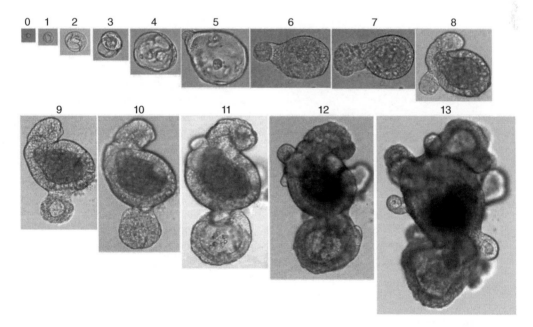

Figure 10.3 Intestinal organoid growing in Matrigel. This colony was founded by a single Lgr5+ stem cell, and growth was recorded each day. The numbers of days of development are shown. (From: Sato, T., Vries, R. G., Snippert, H. J., van de Wetering, M., et al. (2009) Single Lgr5 stem cells build crypt–villus structures in vitro without a mesenchymal niche. Nature **459**, 262–265. Reproduced with the permission of Nature Publishing Group.)

histochemical methods. A striking finding was that intestinal epithelial crypts were almost always of a single genotype, indicating that they were the clonal progeny of a single cell. The same applies to humans, an example from human colon being shown in Figure 10.4. This specimen was derived from a female heterozygous for a null allele of glucose-6-phosphate dehydrogenase, whose gene lies on the X-chromosome. Due to the process of X-inactivation about half the cells in her body do not express the normal allele and appear negative in histochemical staining. In the

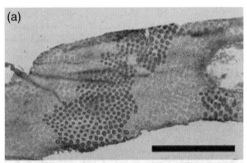

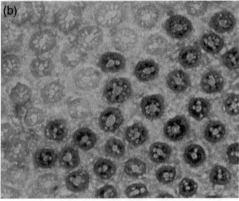

Figure 10.4 X-inactivation clones visualized in the human colon. The specimen is a frozen section from a female heterozygous for the X-linked gene encoding glucose-6-phosphate dehydrogenase, and is stained to reveal presence of the enzyme. (a) Low power, scale bar 2 mm. (b) High power. None of the crypts show mixed clonality. (From: Novelli, M., Cossu, A., Oukrif, D., Quaglia, A., et al. (2003) X-inactivation patch size in human female tissue confounds the assessment of tumor clonality. Proceedings of the National Academy of Sciences of the United States of America 100, 3311–3314. Reproduced with the permission of The National Academy of Sciences.)

specimen the enzyme appears in contiguous patches containing many crypts. All labeled crypts are fully labeled, and there are no partly labeled crypts.

The obvious conclusion from such data was that each crypt contained just a single stem cell. But this was out of line with contemporary cell kinetic studies which indicated 6–8 stem cells/crypt, and even more out of line with recent counts of LGR5-positive cells at the crypt base (12–14 cells). How can such divergent findings be reconciled? The answer lies in the mode of division of the stem cells. If individual stem cells are labeled it is possible to observe directly how many of them undergo equal or unequal divisions. For example, stem cells can be labeled using *Lgr5-CreER x R26R-Confetti* (Figure 10.C.1c). In this system the Confetti reporter expresses one of several different colored fluorescent proteins following CRE-induced recombination (see "Brainbow" techniques, Chapter 3). If a tamoxifen dose is given suitable to induce recombination in about 1 crypt in six, then most crypts containing any labeled cells at all have only a single one. After one cell division only about 10% of labeled clones still contain a single LGR5-positive cell, meaning that the original cell divided into one stem cell and one transit amplifying cell. In about 45% of cases the progeny are both LGR5-positive, indicating division to two stem cells, and in the other 45% neither are LGR5-positive, indicating division to two transit amplifying cells. Because only stem cells persist long term, this stochastic mode of division guarantees that, if a crypt remains labeled, the number of labeled cells must progressively increase until the progeny of only one stem cell remains, although the total number of stem cells in the crypt is still the same. In fact after about two weeks, some crypts are becoming monoclonal with respect to the Confetti label, and after 2 months most of them are. A similar situation exists in the newborn mouse, studied using aggregation chimeras or X-inactivation mosaics. Shortly after formation of the crypts they are all polyclonal, but after about 2 weeks, most are monoclonal. In the newborn mouse the gut is growing rapidly and so crypts are frequently dividing into two. Since the available stem cells will be partitioned at random between the two crypts this accelerates the process of clonal reduction. The mechanism of crypt fission is still not well understood but it is an essential process for normal growth and for regeneration following damage.

The example of intestinal crypt clonality underlines the need for clear thinking about the interpretation of cell labeling data. The fact that a crypt is monoclonal means that it derives from a single cell present at or after the time of labeling. It does not mean that there is any particular difference between crypts, apart from the markers used for labeling. Nor does it mean that the ancestor cell of a particular crypt is any different from the other stem cells in the same crypt before the process of clonal selection. Nor, of course, does it mean that crypts contain only a single stem cell. Cell kinetic calculations indicate that there are about 6–8 stem cells per crypt in the steady state. As mentioned above, this is many fewer than the number of potentially regenerative cells per crypt (30–40); it is also fewer than the number of LGR5-positive cells (12–14); but it is more than the single ancestral cell indicated by clonal marking techniques followed by a period of clonal selection. So at any one time the stem cell population is a subset of the LGR5 positive population and exactly which cells are behaving as stem cells depends on their local environment and interactions. The idea, embedded in the definition of tissue-specific stem cells, that they should persist for life, must of course be taken to refer to the population of stem cells rather than to any one specific cell lineage. This was emphasized in the discussion of in vivo cell labeling presented in Chapter 1.

The Epidermis

In the mouse the epidermis becomes specified at about 9 days of gestation and fully formed shortly before birth. It is a familiar fact that the epidermis of the skin is in a state of continuous cell turnover. It has been estimated that complete renewal occurs in about 9 days in the mouse and about one month in humans, although there is variation in different parts of the body. The epidermis is a squamous epithelium composed of cells called keratinocytes. This means it is multi-layered, with cuboidal cells in the basal layer and progressively more flattened cells towards the external surface (Figure 10.5). Histologically the epidermis consists of a stratum germinativum, the basal layer where all cell division occurs; and above this, from inside to outside, the stratum spinosum, stratum granulosum and stratum corneum. In the stratum corneum the cells are dead and consist mostly of flattened sacs of crosslinked cytokeratin proteins. The epidermis lies on a basement membrane and the underlying dermis contains loose connective tissue, adipose tissue, nerves and blood vessels. In humans there is a characteristic pattern of undulations called dermal papillae, a name also given to the dermal component of the hair follicle in mouse and human.

As keratinocytes move up through the layers, the gene expression pattern changes. Basal layer cells express keratin 5 and 14 (K5, K14), while the higher, post-mitotic, layers express K1 and K10. The *K14* promoter is often used in genetic experiments in mouse to drive gene expression in just the basal layer. In the granular layer are found the proteins involucrin and loricrin. These become cross-linked by transglutaminase to form a cornified envelope around the cell. In the stratum corneum the nuclei and other cell organelles are lost. The key gene determining the properties of the epidermis, as well as all other squamous epithelia, is *p63*, encoding a transcription factor structurally related to the famous p53, important for its involvement in human cancer. The direct targets of p63 include the *K14* gene. In mice, the knockout of *p63* prevents formation of squamous epithelia and is lethal.

Because of the continuous cell turnover it is evident that the basal layer must contain

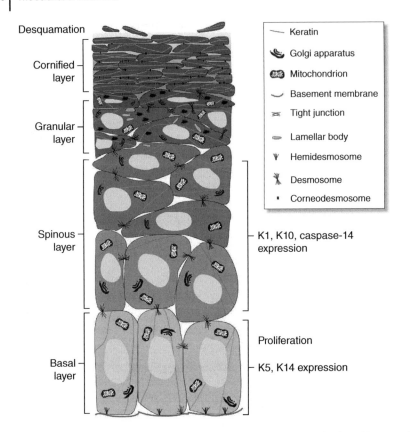

Figure 10.5 Structure of the epidermis. Cell division occurs in the basal layer and progeny are displaced upward. As they move up the keratinocytes differentiate and eventually die and are shed. (From: Eckhart, L., Lippens, S., Tschachler, E. and Declercq, W. (2013) Cell death by cornification. Biochimica et Biophysica Acta (BBA) – Molecular Cell Research 1833, 3471–3480. Reproduced with the permission of Elsevier.)

stem cells. There has been some controversy about whether the stem cells are a small subset of the dividing cells, or whether the basal layer forms a single stem cell compartment with any individual cell having some probability of persisting long term. The basal layer cells express AXIN2, a Wnt signaling target. Wnt signaling does seem to be necessary for normal epidermal homeostasis, as knockouts of pathway components lead to a thin epidermis. The main source of Wnt factors appears to be the basal layer itself, however it is probable that additional factors from the dermis are also needed. Labeling of individual basal layer cells with *K14-CreER* or *Axin2-CreER* shows the expected behavior of stem cells

(Figure 10.C.2). Shortly after the tamoxifen dose, there are many small labeled clones in the basal layer. In the long term a few labeled clones persist, spanning the full thickness of the epidermis. Observation of cells shortly after labeling indicate that both symmetric and asymmetric divisions take place, as for intestinal stem cells. There has been disagreement about whether a minority of cells in the basal layer are true stem cells and the remainder transit amplifying cells, or whether they are all equivalent. Mathematical analysis of the labeling studies shows progressive increase of clone size, reduction of clone number, and scaling behavior. As discussed in Chapter 1 this is consistent with the stochastic model in which all dividing cells are equivalent and they exhibit stem cell behavior as a population.

Epidermal cells can be cultured in vitro. It was discovered in the 1970s that they would form a multilayer tissue similar to the normal epidermis if cultured in the presence of feeder cells. However, like all primary cell cultures these do not grow without limit. It is possible to grow large sheets of cells from a small biopsy of human skin and these have been used on a small scale as autologous grafts to treat severe burns (see online supplement). The grafted skin appears to persist indefinitely but it does not form hair follicles or sweat glands.

In vitro culture also indicates the existence of cells with different propensities for division. Some cells form very large clones while others form only small clones (Figure 1.4). This is one argument for the existence of a hierarchy of dividing cells in the basal layer, as opposed to the stochastic model. It has been considered that the former are the true stem cells while the latter are transit amplifying cells, and in fact the large clone-forming cells do have a higher level of p63 than the others.

Although the keratinocyte is the most abundant cell type in the epidermis, there are some others. Melanocytes are pigment cells derived from the neural crest. They export pigment granules to epidermal cells and to hairs and hence give them their characteristic colors. Langerhans cells are dendritic antigen-presenting cells and are derived from the hematopoietic stem cells of the bone marrow. The Merkel cells are small oval mechanoreceptor cells with connections to the sensory system. Cell lineage labeling using *K14-CreER* suggests that Merkel cells are derived from the epidermal stem cells. Moreover, removal of the transcription factor ATOH1 using *K14-Cre × floxed Atoh1* prevents their formation. So although the epidermal stem cell is generally considered as unipotent, forming only keratinocytes, it does actually generate more than one type of differentiated cell. Moreover the need for ATOH1 indicates a possible similarity in cell

differentiation mechanism to that of the intestine (see Chapter 9).

Hair Follicles

In addition to the squamous epithelium covering the body, the mammalian epidermis gives rise to various appendages: the hairs, nails, sweat glands and mammary glands. In each case the arrangement of stem cells sustaining these structures is somewhat different from the rest of the epidermis (usually called interfollicular epidermis). Hair follicles have been studied very intensively in recent years although a complete understanding of their component cell types and the relationship between them is still lacking.

Hair follicles are formed from ingrowths of the epidermis during embryonic life, as described in Chapter 8. At the base of the mature hair follicle the epithelium surrounds a specialized clump of dermal cells: the dermal papilla, to form a hair bulb. Within this is the hair germ, a region of dividing epidermal cells that generates the multiple layers of keratinized cells within the hair shaft as well as an inner root sheath around the hair shaft and an outer root sheath lining the cavity (Figure 10.6). Some way up the follicle is a sebaceous gland, continuous with the outer root sheath. Each hair follicle is innervated by a sensory neuron and has a small muscle, the arrector pili (sometimes written erector pili) muscle, which can raise the external hair shaft on contraction.

The dermal papilla is critical as a provider of various extracellular signals needed for hair growth and differentiation. These include FGF7, BMPs and Wnts. New follicles can be induced in interfollicular epidermis by implantation of dermal papillae. They can also be induced by overexpression of Wnt signaling components. It might seem obvious that the epidermal stem cells that sustain hair growth should be found at the base of the hair shaft in contact with the dermal papilla. However there is good evidence that the principal reservoir of stem cells lies further up the follicle in a region known as

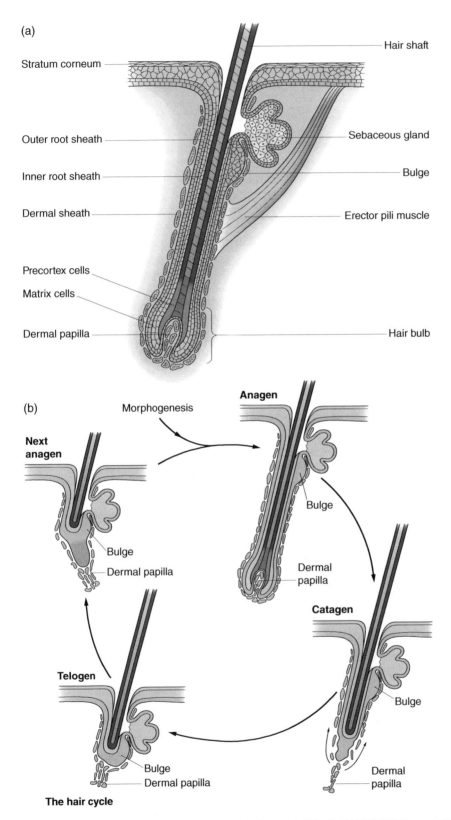

Figure 10.6 (a) Structure of the hair follicle. (b) The hair cycle. (Slack, J.M.W. (2013) Essential Developmental Biology, 3rd edn. Reproduced with the permission of John Wiley and Sons.)

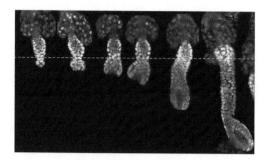

Figure 10.7 Growth of a hair follicle during the hair cycle. These are follicles viewed in vivo by multiphoton microscopy in a mouse expressing a *K14-H2BGFP* reporter. The bulge region lies above the dashed line. As growth commences cells from the bulge move down to surround the dermal papilla. They form a new hair bulb which generates all layers of the hair shaft and the outer and inner root sheaths. (From: Rompolas, P. and Greco, V. (2014) Stem cell dynamics in the hair follicle niche. Seminars in Cell and Developmental Biology 25–26, 34–42. Reproduced with the permission of Elsevier.)

the bulge (which may or may not be an actual physical bulge). This does make sense given that the process of hair growth is cyclical. Individual follicles go through successive phases of growth (anagen), regression (catagen) and dormancy (telogen) (Figures 10.6b and 10.7). In the mouse the first anagen lasts from about postnatal day 6–16, followed by a short telogen and a second anagen to about day 42. There is some regional variation in these timings. During catagen, the cells of the lower part of the follicle are destroyed by apoptosis and the follicle shortens so that the bulge region is brought close to the dermal papilla. The bulge originally attracted attention as a possible location of stem cells because it contained label-retaining cells and because cells derived from the bulge gave rise to long-lived clones in vitro, while those of the hair germ gave rise to shorter lived ones. This last experiment depended upon manual microdissection of the whisker follicles found in rodents which are much larger than other follicles.

Cells of the bulge and the hair germ share various markers including K14, K19, SOX9 and LGR5. However they do also differ in some respects: for example the bulge

expresses the cell surface glycoprotein CD34 while the hair germ expresses the cell adhesion molecule P-cadherin. *LGR5-CreER* labeling during telogen indicates that descendants of *Lgr5*-positive cells populate all parts of the follicle below the sebaceous gland in the course of the next anagen (Figure 10.C.3). So it is probable that the bulge is the permanent stem cell niche, and that it provides cells to populate the hair germ during anagen.

Located as they are in the bulge, the hair follicle stem cells are distinct in location and properties from those of the interfollicular epidermis. However, following wounding, cells from the hair follicles can repopulate the interfollicular epidermis, at least for a period.

Human hair follicles are morphologically similar to those of rodents and their stem cells are probably similar. However the timing of the hair growth cycle is quite different: it is asynchronous between follicles, and much longer than in the mouse: anagen may last for a few years and telogen a few months. The morphological sub-types of hairs and hair follicles also differ between rodents and humans.

Cornea and Limbus

The cornea is the transparent structure at the front of the eye (Figure 10.8a). It is responsible for most of the focusing of incoming light onto the retina, although the lens also contributes to focusing and through its shape changes provides accommodation for different distances. The cornea is composed of three cell layers: an outer epithelium, which is a derivative of the epidermis; a central stroma, mostly composed of collagen fibers; and an inner endothelium. Of these components, the outer epithelium is a renewal tissue maintained by a population of stem cells. The stem cells reside in an annular region around the periphery of the cornea called the limbus. This contains cells high in p63, expresses the transporter protein ABCB5, and, like the basal layer of the epidermis,

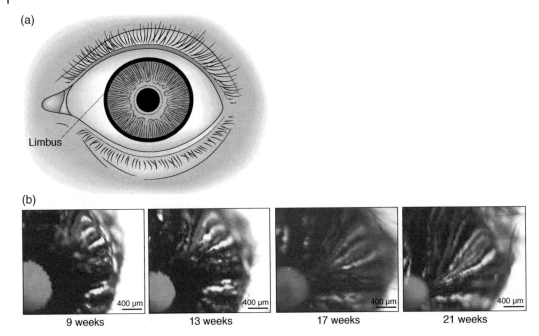

Figure 10.8 (a) Location of the limbus at the periphery of the cornea. (b) Lineage tracing of stem cells in the limbus. Mice were *K14-Cre-ER x R26R Confetti*. Tamoxifen was administered at 6 weeks of age and labeled clones in the cornea visualized after the indicated number of weeks. By 21 weeks clones have reached the center of the cornea. (*Sources:* (a) Modified from Pellegrini, G., Golisano, O., Paterna, P., Lambiase, A., et al. (1999) Location and clonal analysis of stem cells and their differentiated progeny in the human ocular surface. Journal of Cell Biology. 145, 769–782. Reproduced with the permission of The Rockefeller University Press. (b) From Di Girolamo, N., Bobba, S., Raviraj, V., Delic, N.C., et al. (2015) Tracing the fate of limbal epithelial progenitor cells in the murine cornea. Stem Cells 33, 157–169. Reproduced with the permission of John Wiley and Sons.)

expresses cytokeratins 5 and 14. The central cornea does not express these cytokeratins, nor the characteristic K1/K10 found in the upper layers of the epidermis. Instead the central cornea expresses K3 and K12. When a *K14-CreER × R26R-Brainbow* mouse is labeled with a low dose of tamoxifen some limbal stem cells become labeled. These subsequently generate radial streaks of progeny from periphery to the center of the cornea, which is typical stem cell behavior (Figure 10.8b). Limbal stem cells can also be cultured in vitro and are capable of generating large clones.

When a corneal graft is carried out, the epithelium becomes renewed from the limbus of the host. If this does not occur the cornea will become covered with connective tissue and blood vessels and cease to be transparent. Limbal cells can be expanded in

vitro and grafts of limbal cells from one eye to the other have been used to support and maintain corneal grafts. The indication for this is usually in cases where the whole front of one eye, including the limbus, has been destroyed (see online supplement).

Mammary Glands

Another derivative of the epidermis is the epithelium of the mammary glands. As described in Chapter 8, the post-pubertal, but virgin, mammary gland consists of a branched duct system connected to the nipple and embedded in a connective tissue stroma. The ducts have a layer of myoepithelial cells around the outside and luminal cells on the inside (Figure 10.9). Although the basic cell types are the same, there are some differences between humans and mice

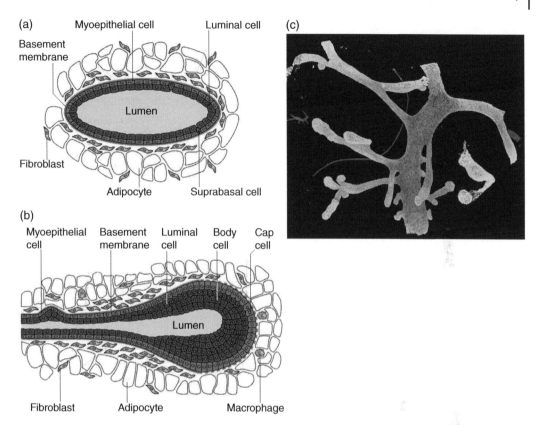

Figure 10.9 Structure of the mammary epithelium in the mouse. (a) Duct showing myoepithelial and luminal cell layers. (b) A terminal end bud. (c) Whole mount of a 3 week old mouse mammary tree, immunostained for keratin 5. (*Sources:* (a&b) from Visvader, J.E. (2009) Keeping abreast of the mammary epithelial hierarchy and breast tumorigenesis. Genes and Development 23, 2563–2577. Cold Spring Harbor. (c) from Rios, A.C., Fu, N.Y., Lindeman, G.J. and Visvader, J.E. (2014) In situ identification of bipotent stem cells in the mammary gland. Nature 506, 322–327. Reproduced with the permission of Nature Publishing Group.)

(Figure 10.10). Human females have just one pair of mammary glands while mice have five pairs. In mice the termini of the duct system form terminal end buds which are growing points, in humans the termini are clusters of small lobules called terminal ductal lobular units. Moreover the mouse stroma is mostly adipose tissue while in humans there is also considerable fibrous connective tissue.

After the post-pubertal maturation of the system, there a limited growth and regression with each estrus/menstrual cycle. But the main developmental changes occur during pregnancy and lactation (Figure 10.11). During pregnancy the epithelium proliferates considerably and generates a large number of

secretory alveoli. The alveoli differentiate during mid-late pregnancy and in mice the transcription factor ELF5, specifically expressed in luminal cells, is required for differentiation. After birth of the offspring, the alveoli secrete milk by an apocrine mechanism, meaning that distal portions of cells containing the components of the milk become detached from the cells and enter the duct system. Contraction of the myoepithelial cells helps drive the milk towards the nipple. High level milk secretion continues during lactation. After lactation is finished there is an involution of the organ involving massive cell death, collapse of the alveoli, and remodeling of the epithelium back to the pre-pregnancy appearance.

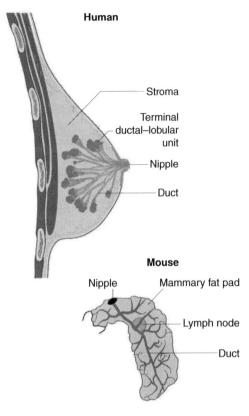

Human

Stroma

Terminal ductal–lobular unit

Nipple

Duct

Mouse

Nipple

Mammary fat pad

Lymph node

Duct

Figure 10.10 Differences between human and mouse mammary glands. (From: Visvader, J.E. (2009) Keeping abreast of the mammary epithelial hierarchy and breast tumorigenesis. Genes and Development 23, 2563–2577. Cold Spring Harbor Laboratory Press.)

The main hormonal stimulus for these changes is prolactin from the anterior pituitary gland. This is a cytokine-type hormone with a cell surface receptor connecting to the JAK-STAT signal transduction pathway (similar to LIF, see Figure 6.1b). In late pregnancy the related hormone placental lactogen, secreted by the placenta, plays a similar role, but after birth prolactin again becomes predominant. STAT5 is the critical intracellular mediator. It is essential for pregnancy in mice and its direct targets include the genes encoding milk proteins and junctional complexes for the secretory cells of the alveoli. Other important signal molecules, which are locally produced in response to the systemic hormones, are RANK-L and members of the epidermal growth factor family.

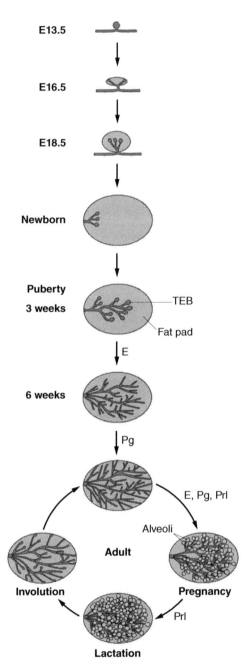

E13.5

E16.5

E18.5

Newborn

Puberty 3 weeks — TEB, Fat pad

E

6 weeks

Pg

Adult

E, Pg, Prl

Alveoli

Pregnancy

Prl

Lactation

Involution

Figure 10.11 The complete development of the mouse mammary gland. At the bottom are shown the phases of growth, secretion and involution accompanying pregnancy and lactation. TEB: terminal end bud; E: estrogen; Pg: progesterone; Prl: prolactin. (From: Visvader, J.E. and Stingl, J. (2014) Mammary stem cells and the differentiation hierarchy: current status and perspectives. Genes and Development 28, 1143–1158. Cold Spring Harbor Laboratory Press.)

Mammary Stem Cells

Although the mammary epithelium is not continuously renewing like the intestinal epithelium or the epidermis, it does obviously require considerable cell renewal to enable the cycles of growth and regression accompanying pregnancy. Modern human females now have few pregnancies in a lifetime, but wild animals such as mice have very many and we should therefore expect to find stem cells in their mammary glands.

The study of mammary stem cells and cell lineage has largely depended on transplantation assays. Because cells may behave in a different way on transplantation, or on in vitro culture, compared to their normal situation in the body, the interpretation of these data requires some care. The usual transplantation system is the "cleared fat pad" of mouse mammary gland. Three week old mice have the nipple region and associated duct system removed from one mammary gland (usually number 4) and this leaves a "fat pad" consisting of adipose tissue, loose connective tissue and blood vessels without an epithelial component. In this environment grafts of pieces of duct system, or even single cells, from genetically compatible mice can develop into a complete duct system (Figure 10.12a). The fact that a single cell from an adult gland can support development of the whole duct system, and undergo differentiation during pregnancy, is considered evidence for the presence of cells with multipotent potential. Serial transplants have been carried out with explants of tissue for as many as seven transplant generations and this indicates that the stem cells are capable of sustaining growth of a mammary duct system for more than a normal mouse lifespan, as is also found for the epidermis and the hematopoietic system. The cleared fat pad system can be modified for the growth of human mammary epithelium by using NOD-SCID (immunodeficient) mice and also by injecting some human fibroblasts into the fat pad to "humanize" the local environment.

There is also a cell culture system in which dissociated mammary cells are cultured in non-adherent dishes in the presence of EGF or FGF. This gives rise to "mammospheres",

small cell clusters which can differentiate into ductal and myoepithelial cells and can be serially propagated. These are presumed to contain mammary stem cells, and have an obvious resemblance to the cultured neurospheres which contain neural stem cells (see Chapter 9).

The mouse cells which give good efficiency in transplantation assays are high in α6 and β1 integrins (also called CD49f and CD29 respectively), in the P-selectin ligand (CD24), and they lack Sca1, a GPI-linked cell surface protein found on hematopoietic stem cells. In humans, various cell types have been reported as being enriched in their ability to form mammospheres. These include high α6 integrin cells, aldehyde dehydrogenase 1 expressing cells, "side population" cells which exclude the dye Hoechst 33342, or slow dividing cells which retain the vital dye PKH26.

Because of the uncertain relationship between transplantation, or in vitro culture behavior, and cell behavior in vivo it is also reassuring to have some lineage label results in vivo. These have been hampered by the fact that tamoxifen, used to activate CreER, has deleterious effects on mammary development. However, it is possible to use doxycycline to induce expression with transgenic mice of the type: *promoter-rtTA; TRE-Cre; R26R*. The promoter of interest drives expression of the Tet activator rtTA. In the presence of doxycycline, this binds to the Tet response element (*TRE*) and causes transcription of *Cre*. This excises the stop sequence from the R26R reporter and causes permanent expression of a marker gene. The *keratin 5* promoter is active in basal mammary cells and can be used to drive the expression of *rtTA*. If a doxycycline pulse is administered in adult life the labeled clones are initially basal but extend to the luminal surface after a few weeks (Figure 10.C.4). These multilayer clones can be quite large and they can also label the alveoli that form during pregnancy. If the doxycycline is administered at the start of puberty, small multilayer clones become labeled all over the duct system. If the rtTA is driven by the *Elf5* promoter, active in luminal cells, then clones

(a)

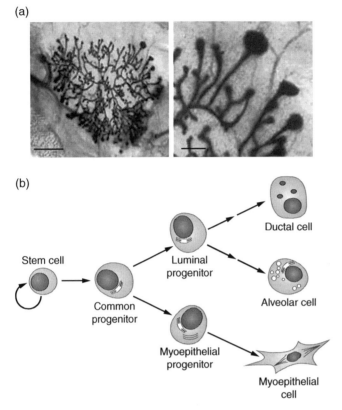

(b)

Figure 10.12 Mammary stem cells. (a) Growth of a whole mammary tree from a single stem cell transplanted to a cleared mammary fat pad. Scale bars: 250 μm for low power, 50 μm for high power. (b) Putative cell lineage for mammary stem cells. (*Sources:* (a) from Shackleton, M., Vaillant, F., Simpson, K.J., Stingl, J., et al. (2006) Generation of a functional mammary gland from a single stem cell. Nature 439, 84–88. Reproduced with the permission of Nature Publishing Group. (b) from Visvader, J.E. (2009) Keeping abreast of the mammary epithelial hierarchy and breast tumorigenesis. Genes and Development 23, 2563–2577. Cold Spring Harbor Laboratory Press.)

are confined to the luminal surface and disappear after 20 weeks, indicating that this promoter is active only in progenitor cells and not in stem cells.

The evidence is good from both transplantation and cell labeling studies for the existence of multipotent stem cells that sustain the growth, differentiation, and involution of the mammary epithelium. But at present the system is not as well characterized in terms of cell lineage and cell kinetics as the intestinal epithelium or the epidermis. A plausible model for the cell lineage is shown in Figure 10.12b.

The Hematopoietic System

The hematopoietic system generates all of the cells found in the blood and the immune system. These comprise erythrocytes (red blood cells), granulocytes (neutrophils, eosinophils and basophils), monocytes and macrophages, megakaryocytes (which generate blood platelets), and lymphocytes. The non-lymphocytes are referred to as myeloid cells. The lymphocytes comprise B (antibody secreting), T (thymus derived) and NK (natural killer) cells.

One often reads in stem cell literature that the hematopoietic system is the "best understood" system and stands as a model for investigation of the others. This may have been true before the advent of CreER labeling, but it is not so true today. This is because the hematopoietic system has been investigated mostly by transplantation and by in vitro culture, so a lot is known about these aspects, but rather less about the behavior of the stem cells in the situation of normal steady state tissue turnover, or homeostasis.

Analysis by Transplantation and in Vitro Culture

The standard transplantation assay for hematopoietic stem cells (HSC) was discovered as an offshoot of work on the atomic bomb in the 1940s. It was found that animals exposed to high doses of radiation would die of bone marrow failure, but that they could be saved with a graft of bone marrow from a healthy donor. By the 1950s it was known that the factor in the marrow consisted of cells rather than a hormone-like substance. It was later found that the recipients of bone marrow transplants developed colonies in the spleen. These consisted of various types of hematopoietic cells, sometimes of one type and sometimes of mixed type. It was concluded that each spleen colony-forming unit was a single cell and that the mixed colonies arose from multipotent cells, considered to be stem cells. In fact the spleen colony assay does not capture the most primitive stem cells, as lymphoid tissues are not formed, but it was important for establishing the concept of a single type of hematopoietic stem cell, resident in the bone marrow, which generated all of the cells of the blood and the immune system throughout life.

The system was later dissected using fluorescence activated cell sorting (FACS) to separate subpopulations of bone marrow cells on the basis of their surface antigens. The identity of the resulting cells could then be established by two methods. One was transplantation into lethally irradiated mice. The radiation destroys the stem cells of the host, and if the graft contains stem cells, they can home to the vacant niches in the bone marrow. All components of the blood and immune system are then restored from the stem cells of the graft and the mice would survive long term. If the graft contained only progenitor cells then a partial spectrum of blood cell types would regenerate, but this would be a temporary effect and the mice would soon die. The other method for characterizing sub-fractions of marrow cells was in vitro culture in soft agar or methyl cellulose supplemented with hematopoietic growth factors, such as GM-CSF (granulocyte-macrophage colony stimulating factor), G-CSF (granulocyte colony stimulating factor), M-CSF (macrophage colony stimulating factor) and erythropoietin. The types of colony formed indicate the nature of the progenitor

cells, and in the case of the more primitive progenitors, or of the stem cells themselves, mixed colonies would be formed. In addition to cell isolation and characterization, information about the hematopoietic system came from mouse knockouts for particular genes needed for blood formation. Some of these give rise to animals lacking specific sub-lineages of cells. Putting together the information from these three sources gave rise to the familiar diagrams found in textbooks. The details of these diagrams vary slightly, but they all show a single type of long-lived hematopoietic stem cell giving rise to progenitors of restricted potency and eventually to the various types of blood cell and lymphocyte (Figure 10.13).

The transplantation assay depends absolutely on the radiation given to kill the stem cells of the host. Without this the niches remain occupied and graft HSCs have nowhere to go. A period of 4 months after transplantation is generally considered to be long enough for the progeny of committed progenitors to disappear, so donor-derived cells persisting longer than this time are considered to arise from HSC. Donor cells forming all types of blood and immune cells, but not persisting more than 4 months are considered to be "short-term hematopoietic stem cells" (ST-HSC), a commonly used designation which is rather an oxymoron if the definition of a stem cell is considered to include long term persistence. Limiting dilution studies indicate that there are about 2–8 long-term HSC per 10^5 bone marrow cells. In humans, using limiting dilution transplants into immunodeficient mice, which may be a more demanding criterion, the estimate is 1–4 long term HSC per 10^7 mononuclear cells. If the HSC in a graft are progressively diluted with non-repopulating cells, it is possible to show that just a single HSC is capable of bringing about full repopulation. It is possible to do serial transplants from the recipient of a transplant into further irradiated mice, which makes it possible to investigate the ultimate longevity of stem cells. Serial transplantation of unfractionated bone marrow can be carried out about five times,

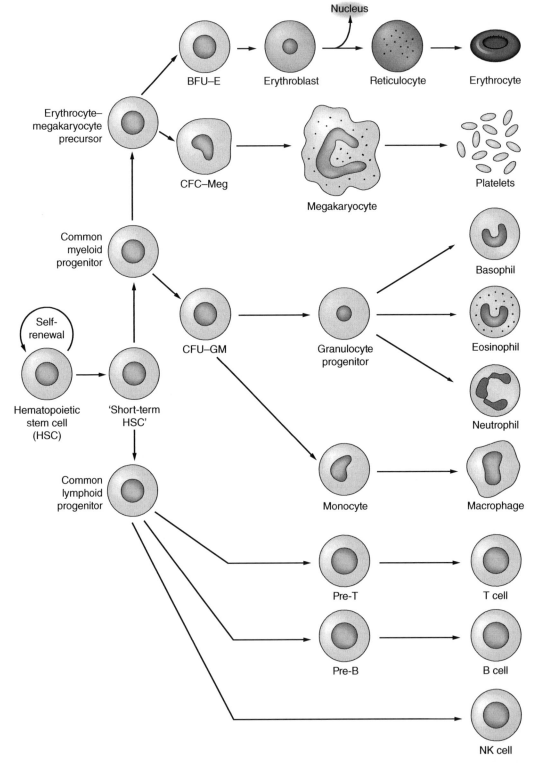

Figure 10.13 Putative hematopoietic lineage. This is a consensus diagram derived from the sort of data described in the text. Other published diagrams may differ in detail. However all have a hierarchical, rather than a reticulate, pattern and a fully multipotent long lived hematopoietic stem cell at the start. (Slack, J.M.W. (2013) *Essential Developmental Biology*, 3rd edn. Reproduced with the permission of John Wiley and Sons.)

beyond which it fails, at least partly because of telomere erosion. Serial transplantation does show that the HSC population can self-renew for much longer than the lifetime of a single mouse.

It is possible to compare different cell populations for stem cell content or vigor by doing competitive reconstitutions, where two populations of cells are introduced into the same irradiated host. In the blood eventually formed by the grafts, the more vigorous strain will prevail over the less vigorous. For competitive reconstitution assays the two donor strains need to be genetically distinguishable but also equally tolerated by the host.

The hematopoietic stem cells themselves can be isolated by FACS. That of the mouse was first isolated as cells of the constitution Lin^-, $Thy-1^{lo}$, $Sca-1^+$. Lin^- indicates the absence of various markers for differentiated blood cells. Thy-1 (= CD90) and Sca-1 are both cell surface GPI-linked glycoproteins, Thy-1 being found at high levels on T lymphocytes. Nowadays mouse HSCs can be isolated by a variety of other criteria, a popular combination being: Lin^-, $Sca-1^+$, $c-Kit^+$, $CD150^+$, $CD48^-$. c-Kit (= CD117) is the receptor for the steel growth factor (= stem cell factor, SCF); CD150 (= SLAM) is a cell surface glycoprotein also found on mature T cells; and CD48 (= BLAST-1) is found on mature B lymphocytes. Also the "side population" of cells that rapidly exclude the dye Hoechst 33342, due to the presence of transporter molecules in the membrane, is enriched in HSC (see Figure 1.2 in Chapter 1).

The fact that a single stem cell can repopulate the entire blood and immune system of an irradiated host indicated that the basic hypothesis of the multipotent HSC was correct. The corresponding hematopoietic stem cell in humans is found in the $CD34^+$ fraction. CD34 is a cell surface sialomucin also found on endothelial cells. Not all $CD34^+$ cells are HSCs and human HSCs may be further purified as $CD38^-$, $CD45RA^-$, $Thy-1^-$. These three components are all found on mature lymphocytes. CD49f (= integrin $\alpha 6$)

has also recently been shown to mark human HSC. It may be noted that although the biological properties of mouse and human HSCs are similar, the cell surface markers by which they are identified differ. For example, CD34 is not found on mouse long term HSC. This is because the markers in question, though useful for isolation by FACS, are probably not necessary for the actual stem cell behavior shown by the cells. In vitro experiments with human HSC use soft agar/methyl cellulose culture techniques similar to mouse. In vivo experiments use lethally irradiated immuno-deficient mice as hosts. The NOG mouse strain combines the defects of NOD and SCID mice with deficiency of the IL2R γ chain (see Chapter 4). It is highly immunodeficient, lacking B, T and NK lymphocytes, and is currently favored for human grafts.

Recently it has been possible to reprogram committed progenitors and even differentiated blood cells back to HSCs, by introducing a cocktail of transcription factors normally expressed in HSCs. The factors were screened by introducing a large number into the target cells using doxycycline-inducible lentivirus and then grafting them into irradiated hosts. Only if cells have been reprogrammed to HSC status can long term repopulation occur. The transcription factors were identified from those present in the successful grafts and were: HLF, RUNX1t1, PBX1, LMO2, ZPF37 and PRDM5. In cases where the target cells were pre-B cells, their immunoglobulin DNA rearrangements were still present in the donor cells populating the new host, indicating a genuine permanent reprogramming of the pre-B cells arising from the transient expression of the active factors. Furthermore, as with normal HSC, secondary transplantation is possible using grafts from the first hosts, indicating their capacity for long term self-renewal.

Hematopoiesis in the Steady State

There are some difficulties with the standard model for hematopoiesis shown in Figure 10.13, particularly in relation to the immune system. For example, if a skin graft is given to a newborn mouse, it can render the mouse tolerant to grafts from the same donor strain in adult life. Likewise, if an animal is immunized then antibody producing B-cells and memory T-cells specific to the immunogen may persist for life. Because the functional cells need to undergo specific DNA rearrangements of antibody genes, or of T cell receptor genes, in the course of their maturation, it seems unlikely that they could be continuously renewed from HSCs. It is much more likely that cells which have experienced those specific DNA rearrangements can persist for life. Such examples indicate the existence of permanent committed populations of memory cells within the hematopoietic lineage. Either such populations do not turn over at all, or they possess their own stem cells which remain active for life, even if they were originally derived from HSC.

Labeling of bone marrow for DNA synthesis using BrdU has been extensively carried out to try to estimate the replication rates of various cell subsets in the hematopoietic system. But there are some toxic effects of BrdU that complicate interpretation. Less subject to toxic effects is the H2B-GFP dilution method in which cells are labeled by a burst of synthesis of fluorescent histone 2B (see Chapter 2). This has recently been applied using a human CD34 promoter to drive the Tet activator (this promoter is active in mouse HSC even though the endogenous mouse CD34 is not expressed). In the absence of doxycycline, H2B-GFP is produced, and when Doxycycline is restored the production stops (Tet-Off system). It is then possible to isolate different cell populations by FACS and ask how much GFP label they retain. This shows highest retention in the HSC themselves, less in the "short-term-HSC", the multipotent progenitors and the common lymphocyte progenitors, and none in the common myeloid progenitors, granulocyte-macrophage progenitors and megakaryocyte-erythroid progenitors (Figure 10.14). Modeling of the data plus other studies of HSC cell renewal indicate that the HSC pool is heterogeneous with some fast and some slow dividing cells.

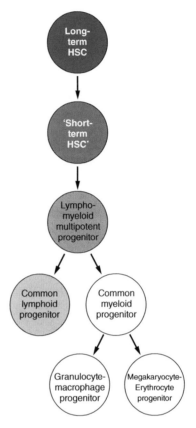

Figure 10.14 Cell division during steady state hematopoiesis. Mouse HSC were labeled by expression of H2B-GFP under the control of a Tet inducible system. The rate of loss of label in each compartment, due to cell division, is indicated by the loss of shading. (From: Schaniel, C. and Moore, K.A. (2009) Genetic models to study quiescent stem cells and their niches. Annals of the New York Academy of Sciences 1176, 26–35. Reproduced with the permission of John Wiley and Sons.)

Until recently there was no suitable promoter known for use in CreER experiments. So genetic labeling of HSC had to rely on induced DNA changes such as insertion sites of applied retroviruses. The most recent such studies use lentiviral libraries of genetic "barcodes" which can be sequenced directly by high throughput machines. Such methods have been used to study the clonal composition of the blood both in a normal turnover situation, and following transplantation. The results tend to indicate many more clones of HSC support normal steady state (homeostatic) blood formation than do so following transplantation. However the results have been rather variable and it remains uncertain whether clones of HSC remain continuously active or whether one clone succeeds another. It has recently become possible to do CreER experiments using the *Tie2* promoter. *Tie2* encodes an angiopoietin receptor which is present on HSC and also on endothelial cells. A dose of tamoxifen causes irreversible labeling of a fraction of the HSCs and it is possible to trace the formation of their progeny by isolating different cell populations using FACS at different times and determining the percentage of label they contain. The results indicate that the label is slow to enter the system. After 4 weeks some "short-term-HSC" and multipotent progenitors are labeled. After 16 weeks all blood cell types are labeled. When the data are modeled it indicates that the "short-term-HSC" compartment in particular shows a very long residence time and does not come to equilibrium during the lifetime of the mouse.

In conclusion, despite the much vaunted status of the hematopoietic system as a model for all others, it seems that the cell lineage is not really known. The pathway shown in Figure 10.13 may not be entirely correct, as some data suggest the existence of precursors able to produce more than one type of offspring and their presence would make the scheme into a reticular network rather than a bifurcating tree, which would be a significant change to the model. Moreover mouse and human may not be exactly the same, and in each species fetal and adult may not be the same either. However the current evidence does support the idea that in homeostasis a large number of HSC clones feed the system, while after transplantation only a small number does so. Furthermore it seems inescapable that some of the precursor populations derived from the HSC are very long-lived and self-renewing and may reasonably qualify to be regarded as stem cells in their own right.

The Hematopoietic Niche

There has been much controversy about the nature of the niche within the bone marrow occupied by the HSCs. This issue is still not settled although a lot of evidence has been gathered in favor of various possibilities. It is likely that many components of the bone marrow are involved and there may be sub-niches for different subsets of HSC, such as quiescent cells, rapidly dividing cells, or cells mobilizable into the circulation.

The bone marrow is present in the cavities of the axial and the long bones (Figure 10.15). The endosteal surface of the bone is covered with flat bone-lining cells together with osteoblasts (bone-forming cells, derived from the lateral plate mesoderm) and osteoclasts (multinucleate bone-resorbing cells derived from the hematopoietic system). This lining exists both on the inner surface of the marrow cavity and also around the masses of trabecular bone. The principal artery and vein lie in the center of the cavity and are connected by arterioles and venules with an anastomosing plexus of venous sinusoids. These sinusoids are just one endothelial cell thick and have no basement membrane, hence allow easy entry and egress of cells. There is also a sympathetic innervation to the marrow which may affect hematopoiesis, for instance in terms of circadian variations.

If bone marrow is transplanted to other tissues in the body, hematopoiesis only continues where a bony ossicle is formed, indicating the importance of the marrow environment for the process. Transplantation studies on cells from different parts of the marrow suggest that the sub-endosteal region is richer in transplantable HSCs than the central region. The whole bone marrow has a relatively low oxygen content, about one-third of atmospheric, and low oxygen has been associated with HSC maintenance via stabilization of the hypoxia inducible factor HIF1α.

One problem in locating the HSC niche is the difficulty of identifying HSCs in situ using immunostaining. None of the antibodies used to isolate HSCs are completely

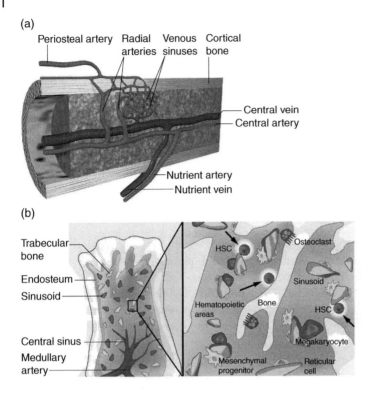

(a)

Periosteal artery Radial Venous Cortical
                   arteries sinuses bone

Central vein
Central artery

Nutrient artery
Nutrient vein

(b)

Trabecular bone
Endosteum
Sinusoid
Central sinus
Medullary artery

HSC
Osteoclast
Sinusoid
Hematopoietic areas
Bone
HSC
Megakaryocyte
Mesenchymal progenitor
Reticular cell

Figure 10.15 Structure of the bone marrow and the putative hematopoietic niches. (a) Overall structure of marrow in a long bone showing the vascular supply. (b) Substructure of bone marrow. The HSC (arrowed) are thought to reside near blood vessels and trabecular bone osteoblasts. (*Sources:* (a) from Travlos, G. (2006) Normal structure, function, and histology of the bone marrow. Toxicologic Pathology 34, 548–565. Adapted from: Abboud, C.N. and Lichtman, M.A. (2001) Structure of the marrow and the hematopoietic microenvironment. In: Williams Hematology, 6th edn. Copyright McGraw-Hill, used with permission. Adaptive drawing by David Sabio. (b) from Morrison, S.J. and Spradling, A.C. (2008) Stem cells and niches: mechanisms that promote stem cell maintenance throughout life. Cell 132, 598–611. Reproduced with the permission of SAGE Publications.)

specific and the FACS method relies on the absence as well as presence of various markers. Studies with CD150 (= SLAM) suggest that the putative HSC may lie adjacent to blood vessels. If HSCs are first purified by FACS they can then be labeled and transplanted to irradiated hosts, whose niches are available for occupation. This method shows that they tend to end up in the sub-endosteal zone adjacent to trabecular bone osteoblasts and probably also to blood vessels (Figure 10.C.5). Mobilization of HSC into the circulation using granulocyte colony-stimulating factor (G-CSF) is associated with the death or flattening of osteoblasts and of marrow macrophages.

Analysis of various mouse knockout strains has indicated certain growth factors and cytokines that are likely to be important components of the niche. These include stem cell factor (SCF), CXCL12 (= SDF1), parathyroid hormone, Notch ligands, and various cell adhesion molecules. Further discrimination has come from the use of transgenic mice in which a promoter active in a specific cell type, such as osteoblasts or subsets of bone marrow stromal cells, are used to ablate or upregulate various factors. Cell ablation studies using Cre-driven diphtheria toxin receptor (see Chapter 2) indicate that subsets of bone marrow stroma expressing nestin or leptin are needed for HSC maintenance.

HSCs have been difficult to expand in vitro. Early methods depended on the use of stroma from bone marrow, which is a complex mixture of fibroblastic cells, blood vessels and adipocytes. More recently some substances have been found that promote division, most notably a compound called stemregenin 1 (SR1), which inhibits the aryl hydrocarbon receptor, a bHLH transcription factor normally resident in the cytoplasm. This has been used to expand the HSCs prior to grafts into human patients.

Spermatogenesis

The process of primordial germ cell migration, spermatogenesis and oogenesis have been described in Chapter 5. But spermatogenesis has another aspect of interest because the dividing spermatogonia constitute a well-characterized tissue-specific stem cell system. As with the hematopoietic system, the stem cells were initially defined by transplantation, but can now be studied in situ using the CreER method of lineage labeling. The transplantation assay uses recipients mice treated with the alkylating agent busulphan. This destroys the endogenous stem cells, making the mice sterile and opening up the niches for repopulation. Injection of spermatogonial stem cells into the testis leads to the formation of permanent donor-derived colonies of cells at all stages of spermatogenesis (Figure 10.16). Like all transplantation, this assay needs to be conducted with immunocompatible mice, or with immunodeficient mice. Using this method, it is possible to do interspecies grafts of rat spermatogonial stem cells into immunodeficient mouse testis and to recover viable rat sperm.

The first transplantable cells arise in the mouse testis at about 4 days after birth. In the mature testis, the type A spermatogonia undergo divisions to form pairs or lines of cells joined by cytoplasmic bridges (Figure 10.17). These become A1 spermatogonia which divide successively to form A2, A3, A4, intermediate and B spermatogonia. These form spermatocytes which undergo meiosis to form four spermatids, each of which becomes a single sperm. The maximum possible yield of 4096 sperm per type A spermatogonium is not achieved because there is some cell death in the A1-B period. The type A spermatogonia are found near the basement membrane of the tubules, adjacent to Sertoli cells. As differentiation progresses the cell types are located further towards the luminal side, still closely associated with Sertoli cells, until eventually

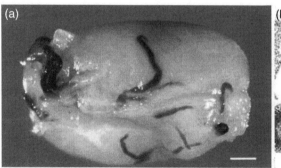
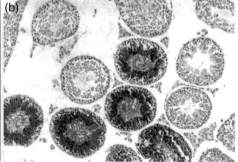

Figure 10.16 Spermatogonial stem cell transplantation. (a) A host testis into which stem cells were injected from a mouse labeled with *lacZ* reporter. The patches of tubule colonized by donor-derived cells are revealed by X-Gal staining (dark). (Slack, J.M.W. (2013) Essential Developmental Biology, 3rd edn. Reproduced with the permission of John Wiley and Sons.) (b) Section through a similar testis. In the colonized patches (dark), all of the germ line, but not the somatic cells, of the tubule are graft-derived. (Reproduced from Brinster and Avarbock (1994) Proceedings of the National Academy of Sciences of the USA 91, 11303–11307, with permission from National Academy of Sciences.)

the mature sperm are released into the lumen (Figure 10.18). It has been known for some time that the spermatogonia depend on Sertoli cells for their survival and multiplication. A key factor is Glial-Derived Neurotrophic

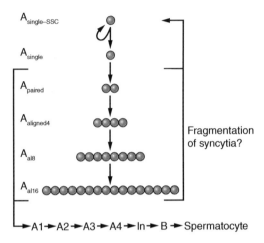

Figure 10.17 Spermatogenesis in the mouse. The spermatogonial stem cells are among the A_{single} population. These divide to form syncytial groups of up to 16 joined cells, which divide through further stages as indicated. The B spermatogonia become spermatocytes each of which undergo meiosis to form four sperm. (From: Griswold, M.D. and Oatley, J.M. (2013) Defining characteristics of mammalian spermatogenic stem cells. Stem Cells 31, 8–11. Reproduced with the permission of John Wiley and Sons.)

Factor (GDNF), which is secreted by Sertoli cells. Its receptors, the tyrosine kinase RET, and the GPI-linked co-receptor GFRα1, are present on the A type spermatogonia. Mice lacking one copy of the *Gdnf* gene have reduced spermatogonial numbers and mice overexpressing *Gdnf* have too many. The Sertoli cell is obviously a key part of the stem cell niche, and it is likely that proximity to blood vessels is also important. The complete process of spermatogenesis takes about 35 days in the mouse and 74 days in human. There is also a "seminiferous epithelial cycle", taking 8.6 days in the mouse and 16 days in human. This is effectively the duration of a cell's residence in any one layer of the tubule. Because division of type A spermatogonia is approximately synchronized, adjacent cells are at about the same developmental stage and this means that there is a particular association of stages for each level from the basement membrane up to the lumen. The phase of this "cycle" varies along the tubule so that different stages are laid out adjacent to each other This arrangement is less apparent in human testis than in those of rat or mouse.

The identification of the stem cells has caused some difficulties. It has generally been thought that the true stem cells lie within the A_{single} (A_s) compartment, although lineage

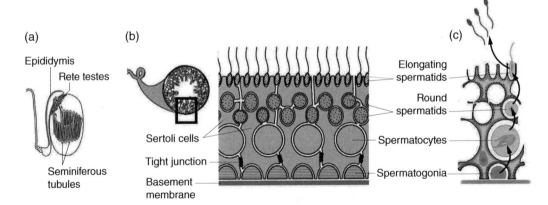

Figure 10.18 Organization of spermatogenesis. (a) Arrangement of the seminiferous tubules in the testis. (b) Structure of a tubule. The spermatogonia lie at the base and cells in different stages of spermatogenesis lie at successively higher levels. (c) Sperm differentiation occurs in very close proximity to the Sertoli cells. (From: Yoshida, S. (2016) From cyst to tubule: innovations in vertebrate spermatogenesis. Wiley Interdisciplinary Reviews:Developmental Biology 5, 119–131. Reproduced with the permission of John Wiley & Sons.)

labeling for *Ngn3*, which is expressed mostly in the syncytial groups, does label some stem cells. Various markers have been proposed for the stem cells, including GFRα1, the GPI-linked receptor for GDNF, and the transcription factors Id4, FOXO, PAX7, BMI1, TAF4b. The cell surface molecule Stra8 is expressed on all the A type spermatogonia and can be used for enrichment of stem cells by FACS. C-KIT, the receptor for the stem cell factor (SCF) is found on all spermatogonia. Cells in the differentiation sequence from A1 spermatogonia onwards are certainly transit amplifying cells. CreER experiments performed with several of these markers yield some clones which are long lived, and include cells at all stages of spermatogenesis, behavior indicative of real stem cells. An example is shown for *Bmi1*, using the multicolor Brainbow reporter system, in Figure 10.C.6. Here, 24 weeks after labeling, each clone covers a whole segment of a tubule.

Considerable analysis has been performed on the stem cells which can be labeled using *Ngn3-CreER*. Most *Ngn3*-expressing cells are transit amplifying cells and yield short lived clones. But some are stem cells and give large, long lived clones. Comparison of the proportion of the testis labeled in the steady state versus the proportion of transplantable cells that are labeled indicates that the transplantable cells are about 30× as abundant as the steady state stem cells. This is similar to the behavior of the intestine and reminds us that transplantation ability, though an interesting and important property, is not the same as being a stem cell in vivo. Furthermore as time goes on the number of *Ngn3* labeled clones decreases, but individual clones get bigger, preserving a roughly equal proportion of the testis labeled. The system shows "scaling behavior" meaning that the frequency distribution of clone sizes remains constant when divided by average clone size. This property can best be explained by the stochastic model in which the individual stem cells may divide to produce two stem cells, one stem cell and one transit amplifying cell, or two transit amplifying cells, such that the overall number of stem cells remains constant (see Chapter 1). This situation resembles that shown in the intestinal epithelium and the skin. As for these systems, it means that the lifetime persistence, which is a key defining feature of a stem cell, is a population property not an individual cell property.

Spermatogonial stem cells can be cultured in vitro. This is achieved by plating enzyme-digested testis, or enriched cell fractions, on a feeder layer of mitotically inactivated cells, with the presence of GDNF in the medium. The stem cells grow as refractile but loose colonies. They resemble gonocytes and are positive for α6- and β1-integrin. They may be frozen and thawed successfully. They can be transplanted to immunologically compatible, busulfan-treated, mice and yield viable sperm. Although they express the pluripotency genes at low levels, they are not the same as ES cells. However, culture of PGC-stage germ cells, or spontaneous transformations of long term stem cell cultures, can give colonies resembling ES cells. Unlike spermatogonial stem cells, these ES-like variants yield teratomas on transplantation.

Further Reading

Intestine

Barker, N. (2013) Adult intestinal stem cells: critical drivers of epithelial homeostasis and regeneration. Nature Reviews Molecular Cell Biology 15, 19–33.

Buczacki, S.J.A., Zecchini, H.I., Nicholson, A.M., Russell, R., Vermeulen, L., Kemp, R. and Winton, D.J. (2013) Intestinal label-retaining cells are secretory precursors expressing Lgr5. Nature 495, 65–69.

Snippert, H.J., van der Flier, L.G., Sato, T., et al. (2010) Intestinal crypt homeostasis results from neutral competition between symmetrically dividing Lgr5 stem cells. Cell 143, 134–144.

van der Flier, L.G. and Clevers, H. (2009) Stem cells, self-renewal, and differentiation in the intestinal epithelium. Annual Review of Physiology 71, 241–260.

Epidermal Tissues

Alonso, L. and Fuchs, E. (2006) The hair cycle. Journal of Cell Science 119, 391–393.

Blanpain, C. and Fuchs, E. (2006) Epidermal stem cells of the skin. Annual Review of Cell and Developmental Biology 22, 339–373.

Doupé, D.P. and Jones, P.H. (2013) Cycling progenitors maintain epithelia while diverse cell types contribute to repair. Bioessays 35, 443–451.

Driskell, R.R., Clavel, C., Rendl, M. and Watt, F.M. (2011) Hair follicle dermal papilla cells at a glance. Journal of Cell Science 124, 1179–1182.

Hennighausen, L. and Robinson, G.W. (2005) Information networks in the mammary gland. Nature Reviews. Molecular Cell Biology 6, 715–725.

Koster, M.I. and Roop, D.R. (2007) Mechanisms regulating epithelial stratification. Annual Review of Cell and Developmental Biology 23, 93–113.

Liu, Y., Lyle, S., Yang, Z. and Cotsarelis, G. (2003) Keratin 15 promoter targets putative epithelial stem cells in the hair follicle bulge. Journal of Investigative Dermatology 121, 963–968.

Rios, A.C., Fu, N.Y., Lindeman, G.J. and Visvader, J.E. (2014) In situ identification of bipotent stem cells in the mammary gland. Nature 506, 322–327.

Rompolas, P. and Greco, V. (2014) Stem cell dynamics in the hair follicle niche. Seminars in Cell and Developmental Biology 25–26, 34–42.

Sennett, R. and Rendl, M. (2012) Mesenchymal–epithelial interactions during hair follicle morphogenesis and cycling. Seminars in Cell and Developmental Biology 23, 917–927.

Van Keymeulen, A., Mascre, G., Youseff, K.K., Harel, I., et al. (2009) Epidermal progenitors give rise to Merkel cells during embryonic development and adult homeostasis. Journal of Cell Biology 187, 91–100.

Visvader, J.E. (2009) Keeping abreast of the mammary epithelial hierarchy and breast tumorigenesis. Genes and Development 23, 2563–2577.

Watt, F.M. (2014) Mammalian skin cell biology: At the interface between laboratory and clinic. Science 346, 937–940.

Hematopoiesis

Busch, K., Klapproth, K., Barile, M., Flossdorf, M., et al. (2015) Fundamental properties of unperturbed haematopoiesis from stem cells in vivo. Nature 518, 542–546.

Bystrykh, L.V., Verovskaya, E., Zwart, E., Broekhuis, M. and de Haan, G. (2012) Counting stem cells: methodological constraints. Nature Methods 9, 567–574.

Hoggatt, J., Kfoury, Y. and Scadden, D.T. (2016) Hematopoietic stem cell niche in health and disease. Annual Review of Pathology: Mechanisms of Disease 11, 555–581.

Kondo, M., Wagers, A.J., Manz, M.G., Prohaska, S.S., et al. (2003) Biology of hematopoietic stem cells and progenitors: implications for clinical application. Annual Review of Immunology 21, 759–806.

Metcalf, D. (2007) Concise review: hematopoietic stem cells and tissue stem cells: current concepts and unanswered questions. Stem Cells 25, 2390–2395.

Morrison, S.J. and Scadden, D.T. (2014) The bone marrow niche for haematopoietic stem cells. Nature 505, 327–334.

Riddell, J., Gazit, R., Garrison, B.S., Guo, G., et al. (2014) Reprogramming committed murine blood cells to induced hematopoietic stem cells with defined factors. Cell 157, 549–564.

Schaniel, C. and Moore, K.A. (2009) Genetic models to study quiescent stem cells and their niches. Annals of the New York Academy of Sciences 1176, 26–35.

Spermatogenesis

de Rooij, D. (2001) Proliferation and differentiation of spermatogonial stem cells. Reproduction 121, 347–354.

Griswold, M.D. and Oatley, J.M. (2013) Defining characteristics of mammalian spermatogenic stem cells. Stem Cells 31, 8–11.

Klein, A.M., Nakagawa, T., Ichikawa, R., Yoshida, S. and Simons, B.D. (2010) Mouse germ line stem cells undergo rapid and stochastic turnover. Cell Stem Cell 7, 214–224.

Oatley, J.M. and Brinster, R.L. (2008) Regulation of spermatogonial stem cell self-renewal in mammals. Annual Review of Cell and Developmental Biology 24, 263–286.

Yoshida, S. (2016) From cyst to tubule: innovations in vertebrate spermatogenesis. Wiley Interdisciplinary Reviews: Developmental Biology 5, 119–131.

11

Regeneration, Wound Healing and Cancer

Some regenerative situations, such as the loss and replacement of cells from the liver, or the regrowth of hair follicles during the hair cycle, have already been mentioned, but these only represent a small part of the range of regenerative phenomena found in animals. There are at least four different types of regeneration that differ considerably in the mechanisms required to underpin them. First, there is single cell regeneration, as for example in the regrowth of severed nerve axons. Second, there is tissue regeneration, for example growth of epidermis covering a wound, or the regeneration of the liver; which restores the organ mass and the histological pattern, but without any macroscopic anatomical pattern. Third, there is the regeneration of limbs and other appendages in vertebrates, and in arthropods, which does involve the formation of new pattern; in the case of urodele limbs a complex arrangement of skeletal elements, muscles and tendons. Most dramatic is whole body regeneration, in which either heads or tails can regenerate from the same body level depending on the orientation of the cut surface. This is best studied in planarian worms, but does occur in some animals considered "higher" in the evolutionary tree, including at least some hemichordates and urochordates, which are phylogenetic neighbors of the vertebrates. These very different levels of regenerative performance must involve different mechanisms. Internal organ regeneration, such as that of the liver, has to be able to sense total organ size and to stop growing when this is

restored. Regeneration of appendages also needs to be able to do this and in addition to generate spatial pattern, including often a pattern of structures that were present in the part removed but are no longer present at the amputation surface. Whole body regeneration needs to be able to do both of these things and also to decide on whether to form a head or a tail before starting the process. This chapter will briefly examine two classic regeneration models: the whole body regeneration of planarian worms and the regeneration of the urodele limb, from the point of view of the role of stem cells in these processes.

Planarian Regeneration

Planarians are small, simple, worms belonging to the phylum Platyhelminthes. Like other animals they have three germ layers: ectoderm, mesoderm and endoderm, and a head-to-tail polarity, with eyes at the head end (Figure 11.C.1). There is a nervous system with a concentration of ganglia at the anterior end, some complex sense organs, muscle layers, and an excretory system composed of a number of protonephridia. They do not possess a through gut. Instead the pharynx, located midway down the body, opens into a blind-ended gut cavity. Also, planarians lack both a respiratory and a circulatory system, so they possess some but not all the attributes of higher animals. Several species have been used for experimental work on regeneration, and mechanisms are not necessarily the same

The Science of Stem Cells, First Edition. Jonathan M. W. Slack.
© 2018 John Wiley & Sons, Inc. Published 2018 by John Wiley & Sons, Inc.
Companion website: www.wiley.com/go/slack/thescienceofstemcells

in all of them. In recent years there has been a tendency to focus effort on one species: *Schmidtea mediterranea*.

Neoblasts

Planarians can reproduce sexually, from sperm and eggs, but they also reproduce asexually by fission. This necessarily involves extensive regeneration of the missing body parts, and is closely related to the regeneration seen following experimental transection of the body. Planarian regeneration can have two outcomes, depending on the polarity of the cut surface. An anterior-facing surface regenerates a new head, while a posterior-facing cut surface regenerates a new tail (Figure 11.1a). As mentioned above, this type of whole body regeneration must involve a control of polarity (head versus tail?) as well as the ability to grow back the missing parts. The only mitotic cells in planaria are the neoblasts. These are small roundish cells with large nuclei and uncondensed chromatin that have often been considered to be stem cells (Figure 11.1b). As a population, they self-renew and their division also provides cells for normal growth and cell differentiation. The neoblasts are responsible for regeneration. The regions of *Schmidtea mediterranea* that lack neoblasts (the pharynx and the region anterior to the photoreceptors) do not regenerate. Neoblasts are sensitive to X-irradiation and worms whose neoblasts have been ablated by radiation continue to move and feed, but they do not grow and they cannot regenerate. Such worms die after a few weeks. Neoblasts can be isolated by fluorescence activated cell sorting (FACS) on the basis of size and granularity, and the transplantation of isolated neoblasts to an irradiated worm can restore regenerative capacity. In contrast to these results indicating the importance of neoblasts for regeneration, the evidence for de-differentiation of functional cell types in planaria is weak and is confined to germ cells. So it is universally agreed that the neoblasts are responsible for regeneration in planarians.

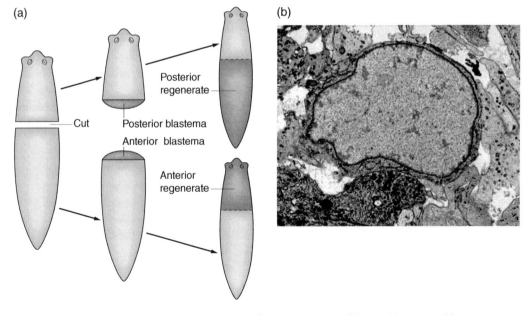

Figure 11.1 (a) Anterior and posterior regeneration of a planarian worm. (b) A neoblast, viewed by transmission electron microscopy. Note the large nucleus containing little heterochromatin, and the small amount of cytoplasm. ((b) Reddien and Alvarado, 2004. Reproduced with the permission of Annual Reviews.)

Since neoblasts comprise all the dividing cells of the worms, it seems unlikely that all of them can be stem cells. More likely some of them are long-lived stem cells and others are transit amplifying cells destined to differentiate into one or a subset of the various functional cell types. There is evidence that at least some of the neoblasts are genuinely pluripotent. One reason for thinking this is that heavily irradiated worms in which the dose is not quite lethal can become reconstituted from a few patches containing dividing cells. Each of these patches is presumed to be a clone derived from one surviving neoblast, and because they contain all the cell types it suggests that the original cell must have been pluripotent. Furthermore, it is possible to transplant single neoblasts from healthy to lethally irradiated worms and, in a

limited number of cases, this leads to complete regeneration from donor cells of all the cell types and structures in the hosts (Figure 11.2). It also restores to the irradiated hosts the ability to regenerate themselves following transection. These experiments do indicate that transplanted neoblasts are, or can be, pluripotent. Of course, as we have seen in several other examples, a transplantable cell population may not comprise exactly the same cells as a stem cell population active in the steady state situation. Based on comparison with other systems such as the mammalian intestine, it is probable that these experiments overestimate the proportion of neoblasts that are normally pluripotent. Nonetheless they do constitute evidence that this type of animal contains some pluripotent stem cells throughout the life cycle. By contrast,

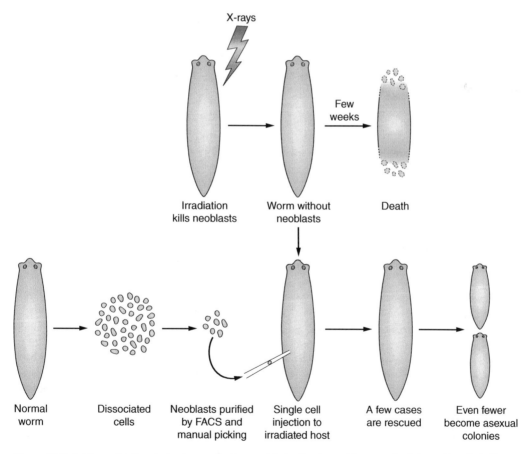

Figure 11.2 Evidence for the pluripotency of some neoblasts. Single neoblasts can be injected into lethally irradiated worms and regenerate a complete worm.

in mammals the only period at which pluripotent cells are definitely present is the short period around implantation of the blastocyst, and the evidence for the presence of pluripotent stem cells after birth is, at best, weak.

Are the neoblasts of planarians really like embryonic stem cells? This is very hard to say. The similarity between them rests mostly on their appearance and their pluripotent behavior. There is probably no direct homolog of the *Oct4* gene in planarians, and comparisons of overall transcriptome between the two types show only a modest similarity. So far it has not been possible to culture neoblasts in vitro. In terms of comparison with mammalian tissue-specific stem cells there is a significant difference in relation to the Wnt pathway. In mammalian systems, Wnt signaling is usually required to maintain proliferation. In planarians on the other hand the role of Wnt signaling seems to be very different, as it is responsible for control of the head-tail polarity of the whole body. Evidence of this is that the expression of Wnt pathway components is elevated in posterior over anterior blastemas. Furthermore, the administration of ds-RNA complementary to β-catenin mRNA can reprogram a posterior blastema to an anterior character, generating a double headed worm.

It is worth noting here a peculiarity of nomenclature in planarians. In general in regeneration studies, the term "blastema", refers to the bud of cells, which appears at the cut surface and gradually differentiates into the missing structures. In animals such as urodeles or annelids the blastema is composed of dividing cells. However in planarians the blastema itself is post-mitotic and the region of excess neoblast division, which generates the cells of the blastema, lies proximal to it.

Amphibian Limb Regeneration

Amphibian limb regeneration is a well-known classic regeneration model that has been studied for over 100 years. Interestingly we are still ignorant of the answer to the simple question: "Why can they regenerate when we cannot?" However we do now have quite good information about cell lineage and about how the pattern of the regenerate is controlled. Among amphibians, newts and salamanders (urodeles) can mostly regenerate amputated limbs while frogs and toads (anurans) cannot, although there is some regeneration of larval stage limbs in anurans. Popular models for regeneration research are the axolotl (*Ambystoma mexicanum*), and the Eastern, or Red-spotted, newt (*Notophthalamus viridescens*).

The Regeneration Blastema

The observable course of events is similar in the two species (Figure 11.3). Following amputation, a thin wound epithelium rapidly covers the cut surface. Then over some days, undifferentiated cells accumulate to form a blastema. The apical epidermis thickens to form a multilayered apical cap and both this and the blastemal proliferate. Once the blastema is about as long as it is wide, the proximal cells start to differentiate into the missing structures. The limb consists of a pattern of skeletal structures: cartilage in young animals progressively replaced by bone as they mature; connective tissues elements comprising dermis, fibrous capsules, tendons and ligaments; and muscles. In addition the regenerate is provided with a system of blood vessels and of nerves.

The blastema forms but does not proliferate if the skin is sutured across the wound, and the properties of the wound epithelium itself seem to be important for proliferation. Another important requirement for growth of the blastema is the presence of a nerve supply. The cut nerve endings emit growth factors including neuregulin (similar to EGF), and there is also a urodele-specific mechanism involving a secreted molecule called "anterior gradient" (AG), which binds to a cell adhesion molecule called Prod1. AG is mitogenic for blastemal cells. Following amputation it is released by Schwann cells of the nerve sheath and is subsequently made and secreted by the wound epithelium and apical

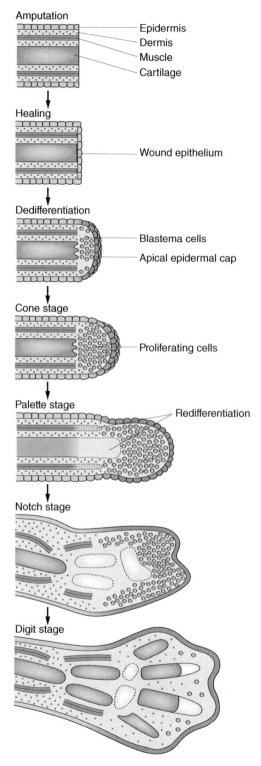

Amputation
— Epidermis
— Dermis
— Muscle
— Cartilage

Healing
— Wound epithelium

Dedifferentiation
— Blastema cells
— Apical epidermal cap

Cone stage
— Proliferating cells

Palette stage
— Redifferentiation

Notch stage

Digit stage

Figure 11.3 The course of urodele limb regeneration. (Slack, J.M.W. (2013) Essential Developmental Biology, 3rd edn. Reproduced with the permission of John Wiley and Sons.)

cap. There have been long debates about the possible role of the immune system in relation to regeneration. In general it seems that inflammation is inhibitory to regeneration and tends to lead instead to fibrotic scar formation. However, there is also a recent report that macrophages in the blastema are necessary for axolotl regeneration.

The cell lineage of the regenerate has been investigated by making transgenic axolotls in which one tissue type is labeled with a genetic marker such as GFP. This is done by suitable operations on the embryo to replace the epidermis, neural tube, somatic mesoderm or lateral plate mesoderm at the body level which will later form the forelimbs. The host embryos are raised to become young animals in whose forelimbs the epidermis, nerve axons, myofibers or connective tissues respectively are labeled. If such limbs are amputated and allowed to regenerate, the GFP should remain present in any tissue types derived from the labeled one. The results indicate that the cell lineage in regeneration is remarkably similar to that in embryonic development. Epidermis generates epidermis, muscle generates muscle, and connective tissue generates connective tissues (Figure 11.C.2). Labeled nerve axons regenerate down the severed tracts, accompanied by Schwann cells from the neural crest.

So, contrary to general belief, the blastemal cells are not pluripotent stem cells. Instead the blastema is a mixture of cells of different potency, each derived from its own tissue type. Even multipotency is confined to those of the blastema cells that are connective tissue-derived. It has been shown that labeled dermis will generate labeled tendons or cartilage and vice versa, so there does seem to be metaplasia between the principal connective tissue types. Beyond this there is no metaplasia in amphibian limb regeneration. There is a curious difference between the two popular experimental species in terms of the origin of skeletal muscle. The newt *Notophthalamus*, shows de-differentiation of multinucleate myofibers, whose nuclei re-enter the cell cycle, and break up into proliferating mononuclear myoblasts, which later re-differentiate into muscle fibers. In the axolotl on the other hand, muscle regeneration follows the mammalian model in occurring from muscle satellite cells, with the damaged fibers degenerating and dying. The reason for this difference is not known.

Pattern Formation in Regeneration

The problem of pattern formation is a very intriguing one to which much attention has been given in recent decades of regeneration research. The limb is inherently asymmetrical in its three dimensions: proximal-distal, anteroposterior and dorsal-ventral. By convention, the anterior of the limb is the digit I (in the human hand the thumb), and the posterior the highest numbered digit (in the human hand, the little finger, digit V). The dorsal side is the back of the hand, and the ventral side the palm. Experiments in which tissue components are rotated or displaced prior to amputation indicate that the pattern is laid down entirely by the connective tissue-derived population of the blastema. The myoblasts, blood vessels and nerves then move in and accommodate themselves to the pre-existing pattern. In order to make a three dimensional pattern from a blastema, it seems that each cells needs three coordinates in order to know what structure to form. In limb development is known that the proximal–distal signal is a gradient of retinoic acid from the flank; the posterior to anterior signal is a gradient of sonic hedgehog from the zone of polarizing activity;

and the dorsal–ventral distinction is made by Wnt7A from the dorsal epidermis. Many of the genes expressed in development are re-expressed in regeneration indicating that at least part of the developmental mechanism is re-deployed. However there are some differences. The proximal–distal pattern must start at the cut surface, not at the flank. The size of the blastema varies with the size of the animal and is generally much larger than a larval limb bud, and this must affect the behavior of gradients of morphogens. The blastema does re-express sonic hedgehog at the posterior margin but the capacity to do this seems to be stored in the mature connective tissue. In the proximal–distal axis there is a gradient of cell adhesivity, which increases from proximal to distal and is necessary for the correct patterning. The GPI-linked cell surface protein called Prod1, which acts to reduce adhesivity, and is expressed at higher levels in proximal than in distal blastemas, has been implicated as the molecular basis of this adhesion gradient. However, for all three pattern axes it seems that the mature connective tissue contains some stable memory of the embryonic situation which can be re-activated following amputation. This is very stable indeed as tissue displacement experiments have shown that the positional memory lasts for at least a year without diminution (Figure 11.4; Figure 11.C.3). Ignorance about its molecular basis explains why the term "positional information" is still used in regeneration research

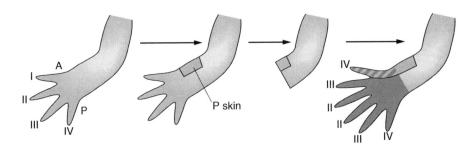

Figure 11.4 Positional memory demonstrated by limb regeneration. A piece of posterior skin is grafted to the anterior. After as much as a year, the limb may be amputated through the graft and it regenerates as a double posterior structure in which part of the respecified pattern comes from host tissue. (Slack, J.M.W. (2013) *Essential Developmental Biology*, 3rd edn. Reproduced with the permission of John Wiley and Sons.)

although has died out in other areas of developmental biology as mechanisms have been discovered and explained. Interesting though the problem of pattern formation in regeneration is, it is not directly relevant to the issue of whether the blastema consists of stem cells, and, if so, what sort of stem cells they are. It is safe to say that the epidermal covering of the blastema contains epidermal stem cells. Also the blastema evidently contains muscle satellite cells, or in the case of the newt, myoblasts derived from myofiber de-differentiation. The key question concerns the nature of the connective tissue-derived blastema cells. Not least because the mechanism of their partial de-differentiation must go to the heart of the unanswered question of why amphibians regenerate limbs while we cannot.

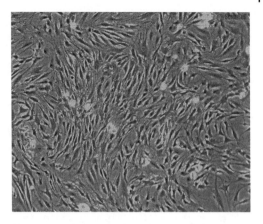

Figure 11.5 Rat mesenchymal stem cells from the bone marrow. (ScienCell Research Laboratories, Inc.)

Mesenchymal Stem Cells

If the regeneration blastema does contain stem cells, then the mammalian counterpart might be the so-called mesenchymal stem cells, or MSCs. These were originally discovered in bone marrow as cells that would adhere to tissue culture plastic and could be cultivated in vitro. They were called CFU-F (colony forming unit – fibroblastic) in line with the various types of colony forming unit (CFU) of the hematopoietic system also found in the bone marrow. They were later called marrow stromal cells or mesenchymal stem cells, both conveniently abbreviating to MSC. CreER labeling of MSC using the inducible *Mx1* promoter indicates that in vivo they serve as precursors for bone. Since they reproduce themselves, generate differentiated cells and persist for the life of the animal they do seem to meet the normal definition of tissue-specific stem cells.

However, the situation is more complex than this. Most work with MSCs has been conducted in vitro (Figure 11.5). It is well established that in vitro, depending on the medium used, they can differentiate not only into bone but also into cartilage and adipocytes. Even a single clone can form all three cell types if its progeny are divided and cultured in the appropriate media. So in vitro, these are multipotent cells. On occasion mesenchymal cells isolated from the bone marrow have been reported to show a much wider potency, approaching pluripotency, although the difficulty of repeating such results indicates that such cells probably arise through somatic mutation combined with selection in culture and are not normally present in vivo. But reports concerning pluripotent cells have confused the literature and have led many to believe that MSCs have a wider potency than the bone, cartilage and adipocytes that is well documented. In fact reports of differentiation of MSC to other cell types are not very reproducible and remain controversial. MSCs also have some negative immunomodulatory effects in various in vitro assays. These are due to a proliferation of secreted and cell contact mediated factors and may account for some of the reported beneficial effects of MSC used in human cell therapy.

MSC-like cells have been isolated from many other tissue types apart from the bone marrow. In particular, human adipose tissue is often available from cosmetic surgery procedures and is frequently used as a source of "stem cells" which have some properties in common with the bone marrow MSCs. Such cell preparations are doubtless very heterogeneous although they inevitably become more homogenous if they are cultured in vitro because of the competition and selection that always occurs in culture. There has been much debate about the nature of MSCs from various tissue sources. The following questions are pertinent:

1) Are the MSCs from different tissues the same or different?
2) What properties can be used to define them to ensure comparability between different research centers?
3) What is the in vivo cell of origin of the MSCs?
4) What is the normal function of this cell or cells in vivo?

The first two questions are hard to answer because of poor specificity of characterization. Adherence to plastic and growth in culture are by no means specific to MSCs. It has been difficult to find specific molecular markers. Those that are often used for human MSCs, such as CD73 (a GPI-linked ecto 5'nucleotidase) and CD105 (endoglin, a TGFβ co-receptor), are not very specific and are also found on ordinary fibroblasts. The ability

to show clonal growth from single cells, and to differentiate into more than one cell type: bone, cartilage, adipocyte and sometimes also smooth muscle and skeletal muscle, does distinguish MSC from fibroblasts. The in vivo counterpart of MSC has often been supposed to be the pericyte, an accessory cell of small capillaries. The histological definition of a pericyte is a mesenchymal cell lodged within the basement membrane of capillaries but not found in larger blood vessels or in lymphatic vessels (Figure 11.6). They are typically elongated in the direction of the capillary with many processes running circumferentially. They occur at different frequencies in different tissues from about 1:1 with endothelial cells in brain down to about 1:10 in skeletal muscle. Their physiological function has been supposed to be concerned with the permeability of capillaries and a possible regulation of blood flow at the capillary level by constriction. A recent study using *Tbx18-CreER* labeling has shown that pericytes do generate MSCs when cultured in vitro. However it also showed that in vivo the pericytes did not become other cell types either during aging or in various pathological states.

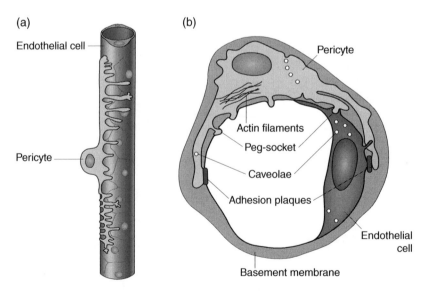

Figure 11.6 The pericyte, a possible cell of origin for the MSC. (a) Diagram of a capillary with an attached pericyte. (b) Cross section through the pericyte. (From: Armulik, A., Genové, G. and Betsholtz, C. (2011) Pericytes: developmental, physiological, and pathological perspectives, problems, and promises. Developmental Cell 21, 193–215. Reproduced with the permission of Elsevier.)

Whether all MSC found in cell culture derive from pericytes is not certain, and they may also derive from fibroblasts or vascular smooth muscle cells, or maybe different cell lines come from different sources. It is prudent to conclude that MSC are rather heterogeneous, based on the tissue of origin and the procedures used for isolation, that their cell of origin is uncertain, and that they have some multipotency of differentiation potential in vitro. Because MSC have been extensively commercialized for use in "stem cell therapy" it is important to note the relative lack of scientific rationale for their use. In particular there is no evidence that injected MSC are likely to contribute to rebuilding of a damaged structure in vivo. Allogeneic MSC are likely to be rejected by the host immune system and to die soon after injection, while autologous ones may persist for some time. In either case any influence they exert is likely to be due to secreted factors (a so-called "paracrine effect") rather than any stem cell differentiation function.

Mammalian Wound Healing

Soft Tissue Wounds

The course of mammalian healing following damage to the surface of the body is shown in Figure 11.7. Following the wound, there is bleeding from ruptured blood vessels. So long as this is not too severe, a blood clot forms by polymerization of fibrinogen to fibrin, brought about by the clotting cascade which climaxes in the activation of the protease thrombin. The clot is very rich in blood platelets which secrete various bioactive substances including platelet-derived growth factor (PDGF) and complement. These substances, together with bacterial debris usually found in the wound, attract abundant neutrophils from the blood. After 2–3 days monocytes are also drawn in from the blood and become macrophages. In the second period, from 2–10 days, new blood vessels grow into the clot, stimulated by vascular endothelial growth factor (VEGF) and FGF from the resident cells. The clot becomes replaced by a mass of capillaries, fibroblasts and macrophages, known as a granulation tissue. Keratinocytes regrow below the surface scab, stimulated especially by the presence of hepatocyte growth factor (HGF), and restore the normal skin barrier to the exterior. Fibroblasts from the wound edge become myofibroblasts and their contraction helps to reduce the wound area. Fibroblasts also secrete large amounts of collagen III. Finally there is a period of remodeling, starting 2–3 weeks after the wounding and extending maybe a year, this involves extensive reworking of the scar by fibroblasts, macrophages and endothelial cells using matrix metalloproteases (MMPs) to digest the matrix, and secreting more matrix. This results in the replacement of the initial collagen III by collagen I together with other extracellular matrix components. During this period the macrophages and myofibroblasts die or depart and leave a scar composed of excessive and disorganized ECM material. The epidermis overlying the scar lacks hairs and sweat glands if the initial wound was deeper than the roots of these appendages in the dermis.

Because of the problems posed by scars there has been some interest in the process of wound healing in mammalian fetuses, in which no scars are formed. The main difference is the lack of an inflammatory response. The closure of the wound occurs by morphogenetic movements of the surrounding epithelium, such as contraction of the wound edge driven by actin cables, as well as by cell proliferation. The matrix formed is softer with a preponderance of hyaluronic acid, fibronectin and tenascin. This once again indicates the dichotomy between inflammation and regeneration, noted above in the context of amphibian limb regeneration.

Healing of Bone Fractures

The events following a bone injury have many similarities to the soft tissue situation,

(a)

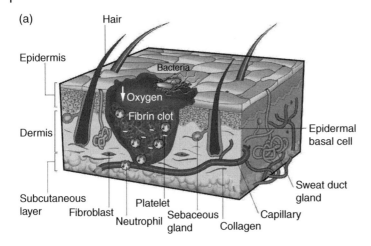

(b)

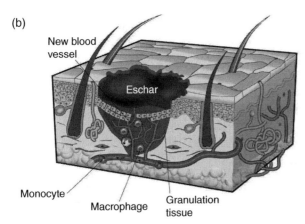

(c)

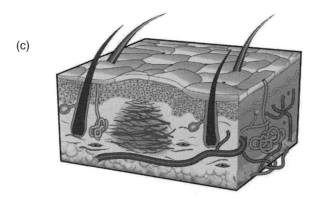

Figure 11.7 Healing of a mammalian skin wound. (a) Formation of the fibrin clot. (b) Formation of the granulation tissue. (c) Following healing, a collagenous scar persists. (From: Gurtner, G.C., Werner, S., Barrandon, Y. and Longaker, M.T. (2008) Wound repair and regeneration. Nature 453, 314–321. Reproduced with the permission of Nature Publishing Group.)

the main difference being that the regenerative events generate new cartilage and bone rather than epidermis. The initial injury will rupture blood vessels and cause bleeding and the formation of a blood clot. This includes many platelets, neutrophils, monocytes and lymphocytes. Macrophages clear away debris and a granulation tissue forms. Promoted by a low oxygen environment, the granulation tissue becomes replaced by cartilage secreted by chondrocytes and containing much collagen II. This is known as a soft callus. There may also be some deposition of fibrous matrix by fibroblasts. The soft callus becomes populated by osteogenic cells. These secrete VEGF, attracting capillaries. The chondrocytes become hypertrophic and mineralize the matrix, then die. The osteoblasts secrete a bony matrix, called hard callus, to replace the soft callus. This depends on a newly formed vascular system and a higher oxygen level, and also on BMPs. Finally there is a prolonged period of remodeling during which osteoclasts, which are multinucleate cells derived from the hematopoietic system, degrade the hard callus and osteoblasts secrete new bone matrix. The remodeling process is similar to the normal bone remodeling which occurs throughout life. In cancellous bone this occurs at the surface of trabeculae, and in compact bone, the osteoclasts ream out the old blood vessel spaces and osteoblasts follow them to rebuild the matrix.

There has been much debate about the origin of the chondrocytes and osteoblasts which participate in bone repair. There is abundant cell division in the periosteum near the injury. The inner layer of the periosteum consists of spindle shaped cells, also found on endosteal surfaces, which can be considered to be MSCs, or skeletal precursor cells, so it is likely that most of the new cells come from the damaged bone. However it is known that new skeletal differentiation can occur in soft tissues following localized treatment with BMPs, so it is possible that some cells may also be recruited from the surroundings.

Spinal Cord Injuries

Spinal injuries tend to arise from compression, contusion (bleeding), or lacerations. Actual transection of the cord is rare in human injuries although may sometimes be examined in animal models. Damage to neurons and axons arises from the indirect effects of inflammation as well as from the initial trauma. Some hemorrhage occurs immediately following the injury. In the following 1–2 days there is an influx of neutrophils, an activation of microglia (CNS macrophages), necrosis of neurons, astrocytes and oligodendrocytes, swelling of damaged axons, and loss of myelin sheaths from axons. In the next 7–10 days the microglia and macrophages from the blood clear debris and secrete various bioactive substances including cytokines, superoxides and nitric oxide. At the edge of the wound, astrocytes become hypertrophic, or reactive, with much larger cytoplasmic processes than normal. They secrete numerous neural growth factors, extracellular matrix and cell adhesion molecules. There is some generation of new astrocytes and oligodendrocytes from the ependymal layer but no generation of new neurons. Angiogenesis increases the capillary density in the damaged area. In the later phase, which may take months or years, there is degeneration of the distal parts of damaged axons and their sheaths. Unlike in the peripheral nervous system, CNS axons do not regenerate down their former tracts. The reactive astrocytes, together with some microglia and macrophages, form a "glial scar" consisting of tightly interwoven cells surrounded by a matrix rich in chondroitin sulfate proteoglycan. In the case of lacerating injuries there is also likely to be a fibroblastic proliferation and secretion of matrix to form a collagenous scar. There may also be a proliferation of Schwann cells from the nerve roots (Schwannosis) and some formation of cysts. The glial scar is considered to have a protective function in the early stages as it limits the spread of excitatory amino acids, free radicals and cytokines which are produced in the damaged area and are liable to cause additional neuronal cell death. It also has an immunomodulatory effect and reseals the blood brain barrier. However the long-term cost of the glial scar is a substantial inhibition of axonal regeneration. The net result of the injury is likely to include loss of motorneurons and of interneurons located in the wound area, and loss of axons from neurons located at higher levels of the CNS. Although human spinal injuries show a great diversity of clinical outcomes, at the cellular level these changes are not reversible.

Regeneration and Repair

It is usual to refer to events that re-create the original tissue structure and cell types with the term "regeneration", while events that enable the animal to continue to function, but without restoration of the original structure is called "repair". Both may be seen in mammalian wound healing. There is genuine regeneration of the epidermis, which can grow to cover significant gaps. However there is not regeneration of the epidermal appendages: hair follicles and sweat glands, which, as we have seen in Chapter 10, arise from their own stem cell populations. Nor is their proper regeneration in the dermis which usually ends up containing a collagenous scar. There is also genuine regeneration of bone, fed by skeletal progenitor cells, which may be the same cell type as the bone marrow mesenchymal stem cells (MSCs). Mammalian bones can therefore heal but they cannot grow distally following an amputation in the manner of the amphibian limb. In the central nervous system there is little regeneration. The main events of wound healing are concerned with sealing off the injury by means of the glial scar, and not with regrowing neurons and axons in the way that is seen in amphibians and fish.

It is still not known why amphibians can regenerate more structures and tissues than mammals. One reason may be a greater ability of cells to dedifferentiate. The ability of connective tissue cell types in the limb to interconvert during regeneration suggests that they may dedifferentiate to something resembling a mammalian MSC. What underlies this ability, for instance a difference of chromatin organization, is not known. In the central nervous system of lower vertebrates there is considerably more continuing neurogenesis from neural stem cells than is found in mammals and this doubtless enables a much higher degree of spinal regeneration. There also seems to be a fundamental incompatibility between an inflammatory response and tissue regeneration. Inflammation seems generally to be linked to scar formation rather than to regeneration, and the ability of the fetal mammals, which have a limited development of the immune system, to heal wounds without scarring, is supportive of this idea.

The zebrafish exhibits some types of regeneration, including that of fins, parts of the heart and the CNS. Maybe the large community of zebrafish workers using the range of new techniques now available, can in due course answer some of the old problems concerning regeneration.

Cancer

What are cancer stem cells? Are they a special form of stem cell that is also cancerous? Are they normal tissue-specific stem cells which are found in cancers? Or maybe they do not exist at all and are an illusion of the genetic variability found in all cancers? In recent years cancer stem cells have been a particularly contentious area in stem cell biology and to evaluate the situation it is necessary very briefly to consider some of the background of the molecular biology of cancer.

Cancer occurs in all animals but the focus is usually on the human disease because it is such a major cause of death and suffering. Cancer is fundamentally a growth of the body's own cells outside of its normal system of regulation. This means that a cancer will expand disproportionately, perhaps causing functional problems in its own organ or mechanical problems in the surroundings. It is likely to grow and invade into surrounding tissues, causing further damage. It may also export cells to distant parts of the body where they grow as secondary cancers or metastases, causing further mechanical damage. In addition cancers often secrete toxic factors that cause systemic illness.

The molecular biology of cancer has been extensively studied in the last 40 years and it is now quite well-known. Various aspects of cell behaviour are altered in cancers and the molecular basis of these is understood to some extent. One important type of change is an upregulation of intracellular signaling pathways leading to increased cell proliferation. This may arise from gain of function mutations in many different components of the pathways concerned. The mutated versions of the genes encoding these components are known as oncogenes. Another common type of change is a loss of negative controls on cell division. The normal genes encoding such factors are called tumor suppressor genes, and, when these are lost by mutation, the cells divide out of control. A tendency toward invasive behavior is acquired by mutations which decrease the normal degree of cell adhesion and promote epithelial to mesenchymal transitions. Cell death mechanisms are also often disabled. Normal cells usually die when they acquire abnormalities but cancer cells frequently inactivate mechanisms that sense damage and then survive and proliferate in their altered form. Normal cells, other than stem cells, are subject to replicative senescence, and can only divide a certain number of times. However, cancers often have upregulated telomerase which enables repair of the telomeres following DNA replication and evades replicative senescence. Finally cancers usually secrete angiogenic factors to encourage the ingrowth of new blood vessels to maintain their supply of nutrients necessary for their proliferation and expansion.

In terms of nomenclature the term "neoplasia" means new growth. Cancers and related growths may be described as neoplasms. The word "tumor" originally just meant any type of swelling but is now applied to neoplasms so tumor and neoplasm are essentially synonymous. "Cancer" normally refers specifically to a malignant neoplasm, that is to say those showing invasive behaviour and metastases. A "benign" neoplasm is confined in its growth, for example by an external capsule, and less likely to cause widespread damage. However "benign" and "malignant" are clinical rather than biological terms and refer more to the outlook for the patient than to the essential biology of the tumor.

The classification of tumors is based on the normal histology of tissues. A carcinoma is a tumor of an epithelial tissue, so when lung, stomach or breast cancer are mentioned these normally refer to epithelial neoplasms of the relevant organ. Carcinomas also contain a certain percentage of connective tissue, or stroma, and this fact needs to be kept in mind when evaluating bulk forms of analysis such as DNA sequencing of tumor samples. A benign tumor of an epithelium is called an adenoma if the epithelium is of glandular structure, and a papilloma if it is of squamous type. The corresponding carcinomas are adenocarcinomas (origin from glandular epithelium) and squamous cell carcinoma (origin from stratified squamous epithelium). Sarcomas are tumors of connective tissues: so an osteosarcoma is derived from bone, chondrosarcoma from cartilage, leiomyosarcoma from smooth muscle and so on. In addition there are leukemias which are cancers of white blood cells disseminated through the blood and bone marrow. Lymphomas are cancers of lymphocytes which tend to be localized as solid growths. A few other tumor types do not fall into this classification, most notably the germ cell tumors including seminomas and teratomas. The latter contain many disorganized differentiated tissues and can also be formed experimentally by implantation of pluripotent stem cells into immunodeficient animal hosts, as described in Chapter 6. Cancers which are very undifferentiated may defy histological classification because of lack of identifying features and be described as "anaplastic" (without form). They are often highly malignant.

Genetic Heterogeneity of Cancer

In the 1950s it was noted that many types of human cancer increase in incidence very steeply with age. This is why the overall cancer death rate continues to rise despite improved treatment: it is simply because the proportion of older people in the population continues to rise, leading to correspondingly more cases of cancer and more deaths caused by cancer. It was found that the age specific incidence curves could be modeled quite well under the assumption that cancer arose in a single cell which had collected 4–6 independently occurring somatic mutations, all of which were necessary to cause the cell to become cancerous. This model particularly fitted carcinomas although breast cancer also shows a negative inflection in the age-incidence curve associated with the menopause. The model did not fit leukemias, for which a smaller number of mutations seemed to be needed, or pediatric tumors, where there is a greater role for inherited predisposition. Because the principal cause is somatic mutation, this model emphasized the following features: the chance nature of cancer; the probable variability between different cancers of the same histological type; and the likelihood of considerable heterogeneity within a single tumor. All of these predictions have turned out to be correct following recent studies of cancer genetics.

Where the model did not fit so well was in accommodating the various well-known precancerous lesions. For example cancer of the colon usually arises in pre-existing adenomatous polyps, and squamous carcinoma of the bronchus usually arises in patches of squamous metaplasia of the normally columnar respiratory epithelium. It seems reasonable to suppose that a precancerous lesion arises because of one somatic mutation, and a cancer arises within it due to a second. This involves the assumption that the precancerous lesion grows faster than the normal tissue so the number of target cells carrying the first mutation is increased, and the likelihood of a cell acquiring the second mutation is correspondingly increased. The other issue was the discovery of substantial genetic damage in cancers, especially chromosome losses, gains, translocations and rearrangements. This led to the idea that the early mutations increase the probability of subsequent ones on a per cell basis. This might occur, for example through mutations which disable parts of the DNA repair mechanisms, leaving more DNA damage unrepaired. Both of these processes, the growth of precancerous lesions and the occurrence of mutations destabilizing the genome, would reduce the number of obligatory mutations below the 4–6 required in the initial model.

The consensus view of cancer that has developed involves putting these various ideas together. First there is initiation by a somatic mutation in a single cell. This causes more rapid expansion than usual and increases the risk of a second mutation by increasing the number of target cells. At a later stage another mutation increases the frequency of mutations per cell above the normal. Many somatic mutations then occur and some combinations of these increase the aggressiveness of the tumor and become selected for by differential growth so that they overgrow the whole of the tumor. Eventually mutations will arise, and be selected for, that lead to complete loss of normal growth control, de-differentiation, invasive behaviour and metastasis. An influential model of the development of colon cancer, including typical mutations causing each step of the progression, was presented by Vogelstein in 1990 (Figure 11.8).

This view leads us to expect that any cancer will be different from others of the same histological type, that the whole of the tumor will have undergone selection for a range of

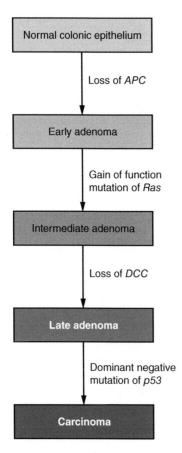

Figure 11.8 The "Vogelgram". A model proposed by Vogelstein to account for the stages in development of colon cancer. In reality there is considerable variation between individual colon cancers. (Slack, J.M.W. (2013) Essential Developmental Biology, 3rd edn. Reproduced with the permission of John Wiley and Sons.)

mutations enabling loss of control of growth, and also that the tumor at any time is likely to contain subclones of cells, defined by different sets of mutations. Recently it has become possible to sequence tumor genomes using modern high-throughput sequencing methods to determine exactly what is the nature of genetic change in human cancer. The results show that all cancers contain vast numbers of somatic mutations. In fact normal somatic tissues also contain vast numbers of somatic mutations but the data for normal cells is more limited because single cell sequencing

is not reliable enough to detect mutations. Only where the tissue contains expanded clones can mutations be detected in normal tissues. The problem then arises of which of the thousands of mutations typically present are actually responsible for the pathology of the tumor. These are called "driver" mutations, while those that do not contribute to the cancer phenotype are called "passengers". Passenger mutations exist because in any cell population subject to positive selection, all the mutations already present in the DNA will be expanded together. So any one driver mutation is likely to carry a large number of passengers with it. The identification of drivers has relied partly on the fact that they tend to recur in different cancers, and partly on pre-existing knowledge of mutations that give rise to oncogenes or unmasking of tumor suppressor genes. At present about 5–600 drivers are known. No single driver mutation is found in all cancers, and it is rare for a specific driver mutation to be found even in all cancers of a particular histological type. However some driver mutations are quite common overall. These include mutations in the genes encoding p53, RAS, PI3K, and RAF. p53 normally inhibits cell division, promotes DNA repair, and promotes apoptosis in cells with unrepairable DNA damage. Loss of p53, or dominant negative mutations, reduce these functions and increase the likelihood of damaged cells surviving and proliferating. RAS is the GDP exchange protein that is involved as an early step in both the ERK and PI3kinase signaling pathways (Figure 7.4c and g), and gain of function mutations activate both pathways. PI3 kinase is the first step of the Akt/PKB signaling pathway, and is normally activated in response to insulin and other receptor tyrosine kinase-activating growth factors. Gain of function mutations continue to signal in the absence of their extracellular factors and enable the cell to grow in the absence of these signals. RAF is a protein kinase in the ERK signaling pathway, also stimulated by many growth factors. Gain of function mutations again promote growth in the absence of the usual extracellular factors. These examples illustrate the fact that most driver mutations are dominant, only one copy of the mutation being required to have an effect.

Tumors also contain many structural chromosome abnormalities, particularly in the more advanced stages. An example of an analysis of a human breast tumor sequenced to a depth of 188 fold is shown in Figure 11.9. The ordinate shows fractions of reads indicating particular mutations. The large cluster of points at 35% represents mutations found in the whole tumor. It is not 100% because most mutations are found on one of the homologous chromosomes, so the maximum frequency in tumor cells would be 50%. It is actually 35% not 50% because in this tumor sample, 30% of the cells were from the stroma and 70% from the tumor itself. The stroma may also contain somatic mutations but, as in other normal tissues, these will be scattered at random through all the cells so the abundance of any one of them will be below detection level.

In order to detect subclones within a cancer it is necessary to sequence to considerable depth, i.e. to have hundreds of reads for any given sequence in the DNA. This is because a mutation present on, say, 10% of the cells will only be present in 5% of the DNA assuming the homologous chromosome is still present. It may even be less than this where some of the DNA sample comes from the stroma. To reliably detect a minority sequence present in a few percent of the DNA requires a very large number of reads. However such analyses have been carried out and have confirmed that cancers do typically contain identifiable subclones. Each is the result of positive selection for a single cell with a specific combination of driver mutations, and it will also display all the passenger mutations present when the selection began. It is even possible to work out "phylogenetic trees" for the subclones present in a tumor. This requires that different mutations are associated with a particular subclone by being found together on single reads, or by being retained together following loss of the homologous chromosome in the subclone. Such phylogenetic trees go back to the time when the tumor was homogeneous. This is called the "last common ancestor" and will itself contain many mutations and have undergone considerable selection compared to the normal tissue. The appearance

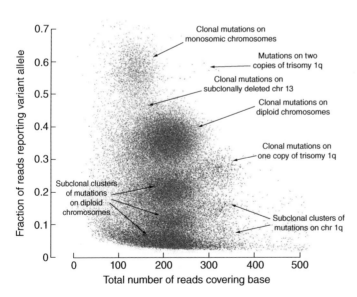

Figure 11.9 Clusters of mutations in a single case of human breast cancer. The sample has been sequenced to a very high depth to enable discovery of subclonal mutations. Each dot represents one sequence containing a mutation. The major cluster at 35% abundance represents mutations found in all tumor (epithelial) cells of the sample. Clusters at 5–20% abundance represent mutations present in subclones of the tumor. (From: Nik-Zainal, S., et al. (2012) The life history of 21 breast cancers. Cell 149, 994–1007. Reproduced with the permission of Elsevier.)

of this last common ancestor genotype will occur probably years before it becomes the only cell type in the tumor, and maybe even before the tumor was clinically detectable. The pattern of subclones, and the driver mutations responsible for their selection are, of course, variable between different individual tumors of the same histological type. A model of human breast cancer development based on these kinds of data is shown in Figure 11.10. All of this genetic complexity and heterogeneity poses considerable challenges for cancer therapy. It is easy to see how a treatment that destroys most of the tumor will aid the selection of subclones that are relatively resistant, and that these are likely to grow back in the form of a tumor recurrence.

Cancer Stem Cells

The above discussion indicates the high degree of variability in cancers derived from somatic mutation and selection at the cellular level. But there are other causes of heterogeneity. The histological classification of cancers is based on their resemblance to the

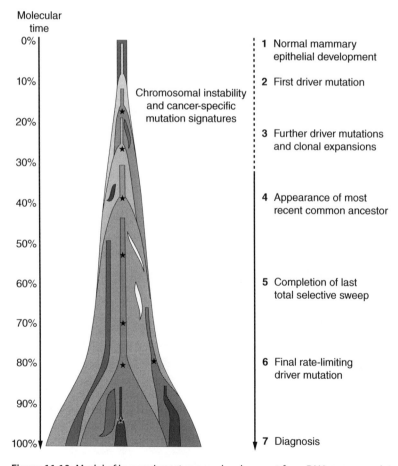

Figure 11.10 Model of human breast cancer development from DNA sequence data similar to that of Figure 11.9. The model indicates the variability of the tumor, which consists of several subclones, and the long time for its development before clinical diagnosis. (From: Nik-Zainal, S., et al. (2012) The life history of 21 breast cancers. Cell 149, 994–1007. Reproduced with the permission of Elsevier.)

normal tissues from which they are derived. If such tissues are maintained by stem cells it seems logical that the cancers should also contain stem cells. Carcinomas also contain stroma that provides a range of microenvironments for the cancer cells, including, presumably, stem cell niches. So we could expect to find stem cells, transit amplifying cells and differentiated cells within a cancer, and in fact pathologists have described possible stem cell populations within cancers for decades.

The more recent evidence for stem cells in cancer has come not so much from the analysis of tissue organization but from a combination of the techniques of cell sorting, in vitro culture, and transplantation into immunodeficient mice (Figure 11.11). The original example was from human acute myeloid leukemia and showed that if $CD34^+CD38^-$ cells were sorted out, they alone could regenerate the tumor when transplanted to immunodeficient mice. CD34 is a cell surface sialomucin found on human hematopoietic stem cells, while CD38 is a cyclic ADP ribose hydrolase found on the surface of many mature immune cells.

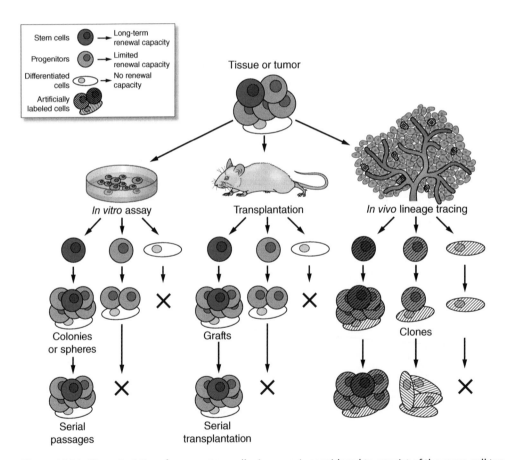

Figure 11.11 Characteristics of cancer stem cells. A cancer is considered to consist of the same cell types as are found in a normal tissue containing stem cells. Only the stem cells themselves will grow in vitro long term, and be capable of serial transplantation. If subject to lineage labeling, only a label in a cancer stem cell will persist and populate the entire tumor. (Modified from Beck, B. and Blanpain, C. (2013) Unravelling cancer stem cell potential. Nature Reviews. Cancer 13, 727–738. Reproduced with the permission of Nature Publishing Group.)

Subsequently it was found that CD44$^+$CD24^{lo} cells from human breast cancer could likewise regenerate the tumor after transplantation. In this case however the antigen combination does not correspond to that found on a normal population of mammary stem cells. Many other examples followed, drawn from various types of solid tumor. In each case one or more cell surface antigens were used to sort a subpopulation of cells which were capable of regenerating the tumor after transplantation into immunodeficient mice. In some cases, in vitro culture of the putative stem cells was also possible, generating "spheres" that contained some stem cells and some differentiated cells.

So the definition of cancer stem cells comprises the following properties. They are a small minority of cells in the original tumor. They can be prospectively isolated using cell surface markers. They can regenerate the entire tumor with all its original organization and cell types, and may be serially transplanted from one immunodeficient mouse host to another. Some might also add the ability of cancer stem cells to be cultivatable as spheres in vitro. The importance of cancer stem cells is potentially that they are the cells that must be destroyed if the cancer is to be eradicated. If they are resistant to therapy then the therapy may shrink the tumor considerably but it will soon grow back from the surviving stem cells.

Stem cells do certainly exist in some cancers, and it has been possible to identify stem cells using CreER lineage labeling in mice. So for example *K14-CreER* labeled cells show stem cell behaviour in mouse papillomas induced experimentally by treatment with a mutagen + phorbol ester (Figure 11.C.4). In human adenomas, clonal patches suggestive of stem cell activity have been identified using the loss of function mutation of mitochondrial cytochrome oxidase (see Chapter 3). However there has been much controversy over the standard methods used to identify cancer stem cells. The xenograft method mostly tests for survival in a mouse environment. If more severely immunocompromised mice are used, then cell fractions may grow which do not grow in less severely immunocompromised mice. Moreover some of the markers used for isolation of cancer stem cells have not proved reliable. Most notably CD133, or prominin, was initially used to identify cancer stem cells from various types of tumor but subsequent results have been inconsistent. Finally, it has always been apparent that a proportion of tumors do not confirm to the stem cell model because in them all the cells are capable of regenerating the tumor following transplantation.

Because of the overreliance on a specific set of techniques to define cancer stem cells there has been some uncertainty about the biological meaning of the concept. Cancers may indeed arise from stem cells, but they can also arise from transit amplifying cells or even from differentiated cells. It may be that multiple mutations will confer a stem cell-like behaviour on a cell that previously did not have stem cell properties. The identification of cancer stem cells with cells resistant to therapy is not universal. And it is clear that most of the heterogeneity evident on tumor progression is due to mutation, as described above, rather than to a normal developmental sequence arising from a stem cell system. A model reconciling the facts about genetic progression and cancer stem cells is shown in Figure 11.12. Here the final step involves formation of a tumor entirely composed of stem-like cells, any of which can propagate the tumor on transplantation.

In conclusion, some cancers do certainly contain stem cells, either derived from actual stem cells or from cells with mutations that confer stem cell behavior. But the "cancer stem cell" concept as a whole has serious limitations as most of the heterogeneity in tumors arises from random and variable somatic mutations.

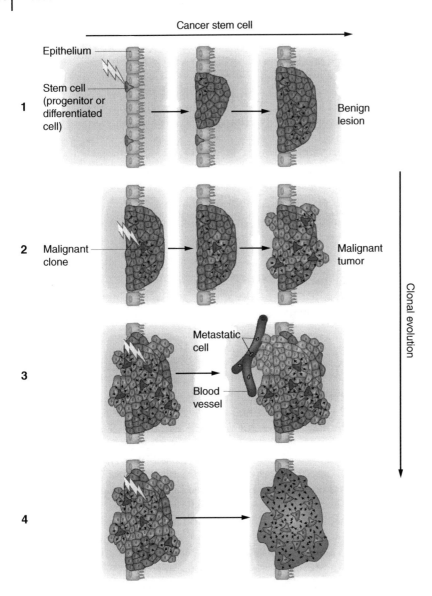

Figure 11.12 A model reconciling the concepts of cancer stem cells with genetic progression of tumors. The first mutation causes growth of a homogeneous benign lesion. The second establishes a malignant subclone. The third enables metastatic behavior in another subclone. At each of these stages the tumor contains stem cell-like cells. The fourth mutation creates highly malignant cells that take over the whole tumor, all of which behave as cancer stem cells. (From: Clevers, H. (2011) The cancer stem cell: premises, promises and challenges. Nature Medicine 17, 313–319. Reproduced with the permission of Nature Publishing Group.)

Further Reading

Regeneration

Bely, A.E. and Nyberg, K.G. (2010) Evolution of animal regeneration: re-emergence of a field. Trends in Ecology and Evolution 25, 161–170.

Elliott, S.A. and Alvarado, A.S. (2013) The history and enduring contributions of planarians to the study of animal regeneration. Wiley Interdisciplinary Reviews – Developmental Biology 2, 301–326.

Kragl, M., Knapp, D., Nacu, E., Khattak, S., Maden, M., Epperlein, H.H. and Tanaka, E.M. (2009) Cells keep a memory of their tissue origin during axolotl limb regeneration. Nature 460, 60–65.

Kumar, A., Godwin, J.W., Gates, P.B., Garza-Garcia, A.A. and Brockes, J.P. (2007) Molecular basis for the nerve dependence of limb regeneration in an adult vertebrate. Science 318, 772–777.

Nacu, E. and Tanaka, E.M. (2011) Limb regeneration: a new development? Annual Review of Cell and Developmental Biology 27, 409–440.

Reddien, P.W. and Alvarado, A.S. (2004) Fundamentals of planarian regeneration. Annual Review of Cell and Developmental Biology 20, 725–757.

Simon, A. and Tanaka, E.M. (2013) Limb regeneration. Wiley Interdisciplinary Reviews – Developmental Biology 2, 291–300.

Wagner, D.E., Wang, I.E. and Reddien, P.W. (2011) Clonogenic neoblasts are pluripotent adult stem cells that underlie planarian regeneration. Science 332, 811–816.

Wound Healing

Armulik, A., Genové, G. and Betsholtz, C. (2011) Pericytes: developmental, physiological, and pathological perspectives, problems, and promises. Developmental Cell 21, 193–215.

Borena, B.M., Martens, A., Broeckx, S.Y., et al. (2015) Regenerative skin wound healing in mammals: state-of-the-art on growth factor and stem cell based treatments. Cellular Physiology and Biochemistry 36, 1–23.

Eming, S., Martin, P. and Tomic Canic, M. (2014) Wound repair and regeneration: mechanisms, signaling and translation. Science Translational. Medicine 6, 265, 265sr6.

Gomez-Barrena, E., Rosset, P., Lozano, D., Stanovici, J., Ermthaller, C. and Gerbhard, F. (2015) Bone fracture healing: Cell therapy in delayed unions and nonunions. Bone 70, 93–101.

Guimarães-Camboa, N., Cattaneo, P., Sun, Y., et al. (2017) Pericytes of multiple organs do not behave as mesenchymal stem cells in vivo. Cell Stem Cell 20, 345–359.

Gurtner, G. C., Werner, S., Barrandon, Y. and Longaker, M.T. (2008) Wound repair and regeneration. Nature 453, 314–321.

Nombela-Arrieta, C., Ritz, J. and Silberstein, L.E. (2011) The elusive nature and function of mesenchymal stem cells. Nature Reviews. Molecular Cell Biology 12, 126–131.

Norenberg, M.D., Smith, J. and Marcillo, A. (2004) The pathology of human spinal cord injury: defining the problems. Journal of Neurotrauma 21, 429–440.

Rolls, A., Shechter, R. and Schwartz, M. (2009) The bright side of the glial scar in CNS repair. Nature Reviews. Neuroscience 10, 235–241.

Schindeler, A., McDonald, M.M., Bokko, P. and Little, D.G. (2008) Bone remodeling during fracture repair: The cellular picture. Seminars in Cell and Developmental Biology 19, 459–466.

Cancer

Beck, B. and Blanpain, C. (2013) Unravelling cancer stem cell potential. Nature Reviews. Cancer 13, 727–738.

Clevers, H. (2011) The cancer stem cell: premises, promises and challenges. Nature Medicine 17, 313–319.

Hanahan, D. and Weinberg, R.A. (2011) Hallmarks of cancer: the next generation. Cell 144, 646–674.

Kreso, A. and Dick, J.E. (2014) Evolution of the cancer stem cell model. Cell Stem Cell 14, 275–291.

Martincorena, I. and Campbell, P.J. (2015) Somatic mutation in cancer and normal cells. Science 349, 1483–1489.

Nik-Zainal, S., et al. (2012) The life history of 21 breast cancers. Cell 149, 994–1007.

Tomasetti, C., Li, L. and Vogelstein, B. (2017) Stem cell divisions, somatic mutations, cancer etiology, and cancer prevention. Science 355, 1330–1334.

Vogelstein, B. and Kinzler, K.W. (1993) The multistep nature of cancer. Trends in Genetics 9, 138–142.

Wong, K.M., Hudson, T.J. and McPherson, J.D. (2011) Unraveling the genetics of cancer: genome sequencing and beyond. Annual Review of Genomics and Human Genetics 12, 407–430.

Index

Page numbers in italic refer to pages where the item is shown in a figure. Entries of the form 2.C.1 refer to figures in the color plates.

2A sequence   *32*, 33
2i medium   80, 93, 95, 100
3D printing of cells   59

a

α cells   176
α fetoprotein   182
absorptive cells, of intestine   189, *190*
acinar glands   *157, 177*
acrosome   76
action potential   164, 169, 172, 175
adeno-associated virus   *32*, 34
adenocarcinoma   229
adenoma   229, 230, 235
adenovirus   *32*, 34
adipocytes   223
adult pluripotent stem cells   103
AG ("anterior gradient")   220
aggregation chimeras   193, 194
AGM (aorta-gonad-mesonephros) region   142
alcian blue stain   15, 17, *2.C.1*
alkaline phosphatase
   endogenous   73, 80, 95, *98*, 100, 101
   as label   17, 18
alizarin stain   17
allantois   *81, 82, 84, 85, 119, 121, 147*
allograft   61
Ambystoma mexicanum (axolotl)   220, 222
amino acids   53
amnion   *81, 84*, 82–85, *121*
amphibian limb   220–223
amphiregulin   137
anaplasia   229

angioblasts   143
angiogenesis   144, 227, 229
anoikis   28
anterior and posterior, definition   111
anterior visceral endoderm (AVE)   116, *117*
antibiotics   54
antibody
   immunity   64
   for immunostaining   17
apical epidermal cap   220, *221*
apoptosis   28–29, 63
arrector pili muscle   197
arterial capillaries, formation   144
ASCL1   183, 192
astrocytes   165–170, 227
asymmetric cell division   6, 8, 78, 134,
      162–164, 196
ATOH1 (=MATH1)   162, 179, 197
atomic bomb   27
atria, of heart   145–146
autograft   61
autophagy   29
AXIN2   196

b

β cells   176–178, 185
banking of cells   55–56
"barcoding"   46–47, 208–209
basal ganglia, of brain   126
basal progenitors, in CNS   166–167, *167*
basement membrane, basal lamina   155, *157*
basophils   204, *206*
benign and malignant neoplasms   229

The Science of Stem Cells, First Edition. Jonathan M. W. Slack.
© 2018 John Wiley & Sons, Inc. Published 2018 by John Wiley & Sons, Inc.
Companion website: www.wiley.com/go/slack/thescienceofstemcells

bile canaliculi   179, *180*
bile ducts   179–182, *180*
biliary epithelial cells (cholangiocytes)
        179–182, *180*
binuclear cells   175, 181–182
bioreactor   51, *52*
biotin/avidin/streptavidin staining   18, 29
"birthday" of neuron   5, 25, 169
bistability, bistable switch   110, *110*, 184, 192
blastema   220–222
blastocyst   38–41, 78–81, *78*, *5.C.3*
blastocyst complementation   104–105, *6.C.4*
blastomeres   78, *78*
blood islands   142
B lymphocytes   62, 99, 204, *206*, 207
BMI1   192, 213
BMP (Bone Morphogenetic Protein)
        signaling   72, *113*, 114, 115, *117*, *7.C.1*,
        125, *131*, 132, 136, 141, 145, 149, 197, 227
body plan *see* General body plan
bone   17, 156, 220, 223–227
bone fracture healing   225–227, 228
bone marrow   49, 59, 64–65, 103, 142, 205,
        208–211, *210*, 223
BRACHYURY (T)   95, 111, 115, *7.C.1*
brain   27–28, 61, 107, 108, 111, 120, 125,
        126–131, *127*, 163, *164*, 165–171, *8.C.1*
brainbow techniques   46, *46*, *3.C.1*, 200, 213
BrdU (bromodeoxyuridine)   4, 24–26, *24*,
        169, *191*, 192, 208
BRN2 (=POU3f2)   183
Brunner's glands   191
bulge, of hair follicle   197–199, *198*
busulphan   211

C
^{14}C dilution   27–28, *27*, 171, 175–176
C2C12 cells   57, 173
cadherins   55, 78, 101, 113, 115, 116, 133,
        139, 155
Caenorhabditis elegans   69, 162–163, *9.C.1*
Cairns hypothesis   25
calcineurin   *63*, 65
calcium ion   63, 65, 76, 77, 172, 175, 178
canals of Hering   *180*
cancer   228–236
cancer stem cell   7, 228–236, *11.C.4*
capillaries   3, 126, 143–145, 172, 173, 224–227
carcinoma   229

cardiac crescent   145, *145*
cardiac muscle (cardiomyocytes)   95, *6.C.1*,
        175, 175–176, 183, 185, *9.C.3*
cardiotrophin   168
cartilage   139, 220, 223, 227
caspases   28
CD *see* Cluster of Differentiation
CD3   61–63, 65
CD4   62
CD8   62
CD24   192
CD34   3, 199, 207, 208, 234
CD133 (prominin)   166, 192, 235
CD150 (SLAM)   207, 210
cdk *see* cyclin-dependent kinase
cdk inhibitors   54
CDX 1&2   79, 112, 119, *7.C.1*, 128–129
C/EBPα,β   151, 180
cell culture *see* tissue culture
cell cycle   21, 22–24, *23*
cell death   28–29, *29*, 229
cell differentiation   5, 56–57, 95–96, 155–185
cell lineage   42–47, *204*, *206*, 232–233
Cerberus-like (Cer-l)   116, *117*, *7.C.1*
cerebellum   *127*, 128
cerebral cortex   126, *127*
chick embryo   119, 125
chimerism   38, 96, 100
cholangiocytes *see* biliary epithelial cells
chorion   *81*, 82, 83, *84*
choroid plexus   126, *128*
chromatid   21, 71
chromatin   73, 87, 94, 158–161, *159*, 183–238
chromosome   21, 71
class 2 safety cabinet   53
cleavage divisions   78
clonal analysis   *43*, 44, 47, 193–195
cloning *see* somatic cell nuclear transfer (SCNT)
clonogenicity   6
clotting   225
Cluster of Differentiation (CD) antigens   20–21
cMYC   97
coelom   *121*, 122, 137, 146
collagen   15, 58, 156–157, 199, 225, 227
columnar epithelium   157, 166, 178
commitment, developmental   107, 111–112
competence, developmental   108–110
conceptus   70, 77
confetti label *see also* brainbow   194

confocal microscopy 16
connective tissue 156–158, *157*, 220, 221
cornea *132*, 199–200
cortical granules 77
cortical plate *167*, 168
COUPTFIII 144, 146
Cre, CreER 7, 35–37, *36*, 41–42, 44–46, *45*, 133, 174, 176, 180, 181, 182–183, 191, 194, 196, 197, 199, 200, *200*, 203, 210, 213, 223, 224, 235
CRISPR-Cas9 *36*, 37–41, *39*, 104
cryopreservation of cells 55–56
cryostat 15
crypts of Lieberkühn 189, *190*
crypt-villus development 149
cumulus cells 76
cup cells 189
cyclin-dependent kinase (cdk) 23
cyclosporin 65, *66*
cynomolgus monkeys 72, 83
cystoblasts 5
cytochrome c oxidase 47, *47*, 192, 235
cytochrome P450 130, 179, 183
cytokeratins *see* keratins

d
δ cells 176
Decapentaplegic 5
decellularization, of organs 60, *4.C.4*
decidua 80
dehydration/rehydration, with ethanol 14
DELTA 162, 176, 179
dendritic cells 64
dense core granules 177
dermal papillae 136, 195, 197–199, *198*
dermis 138, 228
dermomyotome 138–40, 172
desmosomes 156, *157*
determinant, cytoplasmic 72, 163
determination, developmental 107–109, *108*
dHAND 146
diaminobenzidine (DAB) reaction 18
Dickkopf 116, *117*, *7.C.1*, 136
diencephalon 126, *130*
DiI 133, 147
DNA methylation 72, 87, 94, 100, *159*, 160
DNA rearrangement 61, 208
DNA repair 28, 47, 62, 66, 175, 230
DNA replication 21, 70, 175

dominant negative inhibitor 37
dopaminergic neurons 165
dorsal and ventral, definition 111
doxycycline 35, 42, 98–99, 184, 203, 207
DPX (mountant) 15, 17
"driver" mutations 231
Drosophila 69, 162
Drosophila, female germ line 5
drug testing 104
ductal plate 180, *181*

e
ε cells 176
ectoderm 111, 118
EG (embryonic germ) cells 95
egg cylinder 80–82, *81*, 116
eHAND 146
electron microscopy 15–16
electroporation 31
embryoid bodies 95, *95*, 101, *6.C.1*
embryonic stem cells (ES cells), mouse 38–41, *39*, 80, 93–97, *94*, *100*
embryonic stem cells, human *100*, 101
endocrine cells, of pancreas 176–178
endoderm *81*, 82, 111, 118, *7.C.1*, 146–149
enterocytes *see* absorptive cells of intestine
enteroendocrine cells 178, 189, *190*
eosinophils 204, *206*
epaxial and hypaxial myotome 139–140
ependymal cells 126, 166
Eph B4 144
ephrin B2 144
epiblast 80–82, *81*, 95, *7.C.1*
epiblast stem cells (EpiSCs) 96–97, *100*, 101
epidermal placodes 135–136
epidermal stem cells, 195–197, *10.C.2*
epidermis 119, 134–135, *135*, 195–197, *196*, 221, 228
epiphysis (pineal) 127
episomes 97
epithelial-mesenchymal transition (EMT) 116, 133, 139, 229
epithelium 56, 156, 157, 229
epitope 17
epoxy resins 16
ERK signaling pathway 23, *113*, 114, 141, 231
erythrocytes (red blood cells) 142, *8.C.2*, 204, *206*
estrogen 137
ethics, of human embryo/ES cell work 85–86, 102

ethynyl dU   24, 26

exocrine cells, of pancreas   176–177

expression vector   31–33, *32*

extraembryonic membranes   80–85

extraembryonic mesoderm   83

extrahepatic biliary system   179

eye development   131–132, *132*

f

facultative stem cell   2, 174–175

FAH (fumaryl acetoacetate hydrolase) mouse
    model   182, 183, *9.C.5*

fate map   43–44, *43*, 107–109, *108*, 146–147

fat pad, mammary   136, 203

feeder cells   93–95, 101, 197, 213

fertilization   76–77

FGF (fibroblast growth factor) signaling   80,
    93, 113–114, *113*, 119, 128–129, 132, 136,
    139–140, 141, 145, 148–149, 150, 173,
    176, 197, 225

fibroblast(s)   50, 54, 56, 57, 59, 93, 98, 156,
    183, 211, 224, 225–227

first heart field   145

fixation, of specimens   13

FLK-1 (=VEGF receptor 2; =KDR)   143

floor plate   131, *131*, 139, 144

flow cytometry   20–21

FLP-FRT system   37

fluorescence activated cell sorting (FACS)   21,
    22, 205, 207, 218

fluorescence microscopy   16

folding, of embryo body   120–122, *121*

follicle stimulating hormone
    (FSH)   74, 76

forebrain   126

foregut   *121*, 146, *147*

formalin/formaldehyde   13

forward scatter (in flow cytometry)   20

FOXA1–3   111, *7.C.1*, 148, 180, 183

FOXO   115, 213

FUCCI (fluorescent ubiquitin-based cell
    cycle indicator)   26

g

GABA-ergic neurons   183

gall bladder   *147*, 179

gap junctions   155, 175

gastrulation   116–118, *118*

GATA1   142

GATA4-6   80, 95, 111, 145, 148, 183

G-CSF (granulocyte colony stimulating
    factor)   205, 210

GDNF (Glial Derived Neurotrophic
    Factor)   140–141, 212

general body plan   116, *122*

germ cells   3, 5, 70–76, 86, 87, 88, 96, *5.C.1*,
    211–213

germ line *see* germ cells

GFAP (Glial Fibrilliary Acidic Protein)   166

GFP (Green Fluorescent Protein)   16, 26, 31,
    44, 98, 191, 199, 208, 221

glands   156

glia   165–166, *165*

glial scar   227

gliogenesis   168–169

glucose   52

glucose-6-phosphate dehydrogenase   194, *194*

glutaraldehyde   14

glycerol, as mountant   17

GMP (Good Manufacturing Practice)   56, 185

goblet cells   178, 189, *190*

Golgi silver stain   15

gradient, of morphogen   109–110, *109*,
    128–129, 130–131, 138–139, 144, 148,
    168, 180, 222

grafting *see* transplantation

granulation tissue   225–227

granulocytes   204, *206*

granzymes   63

gray matter, of brain   131

growth curve   55, *55*

h

H2B-GFP (histone 2B-green fluorescent
    protein)   26, 208

H-2 complex   61

hair cycle   *198*, 199

hair follicles   135–136, *135*, 197–199, *198*,
    10.C.3, 228

haplotype, in HLA system   64–65

head process   118

heart   *119*, 145–146, *145*

hedgehog signaling   *113*, 115, 129, *131*, 132,
    136, 139–140, 144, 148, 149, 150, 168, 222

hematopoiesis   142–143, 180, 204–211, *206*

hematopoietic growth factors   205

hematopoietic stem cells   142, 183, *206*,
    207–211, *10.C.5*

hematoxylin/eosin stain 15, *2.C.1*
hemogenic endothelium 142–143, *143*
hepatoblasts 151
hepatocytes 179–183, *180*, 183, 185, *9.C.5*
heterochromatin 133, 161, 218
Hex gene 115, *117*
Hes genes/proteins 138, *138*, 151, 179
HGF (hepatocyte growth factor = scatter
    factor) 140, 173, 182, 225
HIF1α 209
hindbrain 126
hindgut *121*, 146, *147*
hippocampus 126, *127*, *128*, 169–171, *169*, *171*
hippo pathway 78–79, *113*, 115
Histoclear 14
histology 13–15
histone 72, 73, 77, 94, 100, 159–160
histone acetylation and methylation 94, *159*, 160
HLA (Human Leukocyte Antigen)
    system 61–63
HNF1α and β 151, 180, 183
HNF4 148, 149, 180, 183
HNF6 180
Hoechst 33342 4
holoclones *7*
homologous recombination *39*, 40
HOPX 192
horseradish peroxidase 17, 18
HOX genes/proteins 111–112, 119, 128–129,
    129, 140, 148
human chorionic gonadotrophin (HCG) 82
human development 82–85, *84*, 119
human embryonic stem cells *see* embryonic
    stem cells, human
human genetic modification 86
human preimplantation conceptus *see*
    preimplantation conceptus, human
Huntingdon's disease 165
hydrogels 58
hyperacute rejection of graft 64, 65
hypoblast 82, *84*, 85
hypophysis 127, 137
hypothalamus 127

i

IGF1&2 (insulin-like growth factors) 87–88,
    113, 115, 137, 179
immune privilege 61
immunodeficient mice 66–67, 203, 205, 207, 211

immunostaining 17–18, *2.C.3*
immunosuppression 65–67
implantation, of conceptus 80, 82
imprinting 87–89, *88*
inbred mice 61, 66, *4.C.5*
induced pluripotent stem cells (iPS cells) 61,
    97–101, *97–100*, *6.C.3*, 161
inducing factors/induction, embryonic
    108–110, 112–115, *113*
inflammation 221, 225, 228
inner cell mass (ICM) 38, 70, 78–80, 82–83,
    93, 95–96, 101–102, 111
in situ hybridization, 18–19, *2.C.3*
insulin 176–178
insulin secretion 177–178
integrins 21, 28, 77, 139, 173, 203, 207, 213
interkinetic migration 166, *167*
interleukin 2 63, 65
interleukin 6 182
intestinal epithelium 189–195, *190*
intestinal stem cells 8, 191–193, *10.C1*
intestine, development *147*, 148–149, *149*,
    178–179
IRES (internal ribosome entry site) *32*, 33
ISLET1 145
islets of Langerhans 176–178, *177*
isograft 61

j

JAGGED 114, 162, 180
JAK-STAT pathway 93, *94*, 202

k

K14 promoter 195–199, 200, 235
keratinocytes 195–197, 225
keratins 135, 195, 200
Ki67 24
kidney development 140–142
KLF4 95, 97
knockout mice 41, 104, 125
Kupffer cells 179, *180*

l

label retention 4–5, 25, 203, 208
lactation 201–202
lamina propria 190
laser capture microdissection 19–20, *20*
lateral inhibition 135, 161–162, *161*
lateral plate 118, *121*, 137, 143

Leblond classification of cell proliferation behaviors   28
left-right asymmetry   146
Lefty   116, *117, 7.C.1*
lentivirus   *32*, 34, 98–99, 207
leukemia   229
Leydig cells   73, 86
LGR5   3, 191–195
LIF (leukemia inhibitory factor)   80, 93, *94*, 95, 100
limbus, limbal stem cells   199–200, *200*
lineage labeling   *7, 8*
liver   142, *147*, 150–151, *150*, 179–183, *9.C.4, 9.C.5*
LMO2   142, 207
lobule, of liver   179–183
lobule, of mammary gland   200–204
LoxP sites   35, *36*
LRIG1   192
luminal cells   200, *201*
Lunatic fringe gene   139
luteinizing hormone (LH)   74, 76
lymphocytes, also *see* B, T, NK lymphocytes   204, *206*, 227
lymphoma   229

m
macrophages   204, *206*, 225–227
MAF A and B   177
Major Histocompatibility Complex (MHC) *see also* HLA   61
mammary glands   136–137, *136*, 200–204, *201, 202, 204, 10.C.4*
mammary stem cells   203–204, *10.C.4*
Masson trichrome stain   15, *2.C.1*
matrigel   58, 193
matrix metalloproteinases   82, 137, 173, 225
M cells   189
mdx (dystrophic) mouse   174
media, for tissue culture   51–53
medium spiny neurons   165
MEF2c   183
megakaryocytes   142, 204, *206*
meiosis   71, *71*
melanocytes   197
memory cells   61, 63, 64, 208
mercuric chloride, as fixative   14
Merkel cells   197

meroclones   *7*
mesenchymal stem cells (MSC)   223–225, 227
mesenchyme   56, *157*
mesoderm   *81*, 82, 111, 118, *7.C.1*
MESP 1&2, 139, *7.C.1*   180
metabolic zonation, in liver   180–181, *9.C.4*
metanephros   140
metaplasia   221, 230
metastases   229
methyl cellulose   205
MEX-5   163
microbial contamination   53–54
microglia   166, 227
microtome   14
midbrain   126
midgut   *121, 147*
mindbomb   163–164
mitochondria   47
mitosis   21, 23, 70–71, *71*
monocytes   204, *206*, 225–227
morphogen *see also* gradient of morphogen 109–110
mosaicism, genetic   41, 193
motor neurons   164–165, 168
MSX1   17
mTOR *see* TOR
mucosa   190
Mullerian duct   86
multipotency, definition   93
muscle *see* skeletal, cardiac, smooth muscle
muscle satellite cells   2, 173–175, *174*
muscular dystrophy   174
muscularis mucosa   190
MYC *see* cMYC
mycophenolate mofetil (MMF)   65, *66*
mycoplasma   54
myelin   164
myeloid cells   204
MYF5   140, 172
myoblasts   57, 140, 172–173, *173*
MYOD   140, 172
myoepithelial cells   200, *201*
myofiber   171–172, *9.C.2*
myofibrils   172
myogenesis   140
myogenin   140
myosins   172, 175
myotome   *139*, 140
MYT1L   183

n

"naïve" embryonic stem cells   96
nanog   72, 79, 80, 94, 95
natural killer (NK) lymphocytes   61, 64, 204, *206*
neoblasts   218–220, *218*
neocortex   126
neoplasia *see* cancer
nephric duct (=Wolffian duct)   140
nephrogenic mesenchyme   140–141, *141*
nephron   141–142
neural crest   *121*, 132–134, *134*
neural plate, neural tube   119, 120, *120*, *121*, *122*, 125
neural stem cells   169–171, *169*, *170*
neuregulin   220
NEUROD   162
neurogenesis   166–168, 169–171, *169*, *170*, 228
neurogenin 12   168
neurogenin 3   162, 176, 212–213
neuromuscular junction   172
neurons   5, 13, 15, 25, 27–28, 126–134, 155, 161–171, *165*, 183, 227–228
neurospheres   171
neurotransmitters   164
neutrophils   204, *206*, 225–227
NFATs (Nuclear Factors of Activated T cells)   63, 65, 173
niche, for stem cells   1, 5–6, 10, 170–171, 192–193, 199, 205, 209–211
NK *see* natural killer
NKX2.2   168, 177
NKX2.5   145
NKX3.2   139
NKX6.1   177
nodal signaling   114, 115, *117*, 117–118, *7.C.1*, 146, 148
node   117–118, *118*, *7.C.1*
NOD-SCID mouse   67, 203
Notch signaling   *113*, 114–115, 139, 141, 144, *161*, 162–164, 174, 176, 179, 180
notochord   118–119, *119*, *121*, 137, 144
Notophthalamus viridescens (Eastern newt)   220, 221
NTBC   182
nude mouse   66

o

OCT4 (=OCT3, POU5F1)   3, 72, 79, 94, 95, 97, 103, 219

OCT (Optimum Cutting Temperature compound)   15
olfactory bulb   126, 170
OLIG1   168–169
oligodendrocyte precursor cells   169
oligodendrocytes   165–166, 227
oncogene   229
oocyte   74, 89, *5.C.2*
oogenesis   74–76, *75*
optic tectum *see* superior colliculus
optic vesicle   127, 131–132, *132*
organ culture   57–58, 125
organ, definition   155
organoids   59, *4.C.3*, 192–193, *193*
osmium tetroxide   16
osteoblasts   209, 227
osteoclasts   209, 227
oval cells   182–183
oviduct (Fallopian tube)   74, 76, 77
oxygen   49, 52, 172, 179, 209, 227

p

p53   231
p63   111, 134–135, 195–197, 199
packaging cell line   33–34
pancreas development   *147*, 150, *150*, 176–178, *177*, *178*
Paneth cells   178, 189, *190*, 191–193
papilloma   229, *11.C.4*
parabiosis   182
paraclones   *7*
paraffin wax   14
parathyroid hormone related hormone (PTHrH)   136
parietal endoderm   *81*, 82
Parkinson's disease   165
PAR proteins   78, 134, 144, 163, *9.C.1*
parthenogenesis   89
PAS (periodic acid-Schiff stain)   15, *2.C.1*
"Passenger" mutations   231
pattern formation   221–223
PAX1   139
PAX3   131, 139, 173
PAX4   177
PAX6   129, 131, 132, *168*, 177
PAX7   131, 173, 213
PAX9   139
PCNA (proliferating cell nuclear antigen)   24
PCR (polymerase chain reaction)   19, 38, 40

PDGF (platelet derived growth factor) 24, 53, 114, 225

PDX1 150, 151, 176–178

perforins 63

pericytes 144, 224, *224*

peroxidase *see* horseradish peroxidase

Peyer's patches 191

phagocytosis 28

pH control (in tissue culture) 52

phosphatidyl serine 28

phosphohistone H3 24

pial (basal) surface, of neuroepithelium 166

PIE-1 163

PI3 kinase *113*, 141, 231

"pioneer" transcription factor 183

pituitary gland *see* hypophysis

placental lactogen 202

planaria 217–220, *218, 11.C.1*

plasmid *see* expression vector

pluripotent stem cells 2, 93–105, 184–185, 219

polar bodies 74, *75*, 76, *78*

polyA addition 31

polyploidy 175, 181–182

portal triad 179

positional information 222

potassium dichromate 14, 15

PP cells 176

prechordal plate 118

preimplantation conceptus
  human 80, *5.C.4*
  mouse 77–80, *78*

"primed" embryonic stem cells 96, 101

primitive endoderm 80, *81*, 82, 95

primitive erythrocytes 142, *8.C.2*

primitive macrophages 142

primitive streak *81*, 82, 116–118, *118*

primordial germ cells (PGCs) 72–73, *72*, 88

prod1 220, 222

progenitor cells 1, 2, 6, 13, 21, 25, 42, 103, 142, 155, 156, 166–168, 170–171, 178, 184, 204, 205–210, 234

progesterone 82

prolactin 202

propidium iodide 20, 21

protamines 73, 77

pseudopregnancy 38

pseudotyping 34

PTF1/p48 150, 176

q

quail marker 133

r

radial glia 166–171, *167*

raf *113*, 231

RAG-1 and-2 deficient mice 62, 64, 67

ras *113*, 231

recombination, genetic 71

regeneration, of limb 220–223, 221, *11.C.2, 11.C.3*

regeneration, of liver 182–183

regeneration, types of 217

regulatory T cells (T-regs) 63

rejection of grafts 64–65

repair, of tissues 228

retinal pigment epithelium (RPE) *132*

retinoblastoma (RB) protein 23–24

retinoic acid 73, *113*, 115, 129–130, *143*, 148, 150, 222

retrovirus *32*, 33–34, 97, 208

rhombomeres 129–130, *130*

RNAi *36*, 37, 220

RNA Scope (enhanced in situ method) 18–19

RNAseq 19

Rosa26 promoter 44

rostral and caudal, definition 111

R-spondin 191, 193

RUNX1 143, 207

s

sarcoma 229

sarcomeres 172

sarcoplasmic reticulum 172, 175

Sca-1 4, 207

scaffolds, for tissue culture 58

scar 225

SCF (stem cell factor) 207, 210, 213

Schmidtea mediterranea 218–220

Schwann cells 133, 134, 164, 220, 221, 227

SCID (Severe Combined Immunodeficiency) mouse 66

SCL 142

scleraxis 139

sclerotome 138–140

SDF1 (stromal cell derived factor 1,=CXCL12) 73, 210

second heart field 145

secretory cells, of intestine   178
seminiferous epithelial cycle   212
seminiferous tubules   211–213
Sendai virus   97
Sertoli cells   73, 211–213
serum (for tissue culture)   24, 53–57, 171, 179
serum-free media   53
sex determination   86
"short term" hematopoietic stem cell   205, 206, 208
side population   4, 203, 207
side scatter, in flow cytometry   20
single cell RNAseq   19
sinusoids   179, *180*, 209, *210*
siRNA *see* RNAi
sirolimus   65, *66*
SIX2   141, 142
skeletal muscle   140, 171–175, 221
small and large intestine, differences   191
smooth muscle   49, 103, 133–134, 144–145, 148, 190, 225, 229
SNAIL   116, 132, 133, 139
soft agar culture   205, 207
soma *see* somatic cells/tissue
somatic cell nuclear transfer (SCNT)   89–90, 102
somatic cells/tissue   70
somatic mesoderm   137
somatic mutations   101, 230–233
somatopleure   137
somites   *119*, *121*, 137–140, 172
SOX1   111
SOX2   72, 94, 97, 111, 183
SOX9   10, 86, 132–133, 136, 169, 176, 180
SOX17   95, 111, 148, 151
specification, developmental   107–109, *108*
sperm   76–77
spermatogenesis   73, *74*, 211–213, *212*
spermatogonia, spermatogonial stem cells, 73, 211–213, *10.C.6*
"spheres", in culture   6, 171, 203
spinal cord injuries   227–228
splanchnic mesoderm   137, 143, 146, 148
spleen colony assay   205
squamous cell carcinoma   229, 230
squamous epithelium   134, 156, *157*, 195
SRY   86
SSEA1 (Stage specific embryonic antigen 1)   95, 100

SSEA-3 &-4   101
stellate cells, in liver   179, *180*
stem cell, definition   1, 9–10
stemregenin   211
sterile technique   53–54
stochastic model, for stem cell renewal   8, *9*, 194, 197, 213
striated muscle *see* skeletal muscle
striatum, of brain   126, 128, 171
stroma   132, 158, 190, 199, 200, 201, 210–211, 229, 232
substantia nigra   128, 165
subventricular zone, of CNS   169–171
Sudan black stain   15, *2.C.1*
superior colliculus   128
sweat glands   136, 228
symmetry breaking   110–111, 116, 135, 144, 163
syncytiotrophoblast   82

t
tacrolimus   65, *66*
tamoxifen   35–37, *36*, 41, *42*, 45, 191, 196, 203
TBX1   145
TBX5   145, 146, 183
T cell receptor (TCR)   61–63, *63*
telencephalon   126, *130*, 166, 168
telomerase   3, 229
telomere   3, 54, 90, 207, 229
teratoma   *95*, 96, 101, *6.C.2*, 183, 213, 229
Tet-On, Tet-Off *see* Tet system
tetraploid complementation   41, 96, 100
Tet system   35, *36*, 98–99, 184, 203, 207
T gene/protein *see* BRACHYURY
TGF β (transforming growth factor β) signaling   180
tight junctions   156, 157
tissue, definition   155
tissue culture (cell culture), 49–60, *50*, *4.C.1&2*
tissue engineering   51, 57–60
tissue-specific promoter   35, *36*, *45*
tissue-specific stem cell, definition   2
T lymphocytes   61–64, *63*, 204, *206*
tolerance, immunological   62
toluidine blue stain   16
TOR (Target of Rapamycin)   65
totipotency, definition   93
TRA 1–60, 1–81 (human ESC markers)   101

transcription factors *see also* the individual factors listed   3, 37, 69, 111–112, *7.C.1*, 158–160
transdifferentiation   6, 183–184
transduction   33–34
transfection   31–33
transgenic mice   38–40
transit amplifying cells   1, 2, 8, 13, 44, 192, 194, 196, 197, 213, 219, 234, 235
transplantation   60–67, 104, 142, 182, 183, 203, 204–205, 209, 211, 221–222, 234–235
tritiated thymidine (^{3}HTdr)   24, 181, 192
trophectoderm/trophoblast   78–80
trypsin   55
T tubules   172, 175
tuft cells   178, 189, *190*
tumor *see* cancer
tumor necrosis factor (TNF)   182
tumor suppressor gene   229
TUNEL (TdP mediated dUTP nick end labeling)   29
turning, of mouse embryo   82, *83*
twinning   78, 85, 86
type A and B spermatogonia   211–212

u
umbilicus   122, 146, *147*
ureteric bud   140–141, *141*
uterus   80

v
vascular endothelial growth factor (VEGF)   143–144, 225, 227
vasculogenesis   143–145
venous capillaries, formation   144

ventricles, of brain   126, *128*
ventricles, of heart   145–146
ventricular (apical) surface of neuroepithelium   166
vertebrae   112, 138, 139–140
villi, intestinal   189, 190
visceral endoderm   *81*, 82, 118, *7.C.1*
VSV-G *see* pseudotyping

w
white matter, of brain   131
wholemounts   17, *2.C.2*
Wnt signaling   3, 4, 5, 10, 93, 112–113, *113*, 115, *117*, 119, *7.C.1*   128–129, 139–140, 141, 148, 149, 181, 191–193, 196, 197, 220, 222
wound epithelium   220, *221*
wound healing   225–228

x
X chromosome   86–87
xenograft   61, 65
Xenopus   69, 119, 125, 138, 140, 142, 162
X inactivation   87, 89–90, 94, 100, 193
Xist   87
xylene   14

y
Y chromosome   86
yolk sacs   *81*, 82, *83*, *84*, 85, *121*, 142

z
zebrafish   69, 119, 125, 138, 140, 142, 228
zona pellucida   74
zygote   77–78

Wiley The manufacturer's authorized representative according to the EU
General Product Safety Regulation is Wiley-VCH GmbH, Boschstr. 12,
69469 Weinheim, Germany, e-mail: Product_Safety@wiley.com.

Printed and bound by CPI Group (UK) Ltd, Croydon, CR0 4YY
24/03/2025
01835786-0001